DOUBLE AWARD

SCIENCE *for* GCSE

GRAHAM HILL

PUPIL'S HANDBOOK

Hodder & Stoughton
A MEMBER OF THE HODDER HEADLINE GROUP

Acknowledgements

The publishers would like to thank the following individuals, institutions and companies for permission to reproduce photographs in this book. Every effort has been made to trace ownership of copyright. The publishers would be happy to make arrangements with any copyright holder whom it has not been possible to contact:

Biophoto Associates (butterfly picture in Life Processes chapters); Science Photo Library (78)/Manfred Cage (electron micrograph of crystals in Materials chapters)/JISAS/Lockheed (X-ray of the Sun in Physical Processes chapters).

We are grateful to the following examining bodies for permission to reproduce examination questions: Assessment and Qualifications Alliance, AQA, which now includes NEAB and SEG, Northern Ireland Council for the Curriculum Examinations and Assessment, NICCEA, Oxford Cambridge and RSA Examinations, OCR, which now includes MEG, Welsh Joint Education Committee, WJEC, and Edexcel.

As author, I am indebted to Charlotte Litt (Senior Desk Editor at Hodder & Stoughton) and my wife Elizabeth for their significant contributions to this book.

Graham Hill
June 1999

Orders: please contact Bookpoint Ltd, 39 Milton Park, Abingdon, Oxon OX14 4TD. Telephone: (44) 01235 400414, Fax: (44) 01235 400454. Lines are open from 9.00–6.00, Monday to Saturday, with a 24 hour message answering service. Email address: orders@bookpoint.co.uk

A catalogue record for this title is available from The British Library

ISBN 0 340 73078 1

First published 1999
Impression number 10 9 8 7 6 5 4 3 2 1
Year 2005 2004 2003 2002 2001 2000 1999

ISBN 0 340 73079 X

First published 1999
Impression number 10 9 8 7 6 5 4 3 2 1
Year 2005 2004 2003 2002 2001 2000 1999

With Answers

Cover photo from Keith Kent, Science Photo Library.
Illustrated by Tom Cross, William Donohoe, Richard Duszczak, Phil Ford, Ian Foulis & Associates, Peters & Zabransky (UK) Ltd.
Typeset by Wearset, Boldon, Tyne and Wear.
Printed in Great Britain for Hodder & Stoughton Educational, a division of Hodder Headline Plc, 338 Euston Road, London NW1 3BH by Redwood Books, Trowbridge, Wilts.

Contents

Life Processes and Living Things

Materials and their Properties

Physical Processes

INTRODUCTION

This book is specially written to provide lots of questions for use in class and for homework, and to help you to prepare for exams.

It is written as a companion book to *Science for GCSE: Double Award*, but can also be used independently.

The book contains a **Summary** of each of the 30 chapters in *Science for GCSE* which correspond with 30 key themes and topics for GCSE. These summaries pick out the important facts and ideas in each topic area.

Each Summary is followed by **Study Questions** related to the Summary. There are three types of study question:

- **Objective Questions** requiring one word or one letter answers. These are used in about one third of double award science courses. They are ideal for quick revision tests and are an ideal preparation for module tests.
- **Short Questions** for four, five or six marks which are used in all the double award courses.
- **Further Examination Questions** taken from past exam papers which provide excellent revision for tests and exams.

In addition to all these there is an *Answers Edition* containing helpful answers and advice on all the study questions.

How to use this book

This book will help you to use homework time effectively, to prepare carefully for class tests and organise your revision thoroughly.

During most of your GCSE course you will be helped and guided by your teacher. He or she will ask you to answer specific questions from different sections of this book as you work through the GCSE course.

At other times, you will be expected to revise for topic tests and prepare for exams. This will require a good deal of self-motivation, particularly as the GCSE exams approach and you have study leave.

In preparing for exams and even for shorter topic tests, it is important to have a schedule. For the GCSE exams, it is crucial to have a schedule and this should be planned from the Easter holiday, allowing time for relaxation as well as time for study.

The following list shows how you could use this book in preparing for a short topic test or a full GCSE exam.

1. Decide which topic you will revise.
2. Read the appropriate chapter in *Science for GCSE* plus your own class notes. Study the Summary in this book.
3. Make your own lists of key words. Learn important definitions and make your own summaries of the key points, possibly using flow charts and diagrams. Keep these notes to jog your memory and help you revise later.
4. Answer the Objective Questions and possibly the Short Questions.
5. Check your answers to the Objective Questions and the Short Questions and make sure you agree with the correct answers. You may need to ask your teacher for the answer if you don't have the *Answers Edition* of this book.
6. Study the topic and the Summary again if your performance on the Objective Questions and Short Questions was below standard.
7. Answer the Further Examination Questions on the topic you are revising. Answering these longer exam-type questions is one of the best ways to prepare for an exam.
8. Check your answers to the Further Examination Questions using the *Answers Edition* of this book. Make sure you agree with the correct answers.
9. Ask your teacher if you don't understand something or you can't get the correct answer. Your teacher is keen for you to do well and will be pleased to see your enthusiasm and commitment.

The following instructions show you how to use [illegible] for a topic test or a full GCSE exam:

1. Decide which topic you will revise.
2. Read the appropriate chapter in [illegible]. Study the summary in this book.
3. Make your own list of key words, formulae, diagrams [illegible] summaries of the key points possible, using flow charts and diagrams [illegible] these notes to help you [illegible] and help you revise later.
4. [illegible] the Objective Questions and then the Short Questions.
5. [illegible] answers to the [illegible] Short Questions and [illegible]

[illegible]

CHAPTER 1

Cells and life

SUMMARY

1 The different places where living things can be found are called **habitats**.

2 The conditions in a habitat make up its **environment**.

3 The environment of habitats can be very different.

4 Living things are often **adapted** to the habitat in which they live. Some adaptations are very obvious like birds to trees. Others are less obvious like Polar bears to the Arctic.

5 Living things are called **organisms**.

6 All organisms carry out seven important **life processes** to stay alive:

Movement
Reproduction
Sensitivity
Growth
Respiration
Excretion
Nutrition

Remember the name
MRS GREN

7 All living things are composed of **cells** which are the basic units of life. Cells can be seen through a microscope.

8 All cells have three important structures.

9 Plant cells have three other structures.

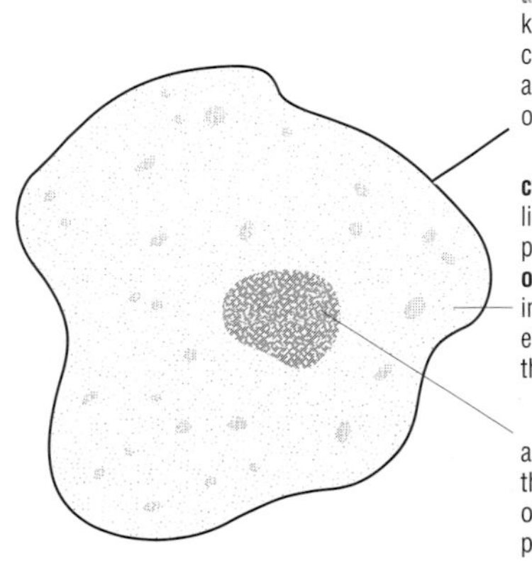

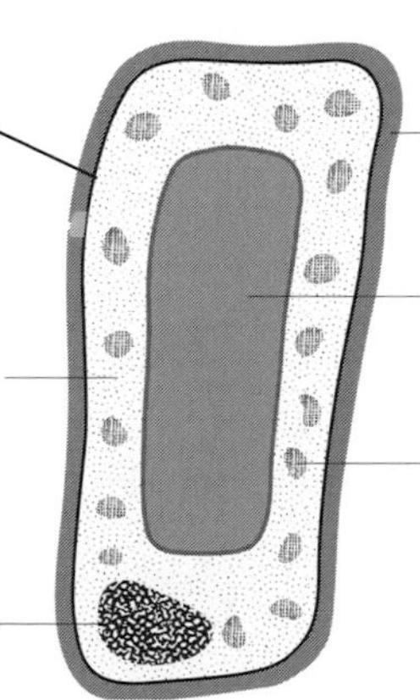

a thin **cell membrane** which keeps the cell together and controls the movement of water and small molecules into and out of the cell

cytoplasm – a jelly-like, watery liquid which contains smaller parts of the cell known as **organelles**. The organelles include mitochondria where the energy producing reactions of the cell occur

a **nucleus** which contains long, thin chromosomes composed of DNA. DNA controls all life processes

a thick **cell wall** made of cellulose outside the cell membrane to support and protect the cell whilst remaining porous and flexible

a **vacuole** in the centre of the cell containing watery cell sap in which are sugars and salts

chloroplasts – small organelles containing the green pigment chlorophyll which absorbs sunlight and uses it to synthesise food for the plant during photosynthesis

10 Organisms are built up from cells which are themselves made up from atoms. In turn, organisms group to form an entire **ecosystem**.

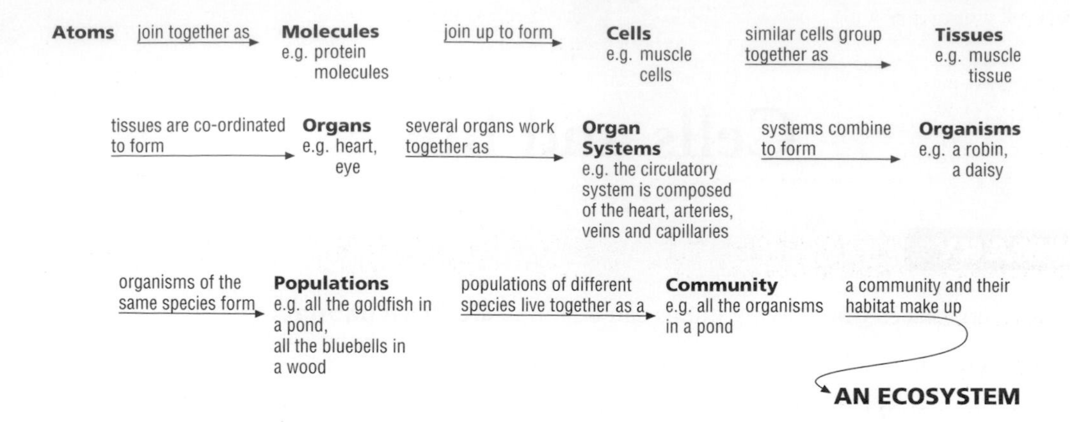
Atoms
join together as
Molecules
e.g. protein molecules
join up to form
Cells
e.g. muscle cells
similar cells group together as
Tissues
e.g. muscle tissue
tissues are co-ordinated to form
Organs
e.g. heart, eye
several organs work together as
Organ Systems
e.g. the circulatory system is composed of the heart, arteries, veins and capillaries
systems combine to form
Organisms
e.g. a robin, a daisy
organisms of the same species form
Populations
e.g. all the goldfish in a pond, all the bluebells in a wood
populations of different species live together as a
Community
e.g. all the organisms in a pond
a community and their habitat make up
AN ECOSYSTEM

11 The cell membrane is **partially permeable**, allowing certain particles (molecules and ions) to pass through it, but blocking the passage of others.

12 Particles enter and leave cells by three processes:

- **diffusion**
- **active transport**
- **osmosis**

13 **Diffusion** involves the movement of particles in a liquid or a gas from a region of higher concentration to a region of lower concentration. Diffusion is just spreading out. It occurs because of the random motion of particles in liquids and gases.

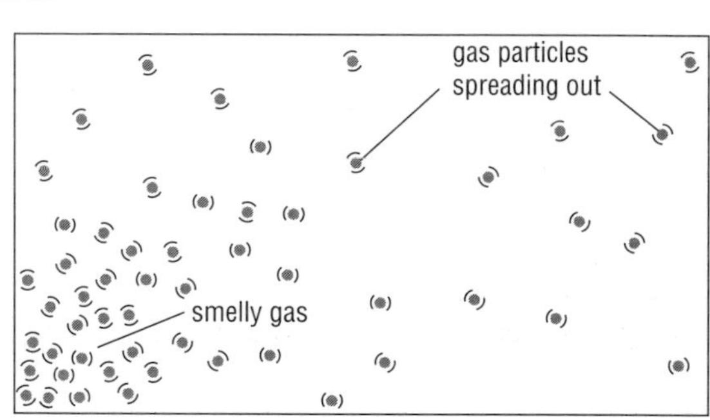
gas particles spreading out
smelly gas

14 **Osmosis** is a special case of diffusion in which particles (molecules or ions) move from a region of higher concentration to one of lower concentration through a partially permeable membrane like a cell membrane.

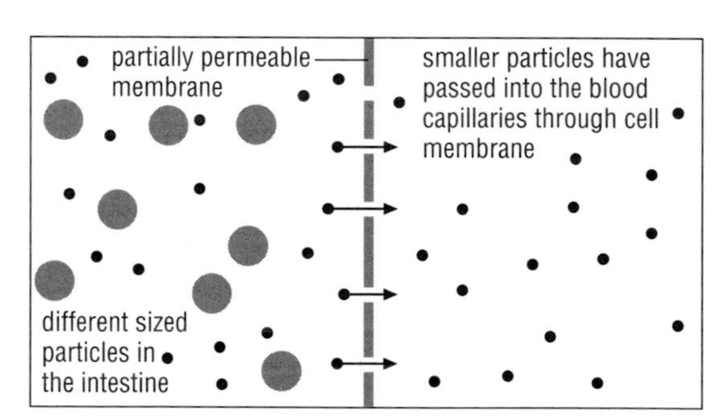
partially permeable membrane
smaller particles have passed into the blood capillaries through cell membrane
different sized particles in the intestine

15 **Active transport** involves the selective movement of dissolved particles from a region of lower concentration to one of higher concentration.

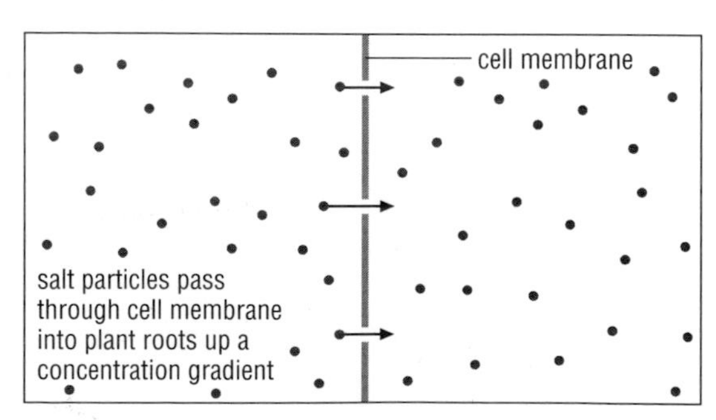
cell membrane
salt particles pass through cell membrane into plant roots up a concentration gradient

16 The table below contrasts diffusion and active transport.

Diffusion	Active transport
Random movement of particles	**Selective** movement of particles
Movement **down** a concentration gradient	Movement **up** a concentration gradient
No energy required	**Energy** required

STUDY QUESTIONS

Objective questions

Questions 1 to 5

The diagram shows a typical plant cell in which parts have been labelled from A to F.

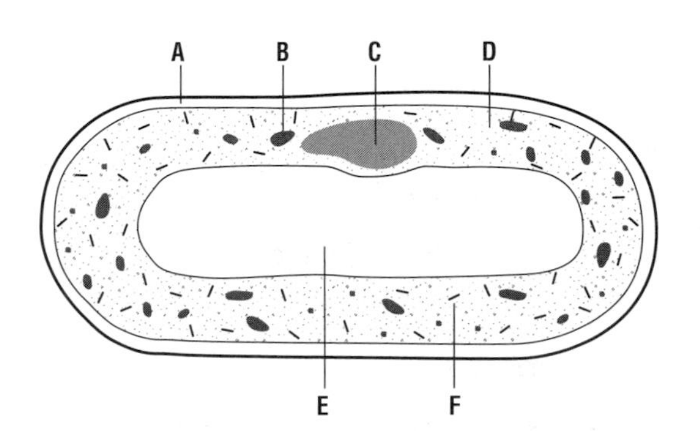

1 Name the part labelled A.

2 What is the function of the part labelled B?

3 Which part contains mitochondria?

4 Which part contains cell sap?

5 Which part controls chemical reactions inside the cell?

Questions 6 to 9

The graph below shows the number of bacteria in the body of a child with whooping cough.

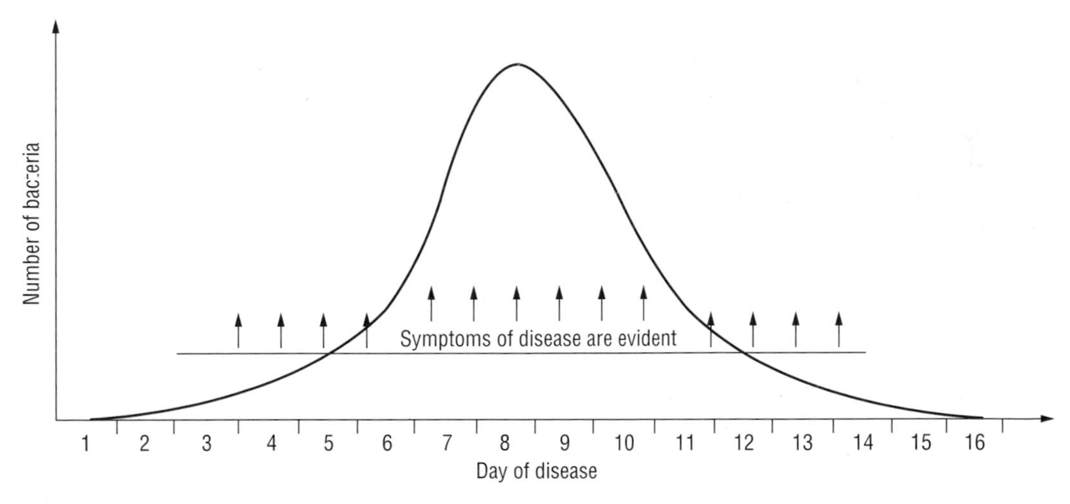

6 On which day did the child first show symptoms of the disease?

7 For how many days did the child show these symptoms?

8 On which day did the bacteria increase the fastest?

9 For how many days did the child retain bacteria after the symptoms were no longer evident?

Questions 10 to 12

The diagram below shows the apparatus that a student set up for an investigation.

In questions 10 to 12, choose from A, B, C or D which is the correct answer.

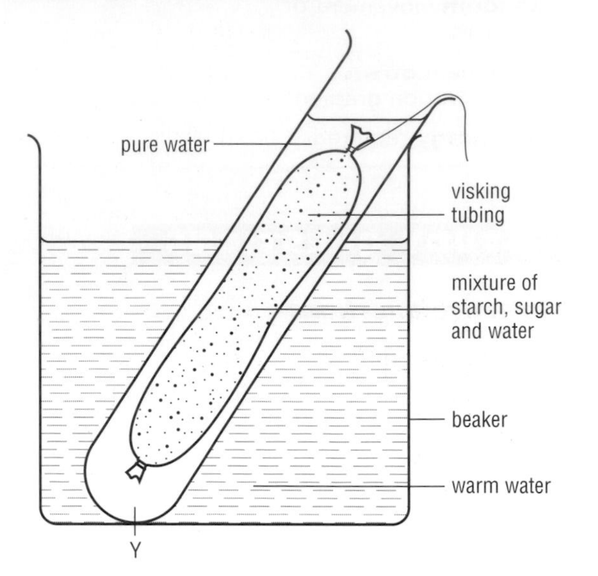

10 Visking tubing is *best* described as
- A impermeable
- B partially-permeable
- C permeable
- D semi-porous

11 After 2 hours, a sample taken from Y will contain
- A water only
- B water and starch
- C water and sugar
- D water, starch and sugar

12 Which of the following processes will *not* occur in the test tube?
- A active transport
- B convection
- C diffusion
- D osmosis

Short questions

13 Copy out and complete the following table for some life processes of a rabbit.

Life process	Description of the life process
Reproduction	The birth of the offspring
a) ______________	Obtaining food by eating plants
Excretion	b) ______________
c) ______________	The response to stimuli

3 marks

d) State *two* life processes of a rabbit not included in the table above. *2 marks*

14 The diagram below shows a sperm cell and an egg cell.

a) Name the parts labelled A.
b) Name the parts labelled B.
c) Name the parts labelled C.
d) Which part of these cells carries information on inherited characteristics?
e) How are sperm cells specially adapted? *5 marks*

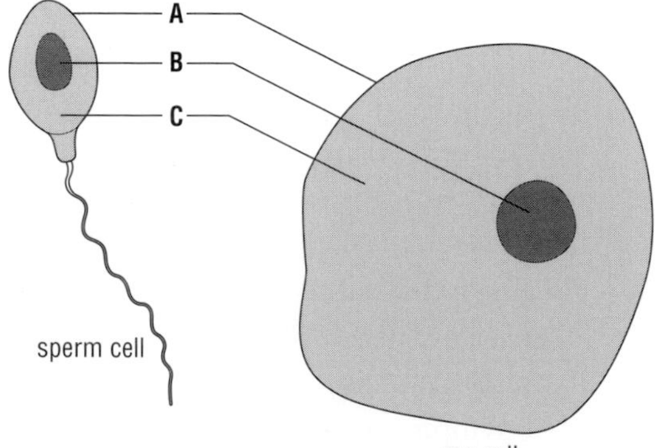

15 Choose words from the following list to replace the letters (a) to (f) in the passage below.

cells cell wall chloroplasts chromosomes genes nucleus organelles organisms organs tissues vacuole

Each cell contains a . . . (a) . . . which determines its characteristics. This contains pairs of . . . (b) . . . which are made up of smaller units of inheritance called . . . (c) . . . . Collections of similar cells working together are called . . . (d) . . . . These make up . . . (e) . . . which work together as systems allowing . . . (f) . . . to survive. *6 marks*

16 Cells are surrounded by a cell membrane. Give *two* functions of the cell membrane. *2 marks*

Further examination questions

17 The diagrams show an animal and a plant cell.

Use the diagrams to answer the following questions.

a) Give *two* similarities between the cells. *2 marks*
b) Name *one* structure present in the plant cell but not in the animal cell. *1 mark*

NICCEA

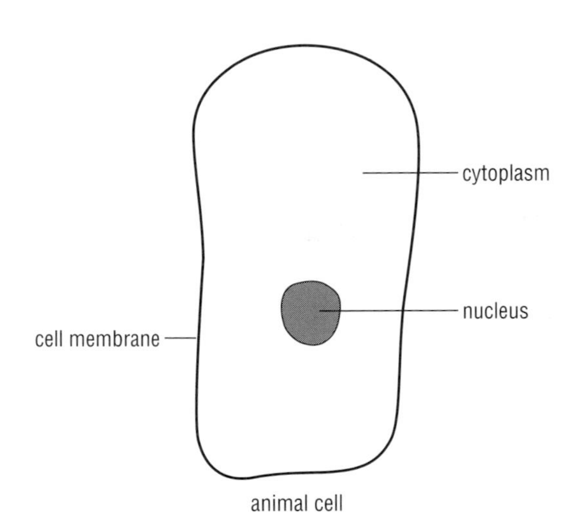

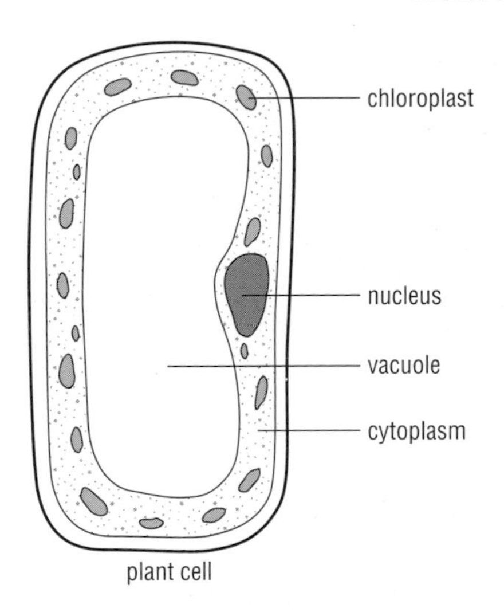

18 The table shows Peter's pulse rate while he was resting.

a) Calculate Peter's average (mean) pulse rate while resting. *1 mark*
b) What would you expect to happen to Peter's pulse rate
(i) if he cycled for 10 minutes
(ii) in the 10 minutes after he stopped cycling? *2 marks*

NICCEA

Time/min	Pulse rate/beats per min
0	73
1	72
2	74

19 The drawing shows a root hair cell from near the tip of a young root.

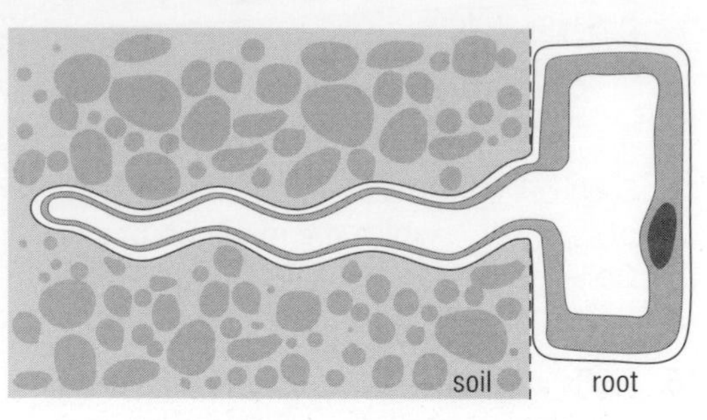

a) This cell needs oxygen. Name the process by which oxygen enters the cell from the air in the soil. *1 mark*

b) Describe the process by which water enters the root hair cells. *3 marks*

c) A seaside garden is flooded by the sea in an exceptionally severe storm. Several of the plants wilt and die. Explain why flooding with sea water caused the plants to wilt. *2 marks* ***AQA***

20 The diagrams show special cells in the skin of a jellyfish.

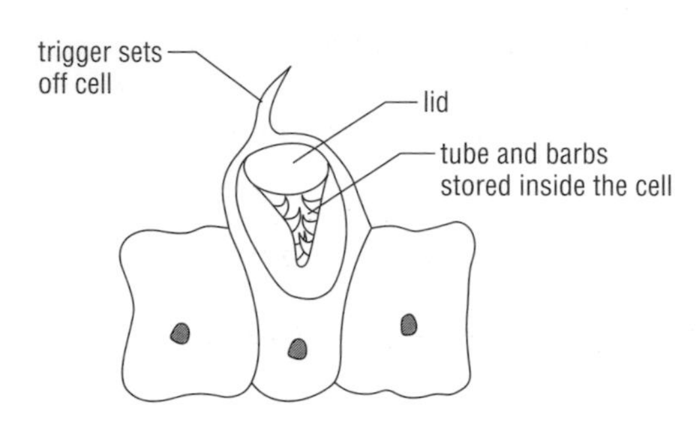

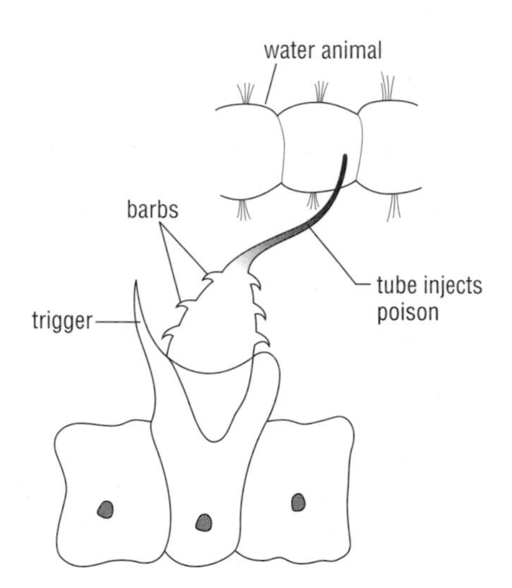

Describe how the special cells help the jellyfish to capture animals for food.
3 marks ***AQA***

21 The drawings show a rubber plant and a pencil rubber.

The rubber plant is living, the pencil rubber is non-living. Give *three* things which the rubber plant can do but the pencil rubber cannot do. *3 marks* ***AQA***

22 a) Chromosomes are present in the nuclei of all cells. Give *two* of their main functions. *2 marks*

b) Organelles are found in the cytoplasm of all cells. They include mitochondria. What important function do mitochondria have? *1 mark*

23 Animals have various cells which are specially adapted to their function.

a) Draw a muscle cell and explain how it is adapted to its function. *4 marks*

b) Draw a nerve cell and explain how it is adapted to its function. *4 marks*

CHAPTER

2

Nutrition and digestion

SUMMARY

1 The process of taking in food for survival is called **nutrition**.

2 A balanced diet contains seven important constituents.
- **carbohydrates** for *energy* in bread, potatoes and rice.
- **fats** for *energy* and for *making cell membranes* in milk, cheese, butter and red meats.
- **proteins** to provide the *chemicals for growth* and the repair of tissues in fish, meat, eggs, peas and beans.
- **vitamins** in fruit and vegetables in very small amounts for vital processes.
- **minerals** such as Fe^{2+} ions for blood, Ca^{2+} ions for bones and Na^+ and K^+ ions in all cells.
- **water** to *maintain cell processes* and keep the concentration of body fluids at a steady level.
- **fibre** as *roughage* to help the movement of food through the gut.

3 **Tests for foods**
- Fats – fatty foods leave a translucent mark on paper.
- Starch – gives a blue/black colour with iodine solution.
- Sugars – simple sugars give an orange/red precipitate on boiling with dilute HCl and Benedict's solution.
- Proteins – give a purple or violet colour with dilute copper(II) sulphate solution + sodium hydroxide solution.

4 **Digestion** is the process in which large insoluble food molecules are broken down into smaller, soluble molecules which the body can absorb and use.

5 Digestion in mammals takes place in the gut or **alimentary canal**. Digestion involves both physical and chemical processes.

6 The physical processes involved in digestion are:
- the chewing and cutting action of teeth
- the churning and mixing by muscles in the wall of the stomach
- the breaking up of fats into an emulsion of tiny droplets by the action of bile.

7 The chemical processes in digestion involve **enzymes** which catalyse the breakdown of individual food molecules into smaller molecules so that they can pass through the walls of the small intestine into the bloodstream.

- Carbohydrase enzymes catalyse the breakdown of carbohydrates.
  - amylase in *saliva* breaks down starch to maltose
  - maltase in the *small intestine* breaks down maltose to glucose.
- Protease enzymes catalyse the breakdown of proteins into peptides and then amino acids.
  - pepsin in *gastric* (stomach) *juices* breaks down proteins to peptides
  - peptidases in the *small intestine* break down peptides to single amino acids.
- Lipase enzymes catalyse the breakdown of fats (lipids) into fatty acids and glycerol in the *small intestine.*

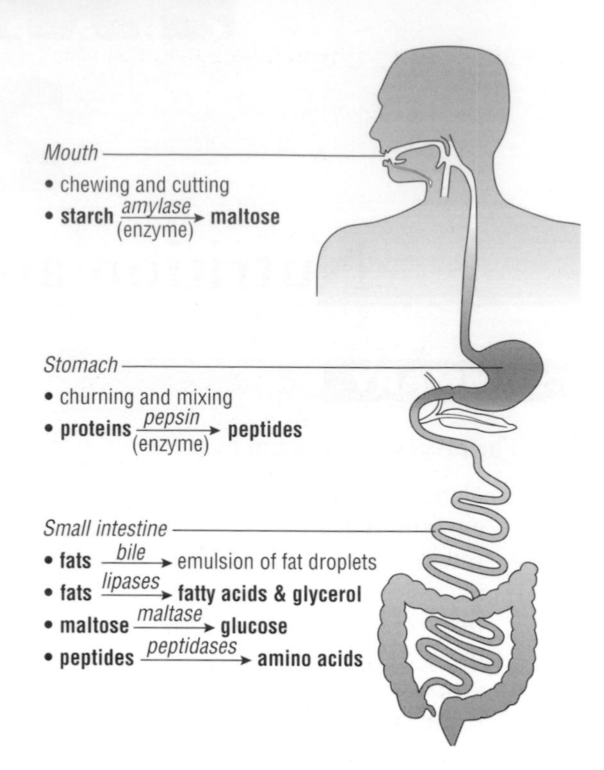

The breakdown of foods during digestion

8 Most of the chemical processes in digestion occur in the small intestine. The small intestine is a very long, thin tube wound round and round in the abdominal cavity. Finger-like villi on its walls have blood capillaries close to their surface. This allows small molecules like glucose and amino acids to seep out of the small intestine and into the bloodstream to supply the rest of the body.

9 Digestion is complete once food reaches the large intestine. Here, water is absorbed by blood capillaries, while semi-solid, indigestible material continues to collect in the rectum. At regular intervals, this is excreted through the anus.

STUDY QUESTIONS

Objective questions

Questions 1 to 4

Choose from the foods labelled A to E below in answering questions 1 to 4.

A cabbage B eggs C dried fish D butter E beans

1 Which food contains the most protein?

2 Which food contains the most carbohydrate?

3 Which food contains the most water?

4 Which food contains the most fat?

Questions 5 to 8

Choose from the parts of the gut labelled A to E below in answering questions 5 to 8.

A stomach B pancreas C gall bladder D large intestine E small intestine

5 Which part of the human gut absorbs water from indigestible material?

6 Which part of the human gut produces enzymes to break down peptides to amino acids?

7 Which part of the human gut contains the enzyme pepsin?

8 Which part of the human gut stores bile?

Questions 9 to 16
Look carefully at the drawing of the human alimentary canal.

9 In which parts of the alimentary canal is starch broken down?

10 In which parts of the alimentary canal are fats broken down?

11 Where is the substance which neutralises acids from the stomach produced?

12 What is the name of the substance which neutralises acids from the stomach?

13 In which parts of the alimentary canal are enzymes produced?

14 Chewing divides food into smaller pieces. How does this help the action of enzymes?

mouth
salivary gland
oesophagus
liver
stomach
bile duct
pancreas
small intestine (duodenum → ileum)
large intestine (colon → rectum)
anus

15 Where does peristalsis occur in the alimentary canal?

16 What is the main constituent of faeces?

Short questions

17 The table below lists information about some important vitamins.

Vitamin	What it does	Disease if vitamin is deficient	Sources of vitamin
A	aids eyesight	poor eyesight in dim light	oily fish, carrots, green vegetables
B	helps to produce energy	beri-beri (stomach complaints, muscle paralysis)	yeast, cereals, rice (mainly in husks)
C	controls oxidation and other processes	scurvy (general ill health, bleeding of gums)	oranges, lemons, green vegetables
D	promotes formation of strong bones	rickets (children grow with deformed bones)	oily fish, eggs, milk – also made by skin when exposed to sunlight

Look carefully at the table.

a) Why are green vegetables an important part of our diet? *1 mark*
b) The disease beri-beri is common in parts of the world where polished rice, rather than whole-grain rice, is the main food. What do you think polished rice is? *1 mark*
c) Why does whole-grain rice prevent beri-beri? *1 mark*
d) Why did Admiral Nelson always insist that his sailors ate citrus fruits when on a long voyage? *1 mark*
e) What is the name for the major oxidation process controlled by vitamin C? *2 marks*
f) It was noted at the beginning of the 20th century that children living in cities where there was heavy smoke pollution were more likely to develop rickets. Why do you think this was so? *2 marks*

18 Peter and Phillipa usually spend their Saturday mornings shopping. After shopping, they have a snack. The table below gives information about the food which Peter usually eats.

Food	Fat measured in kilojoules (kJ) of energy supplied	Nutrients			
		Protein	Fibre	Carbohydrate	Vitamins
hamburger	2500	high	low	high	A, B, D
chips	1200	low	low	high	low
fizzy drink	none	none	none	some	none
chocolate bar	1200	some	none	high	A and D

a) Draw a bar chart using graph paper to show each food and its fat energy. *4 marks*
b) Calculate the total energy Peter gets from the fat in his meal. *1 mark*
c) Peter's body uses the fat from his snack. Write down two uses of this fat. *2 marks*
d) Name one nutrient not present in Peter's snack. *1 mark*

19 Amylase is an enzyme which breaks down starch. The effect of temperature on this reaction was investigated. Two test tubes were set up as shown in the diagram.

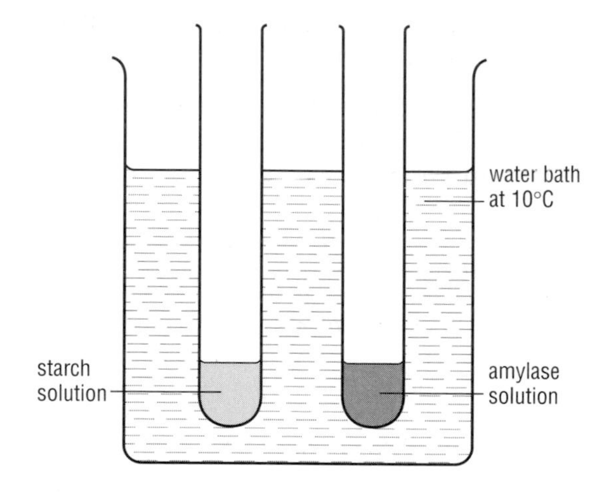

After 5 minutes, the amylase solution was mixed with the starch solution. The mixture was left in the water bath at 10°C. Every minute, a sample of the mixture was tested to find the percentage of starch left. The experiment was repeated with the water bath at 20°C.

The results of the experiment with starch and amylase at 10°C and 20°C are shown in the table.

Time/minute	0	1	2	3	4	5	6	7	8	9	10
% of starch remaining at 10°C	100	98	94	92	89	87	84	81	78	75	72
% of starch remaining at 20°C	100	97	92	85	80	70	62	58	52	48	40

a) Plot the results at 10°C on a graph showing percentage of starch vertically and time in minutes horizontally. *3 marks*
b) Draw the best line through these points. *1 mark*
c) Plot the results at 20°C on the same graph and draw the best line through these points. *2 marks*
d) Use the graphs to describe the effect of temperature on the rate of amylase activity. *2 marks*

Further examination questions

20 In 1882, a young man called Alexis was shot in the stomach by accident. The hole in his stomach would not heal. His doctor carried out a number of experiments on Alexis to get information on the digestion of food.

a) What is meant by digestion? *2 marks*

b) In one experiment, pieces of meat were tied to a silk thread and pushed into Alexis' stomach. Meat is mainly protein.
Explain what happened to the meat.
2 marks

c) Food, especially spicy sauces containing vinegar, can be very acidic. These acids can sometimes affect the enzymes in the digestive system. Explain the role of enzymes in digesting food. *2 marks*

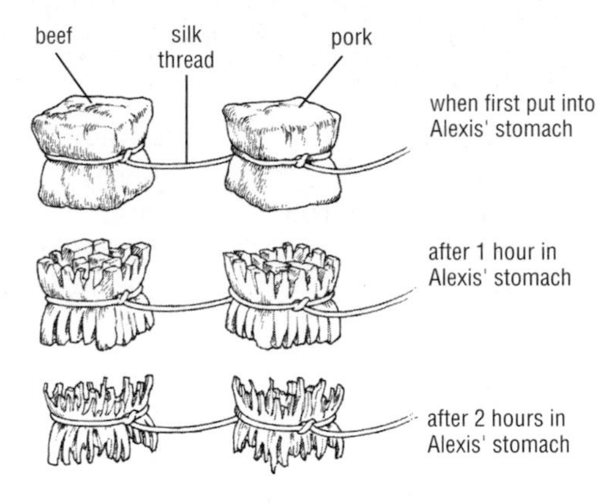

OCR

21 This question is about digestion. Enzymes in the mouth, stomach and small intestine help to digest food.

a) What do enzymes do to food during digestion? *1 mark*

b) What must then happen to the digested food before it can be used by cells in different parts of the body? Explain as fully as you can. *3 marks*

c) Glucose (sugar) is one of the foods used by body cells. What is the glucose used for? Explain as fully as you can. *2 marks*

OCR

22 Read the following passage.

> There is no such thing as bad food.
> There is, however, such a thing as a bad diet.
>
> Experts recommend a balanced diet, containing a mixture of carbohydrates, fats and proteins. The amount of animal fat should be kept low, to reduce the chance of heart disease. The diet should include plenty of starchy foods, such as bread and pasta, and at least one piece of fresh fruit per day. Only small quantities of alcohol should be drunk.

a) As well as carbohydrates, fats and proteins, name *two* other types of substance that a balanced diet should contain. *2 marks*

b) Suggest why 'one piece of fresh fruit' should be eaten each day. *1 mark*

c) Explain why 'only small quantities of alcohol' should be drunk. *2 marks*

d) Keeping the amount of animal fat in the diet low helps people avoid heart problems. Explain, as fully as you can, why this is so. *3 marks*

e) Fat in the diet may eventually be stored under the skin. A person eats a meal containing fat. Describe what happens to the fat from the time it is eaten to when it is stored under the skin. *4 marks*

OCR

23 Lots of people use fast food restaurants.

a) (i) When a beefburger is eaten, what *two* processes occur in the mouth? *2 marks*
(ii) When the burger is swallowed, it passes down the oesophagus (gullet). What is the function of the oesophagus? *1 mark*

b) Match up each food in the left hand list below with one related word on the right hand list.

sugar	protein
butter	energy
white bread roll	fibre
onion	insulation
beefburger	carbohydrate *3 marks*

c) The Government recommends that we eat more bread but less butter and cheese.
(i) Explain why eating too much cheese and butter can lead to heart disease. *3 marks*
(ii) Name *three* types of constituent, other than protein, fat and carbohydrate, which are present in a healthy diet. *3 marks*

24 The diagram shows an investigation into the action of a carbohydrase called amylase. The apparatus was left for 20 minutes.

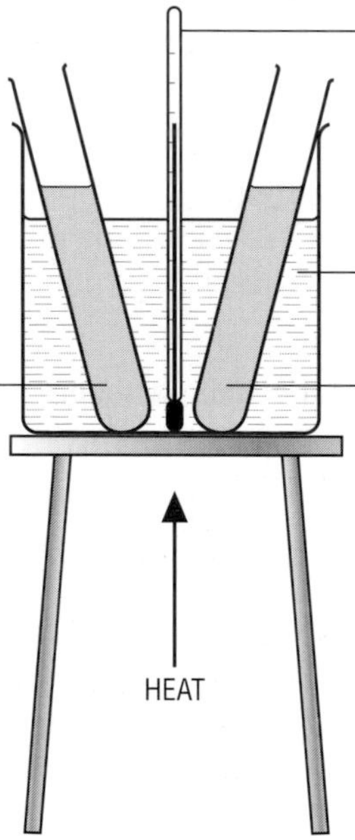

Iodine changes from yellow-brown to black if starch is present. When boiled with a reducing sugar, Benedict's solution changes from clear blue to brick red.

a) After 20 minutes what colour would you expect to see if the contents of tube A were added to
(i) iodine *1 mark*, (ii) Benedict's reagent and boiled? *1 mark*

b) What colour would you expect to see if the contents of tube B were added to (i) iodine, (ii) Benedict's reagent and boiled? *1 mark*

c) What does this investigation tell you about the effect of temperature on enzyme action? *1 mark*

d) Give a definition of an enzyme. *2 marks*

e) What is meant by the optimum pH or temperature at which an enzyme works? *1 mark*

WJEC

3

Respiration and breathing

SUMMARY

1 Respiration involves:
- **Breathing** – chest movements to ensure you inhale and exhale
- **Gas exchange** – in the capillaries surrounding alveoli in the lungs
- **Cellular respiration** – chemical reactions in body cells which produce **ENERGY**

2 Breathing involves:
- **Inhalation** – ribs move up and out, diaphragm flattens → chest cavity (thorax) INCREASES in volume.
- **Exhalation** – ribs lower and move in, diaphragm resumes dome shape → chest cavity (thorax) DECREASES in volume.

3 **Gas exchange** involves:
- diffusion of oxygen from the inside of tiny alveoli (air sacs), through their thin walls (only one cell thick), into the blood capillaries covering the alveoli.
- diffusion of carbon dioxide in the opposite direction from the blood capillaries covering the alveoli through their thin walls and into the lungs.

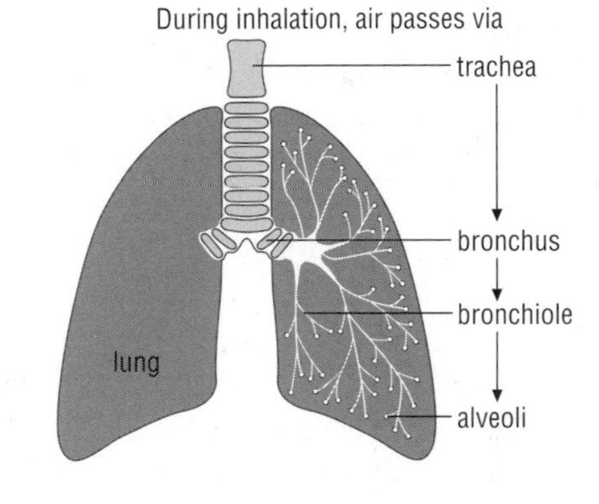

4 **Cellular respiration** involves the reaction of small molecules, such as glucose, with oxygen to produce carbon dioxide, water and energy.

glucose + oxygen → carbon dioxide + water + energy

$$C_6H_{12}O_6 + 6O_2 \rightarrow 6CO_2 + 6H_2O + 2900\ kJ$$

In this process, the small food molecules are **oxidised** and the process is often called **aerobic respiration**.

5 **Smoking**

The major toxic substances in tobacco are:
- carbon monoxide which forms a strong bond with haemoglobin, reducing the capacity of the blood to carry oxygen
- nicotine which is a carcinogen, inducing cancerous growths in the respiratory tract
- tar which paralyses the ciliated cells in the nasal cavity, so that harmful bacteria and viruses are not removed.

Smoking can cause:
- cancers of the throat, bronchus and lungs
- chronic bronchitis and emphysema
- heart disease and arteriosclerosis.

6 Aerobic respiration occurs in the presence of air (oxygen). Respiration can also occur in the absence of oxygen and this is called **anaerobic respiration**. Anaerobic respiration occurs in:

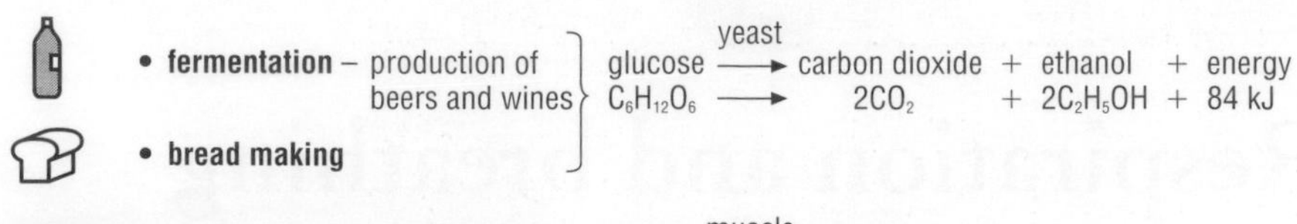

STUDY QUESTIONS

Objective questions

Questions 1 to 3
Choose words from the list below to complete the following sentences.

A alveoli B bronchus C bronchiole D capillaries E trachea

1 Air passes down the windpipe into the lungs. Another name for the windpipe is ....

2 Inside the lungs the windpipe splits into two branches and these branches divide repeatedly into smaller branches. Each smaller branch is a ....

3 At the ends of the smaller branches are large numbers of ....

Questions 4 to 8
In questions 4 to 8, choose from A, B, C or D which is the correct answer.

4 When we breathe in, the rib cage is pulled upwards by
A the diaphragm
B the intercostal muscles
C the lungs
D the ribs

5 Which of the following pairs both contract when we inhale?
A intercostal muscles and trachea
B intercostal muscles and diaphragm
C ribs and diaphragm
D diaphragm and trachea

6 Long distance runners sometimes take in glucose drinks during a race. The glucose is needed to provide
A energy
B food
C oxygen
D water

7 After a long distance race, a runner's muscles hurt and ache because
A his/her body temperature is too high
B he/she is dehydrated due to sweating
C his/her muscles have insufficient oxygen
D he/she is suffering from cramp

8 Yeast causes bread dough to rise by
A increasing the temperature
B allowing expansion to occur
C absorbing oxygen from the air
D producing a warm gas

Short questions

9 Smoking affects the risk of dying from lung cancer. The graph shows how.

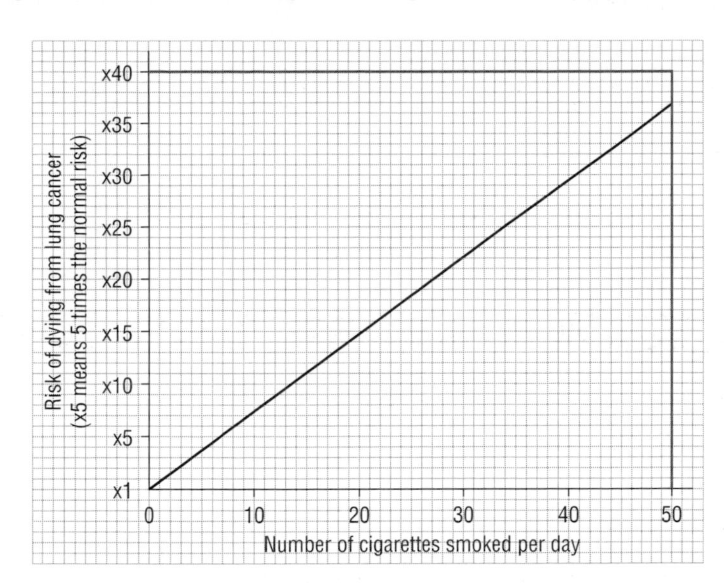

a) What is the risk of dying from lung cancer if you smoke 20 cigarettes a day? *1 mark*

b) What does the graph tell you about the risk of dying from lung cancer? *3 marks*

c) A smoker who gives up smoking halves his/her risk of death from lung cancer. Draw a sketch graph, similar to the graph above, to represent this situation. *2 marks*

10 Oxygenated blood serves all the cells in the body where oxygen combines with glucose to provide energy.

a) Explain what is meant by mechanical energy. *1 mark*

b) In cold weather, our body temperature must be maintained. What form of energy is needed for this? *1 mark*

c) Why do our body cells sometimes respire anaerobically? *2 marks*

11 The equation below shows the reactant and product of anaerobic respiration in our cells.

$$C_6H_{12}O_6 \rightarrow C_3H_6O_3 \quad \Delta H = -120 \text{ kJ}$$

a) Balance the equation and show the names of the reactant and product. *3 marks*

b) What does $\Delta H = -120$ kJ tell you about the reaction? *3 marks*

12 a) What is meant by the term 'anaerobic'? *1 mark*

b) Which form of respiration is the more efficient way of getting energy – aerobic or anaerobic? *1 mark*

c) Why do organisms find it useful to respire anaerobically? *2 marks*

13 a) What happens in bread dough during fermentation? *4 marks*

b) Why does fermentation stop when the bread dough is baked? *1 mark*

c) Why does bread not contain any alcohol even though alcohol forms as the dough rises before it is put into the oven and baked? *1 mark*

Further examination questions

14 Look at the diagram. It shows an alveolus (air sac) in the lungs.
Gas exchange happens in the alveoli (air sacs).

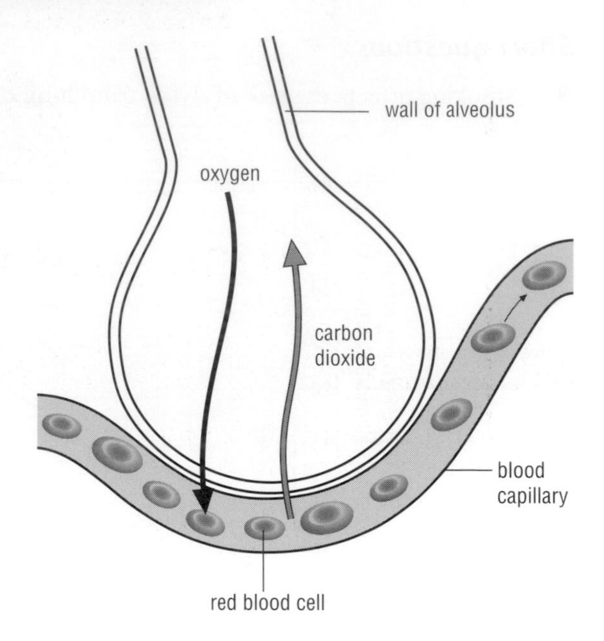

a) Alveoli are adapted for efficient gas exchange. Write down *three* ways in which they are adapted for this.
3 marks

Carbon dioxide passes from the blood into the alveoli.

b) (i) Where in the body is this carbon dioxide made? *1 mark*
(ii) What process produces this carbon dioxide? *1 mark*

c) How is the level of carbon dioxide in the blood controlled? Answer the question by completing these sentences.

High levels of carbon dioxide are detected by .... Signals are sent to the diaphragm and rib muscles to increase the rate of .... This causes the level of carbon dioxide in the blood to .... The system of control is called .... *4 marks*

OCR

15 Read the following passage.

One of the deadliest diseases seems to be making a comeback in Britain. Doctors are alarmed at the rising number of cases of tuberculosis (TB).

TB is caused by microbes called bacteria. When people carrying the TB bacteria cough or sneeze, the TB bacteria get into the air. Other people may then breathe them in.

a) Which organs will be infected first when someone breathes in the TB bacteria? *1 mark*

b) Explain how the TB bacteria may cause disease inside the body. *2 marks*

c) Name *one* other group of microbes that often cause disease. *1 mark*

d) Suggest why people who live in overcrowded areas are much more likely to catch TB than people who live in less crowded areas. *1 mark*

e) People infected with a small number of TB bacteria often do not develop the disease. Explain, as fully as you can, how the body defends itself against TB bacteria. *3 marks*

f) Smoking can cause diseases such as lung cancer. Explain, as fully as you can, why the diseases caused by smoking cigarettes are not infectious.
2 marks

AQA

4

Blood and circulation

SUMMARY

1 The **circulatory system** consists of the heart and blood vessels. It has two main functions:

- transporting oxygen and digested food to all parts of the body
- removing waste products from the body.

2 The diagram illustrates how oxygen and digested food are obtained by the blood and how waste products are removed from it.

1. **Deoxygenated blood** enters the heart via the vena cava.
2. **Deoxygenated blood** is pumped via the pulmonary artery to the lungs.
3. In the **lungs**, blood picks up oxygen and gets rid of carbon dioxide.
4. **Oxygenated blood** travels back to the heart in the pulmonary vein.
5. **Oxygenated blood** is pumped to vital organs and the whole body via the aorta.
6. Digested food is absorbed by blood through walls of the **small intestine**.
7. Digested food is processed and stored in the **liver**.
8. Waste products (urea, water, salts etc.) are removed from the blood in the **kidneys**.
9. Blood supplies the **rest of the body** with food and oxygen. It carries away waste products.

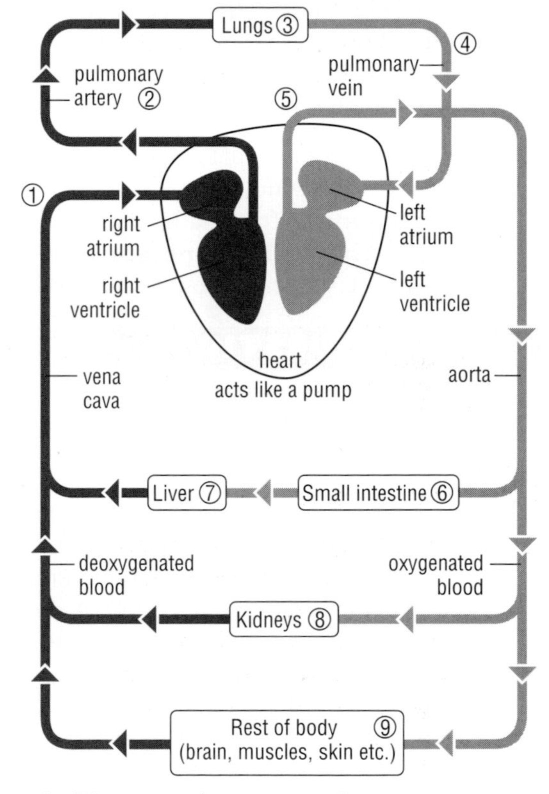

3 Blood circulates around the body in **blood vessels**. There are three types of blood vessel with structures related to their own particular function.

- **Arteries**
  - carry blood **away** from the heart at **high pressure**
  - have thick, muscular walls to withstand the pumping pressure
- **Capillaries**
  - divide from the arteries and then rejoin to form veins
  - are very narrow tubes with thin walls through which digested food, oxygen and waste products can diffuse
- **Veins**
  - carry blood **back** to the heart at **low pressure**
  - have thinner, less muscular walls than arteries
  - contain valves to prevent the backward flow of blood

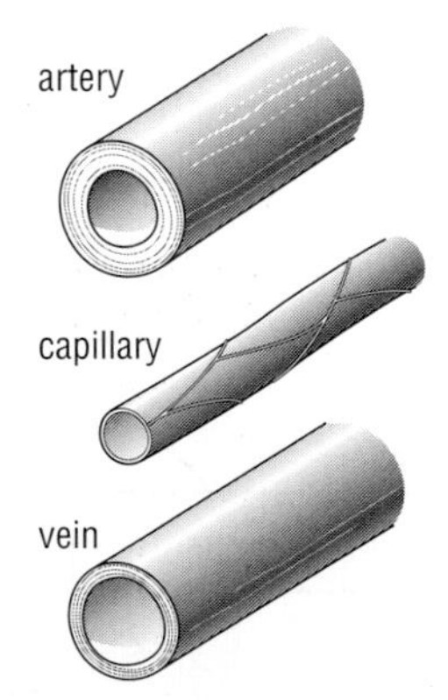

The constituents of the blood are plasma, red blood cells, white blood cells and platelets.

- **Plasma**
  - a straw coloured, watery liquid containing nutrients and waste products
- **Red blood cells**
  - contain haemoglobin which combines with oxygen. They transport oxygen to every part of the body
- **White blood cells**
  - **phagocytes** engulf harmful bacteria and secrete an enzyme to kill them
  - **lymphocytes** produce antibodies to kill bacteria and viruses and neutralise poisonous chemicals
- **Platelets**
  - small fragments of cells which help the blood to clot when the skin is cut

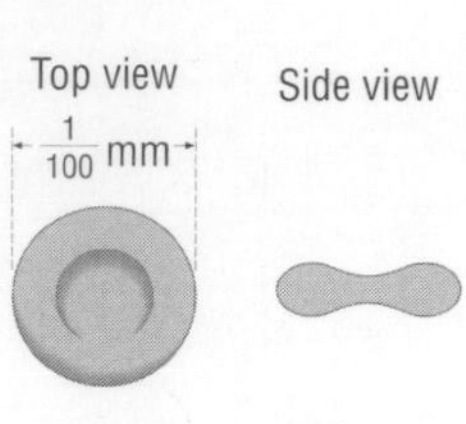

Red blood cells

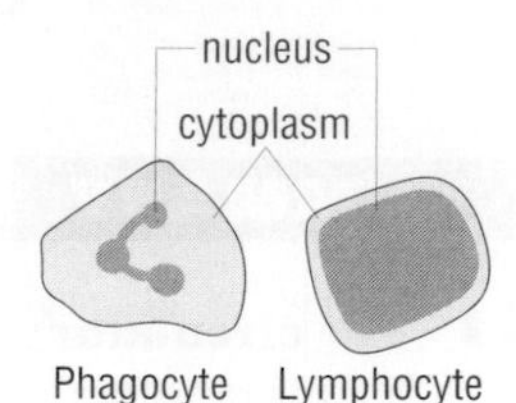

White blood cells

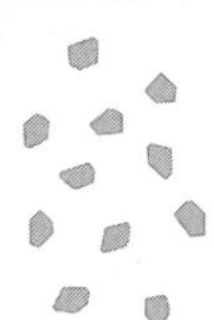

Platelets

5 The blood has three main functions:

- **transporting** nutrients, oxygen and waste products
- **protecting** us from pathogens (bacteria, viruses and chemicals)
- **regulating body temperature** via the blood vessels (mainly capillaries) close to the surface of our skin.

STUDY QUESTIONS

Objective questions

Questions 1 to 5

Fill in the blanks in the following passage.

> The blood flows to the organs through . . . 1 . . . and returns to the heart through . . . 2 . . . . The heart is made mainly of . . . 3 . . . . Blood enters the . . . 4 . . . of the heart and then passes to the . . . 5 . . . before circulation to other organs.

Questions 6 to 12

Choose from the letters A, B, C or D which is the correct answer.

6 The pulse that you can feel in your neck is due to:

A blood pumping along an artery
B blood pumping along a vein
C valves opening and closing
D increased blood flow from exercise.

Questions 7 and 8

A simplified diagram of the human blood circulation is shown below.

7 The order in which oxygen is pumped through the blood vessels is:

A 1, 2, 3, 4
B 1, 2, 4, 3
C 1, 3, 2, 4
D 1, 3, 4, 2

8 Which of the blood vessels carry deoxygenated blood?

A 1 and 2
B 1 and 4
C 2 and 3
D 3 and 4

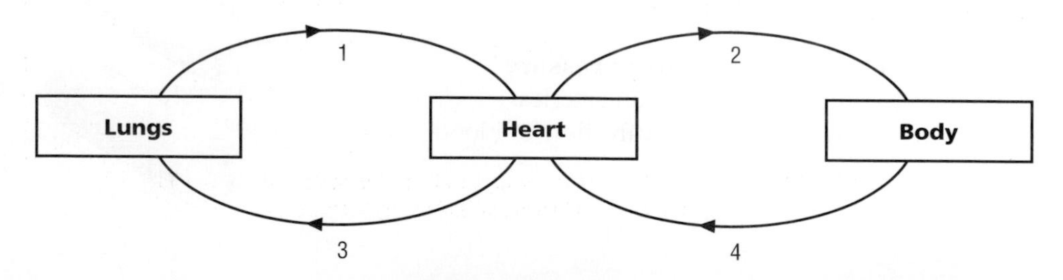

9 Urea is transported to the kidneys by:
A plasma
B platelets
C red blood cells
D white blood cells

10 Which of the following parts of the blood help us to fight infection?
A red and white blood cells
B phagocytes and lymphocytes
C platelets and plasma
D red blood cells and plasma

11 Which chamber of the heart has the thickest walls?
A the left atrium
B the left ventricle
C the right atrium
D the right ventricle

12 A raised body temperature is one symptom of infection. Which of the following helps to prevent the temperature rising to a dangerous level?
A contraction of hair erector muscles
B decreased sweat production
C expansion of skin capillaries
D increased metabolic rate

Short questions

13 Look at these diagrams. They show a blood cell attacking bacteria.

a) Which type of blood cell is shown in the diagrams? *1 mark*

b) Look at the diagrams. Describe how this type of blood cell attacks bacteria and gets rid of them. *2 marks*

c) Look at the list of substances found in the blood.

antibodies fibrin glucose plasma

One of the substances is made by blood cells attacking bacteria. Which one? *1 mark*

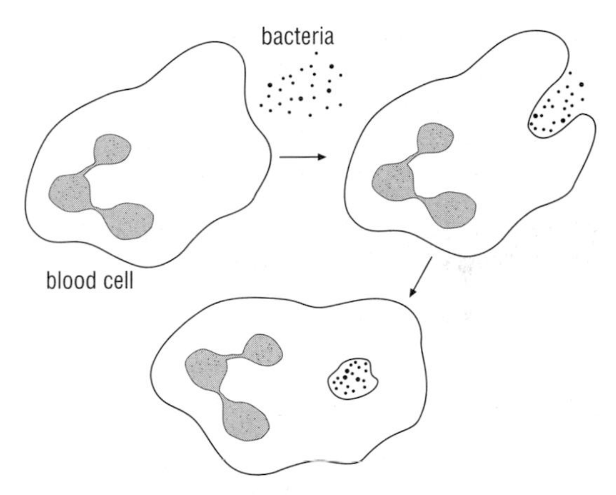

OCR

14 The diagram shows how the circulation of blood changes between rest and exercise.

a) Use the information from the diagram to redraw and then complete the table below.

Parts of the body to be included:

digestive system skin brain bone
arteries of heart muscles of skeleton

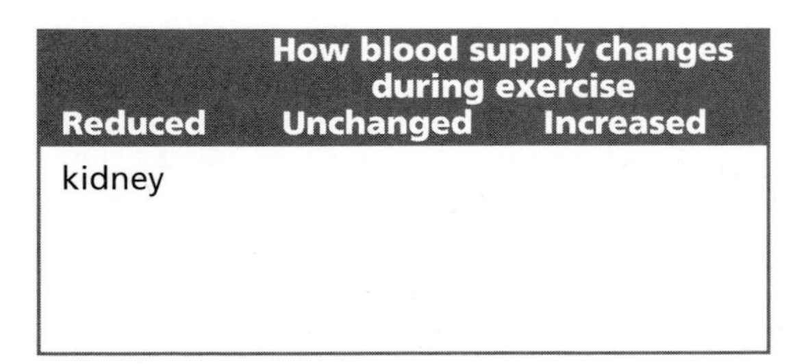

How blood supply changes during exercise		
Reduced	**Unchanged**	**Increased**
kidney		

4 marks

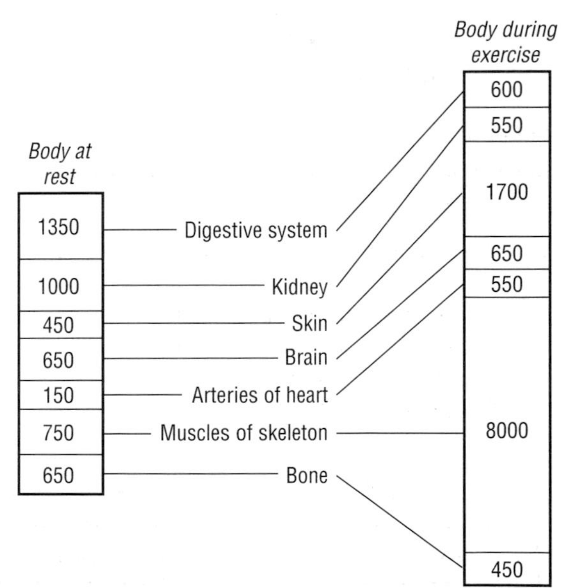

Rate of supply of blood to parts of the body (cm³/min) when at rest and during exercise

b) What happens to the rate of supply of blood to the whole body with exercise? (You should make full use of the information provided.) *2 marks*

AQA

15 The diagram shows a calf foetus and its placenta.

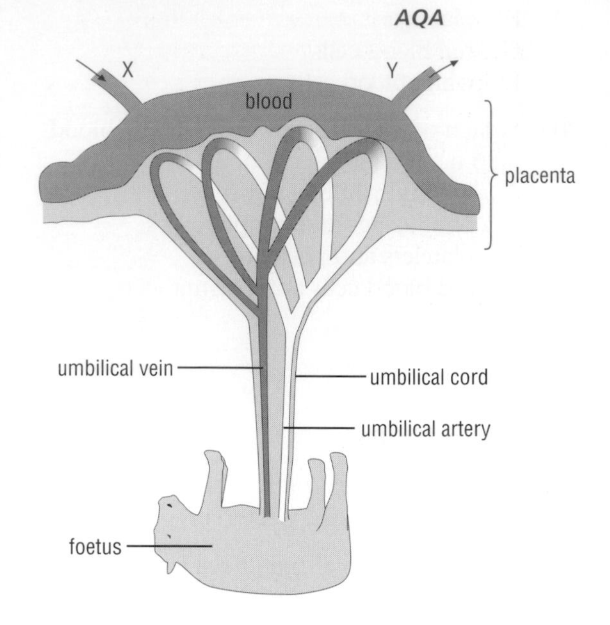

a) Which type of blood vessel serves the placenta at X? *1 mark*
b) Which type of blood vessel leaves the placenta at Y? *1 mark*
c) What does the blood in the umbilical artery contain in order for the foetus to grow? *2 marks*
d) The umbilical vein carries blood containing the waste products urea and carbon dioxide. Why can the foetus not excrete carbon dioxide through its lungs? *1 mark*
e) What do you think may happen if the supply of oxygenated blood to the foetus is reduced or cut off? *2 marks*

16 The bar chart shows the number of successful heart transplants in the United Kingdom since 1983.

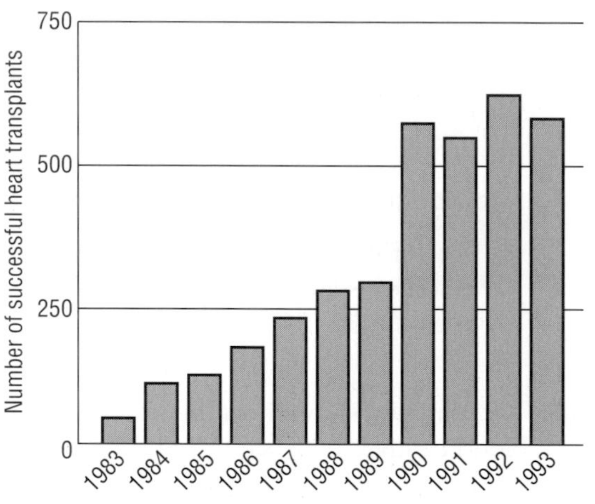

a) How many successful heart transplants were there in the United Kingdom in 1992? *1 mark*
b) What pattern do you see in the number of transplants between 1983 and 1993? *2 marks*
c) The patient's body sometimes rejects the transplanted organ. A new drug was given to reduce this problem. Suggest in which year this new drug was first used. *1 mark*

Further examination questions

17 The diagram shows the stages in a process that occurs in the body.

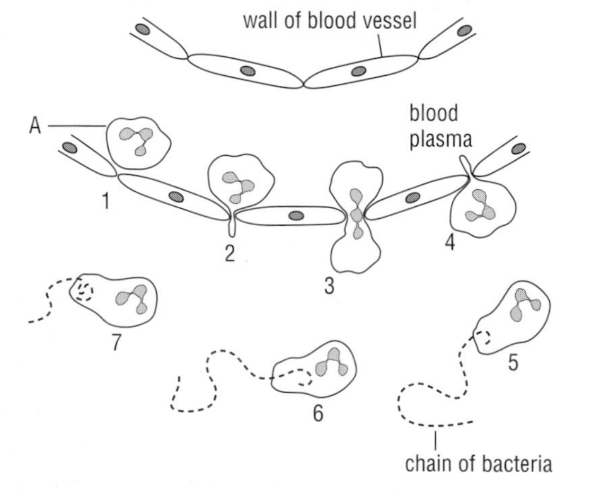

a) Name the type of cell labelled A. *1 mark*
b) Briefly describe what the diagram shows. *2 marks*
c) Copy this table and fill in the blank spaces. *2 marks*

Part of blood	Function (job)
red blood cells	
	transport of urea and soluble food

WJEC

18 Doctors sometimes need to know how much blood a patient has. They can find out by using a radioactive solution. After measuring how radioactive a small syringe-full of the solution is, they inject it into the patient's blood. They then wait for 30 minutes so that the solution has time to become completely mixed into the blood. Finally, they take a syringe-full of blood and measure how radioactive it is.

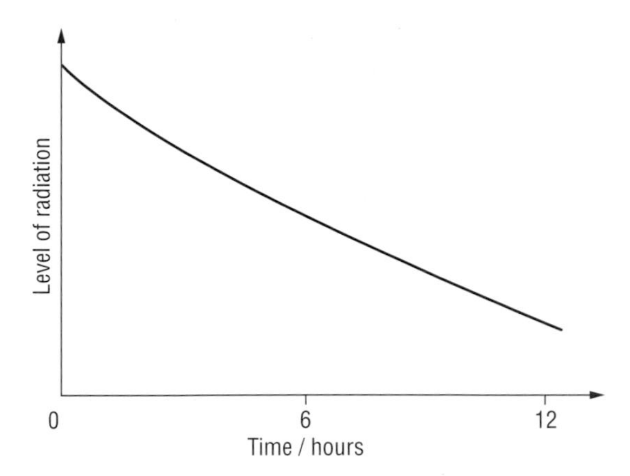

As time passes, the solution becomes less radioactive *on its own*

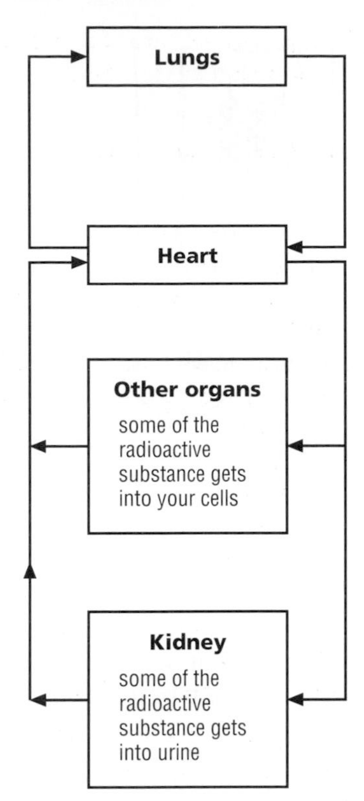

Radiation from radioactive substances can harm your body cells

a) After allowing for background radiation:
- 10 cm^3 of the radioactive solution gives a reading of 7350 counts per minute
- a 10 cm^3 sample of blood gives a reading of 15 counts per minute.

Calculate the volume of the patient's blood. (Show your working.) *4 marks*

b) The doctor's method of measuring blood volume will not be completely accurate.
(i) Write down *three* reasons for this. *3 marks*
(ii) Will the doctor get a figure for the patient's blood volume that is too high or too low? Explain your answer. *2 marks*

c) (i) The doctors use a radioactive substance which loses half of its radioactivity every six hours. Explain why this is a suitable radioactive substance to use. *2 marks*
(ii) Besides half life, suggest *two* other factors which might make the radioactive substance used to measure blood volume less likely to harm the cells in your body. *2 marks*

AQA

19 Polar bears live in areas close to the North Pole where it can be very cold. The diagram over the page shows some of the blood vessels and the flow of blood through a polar bear's foot in cold conditions.

a) Explain how the differences in the diameters of the blood vessels shown in the diagram help the polar bear to survive in cold conditions. *2 marks*

b) Describe *one* way in which the diameters of the blood vessels would change in warmer conditions. *2 marks*

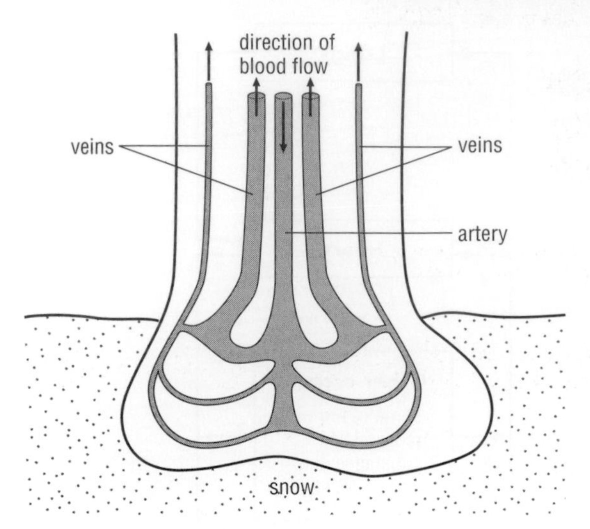

c) It takes several months for young polar bears to grow fur. During this time:
- they keep still
- they move only to feed off the mother's fat rich milk
- they can be found huddling together.

Choose two of these three points and explain how each helps a young polar bear to survive cold conditions. *4 marks*

AQA

20 a) Why can lack of exercise cause the heart pump to be inefficient? *1 mark*
b) The coronary arteries may become diseased and the narrowing of these vessels causes insufficient blood to serve the heart muscle. What medical catastrophe may happen? *1 mark*
c) What steps can we take in our diet to avoid disease of the blood vessels? *1 mark*
d) During infectious diseases such as measles, the pulse rate rises. Why do you think this happens? *2 marks*
e) Give *two* other circumstances which cause the pulse rate to rise? *2 marks*

21 Microbes can invade the body through damaged blood vessels. Blood cells help to defend the body against such microbes.

a) What type of blood cells are able to defend the body against microbes? *1 mark*
b) State *three* ways in which these cells can defend you. *3 marks*
c) Describe *two* ways in which people can be immunised against some diseases. *2 marks*
d) Susan came into contact with a virus on three different occasions. The first time she was infected she showed very strong symptoms of the infection and was very ill. The second time she showed a few mild symptoms but was not so ill, and the third time she showed no symptoms at all. The graph shows Susan's antibody production after the first two infections.

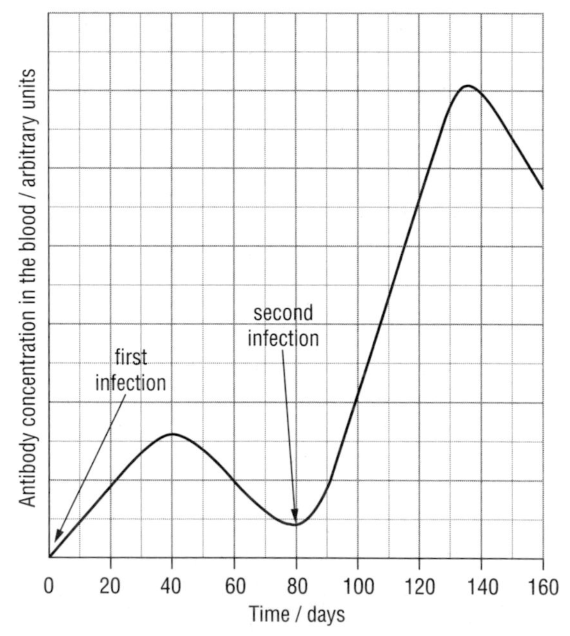

Use all of the information given and your knowledge of antibodies to suggest what happened during
(i) the first infection, (ii) the second infection and (iii) the third infection.
3 marks

WJEC

CHAPTER

5

Photosynthesis

SUMMARY

1 **Photosynthesis** is a chemical process in which all plants make their own food (i.e. glucose) from carbon dioxide in the air and water in the soil.

$$\text{carbon dioxide} + \text{water} \xrightarrow[\textit{chlorophyll}]{\textit{light}} \text{glucose} + \text{oxygen}$$

$$6CO_2 + 6H_2O \qquad C_6H_{12}O_6 + 6O_2$$

2 **Conditions needed for photosynthesis**

Four conditions are essential for photosynthesis: carbon dioxide, water, light and chlorophyll. If one of these is missing, photosynthesis cannot occur.

- Glucose and oxygen are the products of photosynthesis.
- As the glucose is produced, it is either converted into starch and stored in the leaves of the plant or it is transported via the vascular system (see chapter 6) to other parts of the plant.
- As the oxygen is produced, it either passes out of the stomata in the leaves into the atmosphere or is used up during respiration.

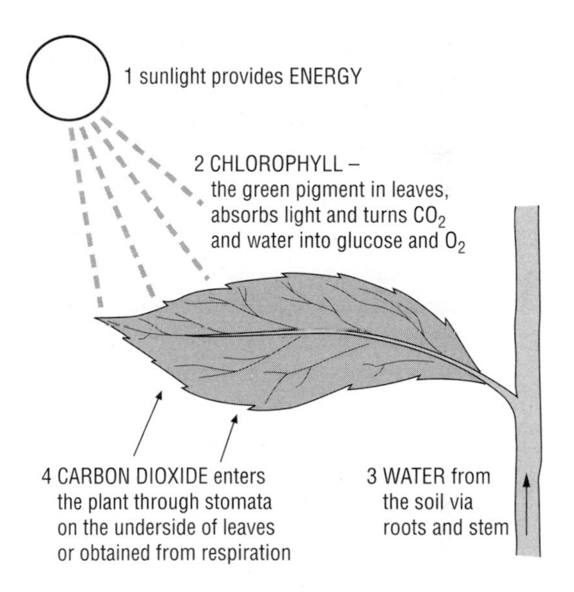
1 sunlight provides ENERGY
2 CHLOROPHYLL – the green pigment in leaves, absorbs light and turns CO_2 and water into glucose and O_2
4 CARBON DIOXIDE enters the plant through stomata on the underside of leaves or obtained from respiration
3 WATER from the soil via roots and stem

3 Testing for starch in plant leaves

(i) Allow a leaf to photosynthesise with black tape covering part of it. Dip the leaf in boiling water for 30 seconds to soften it.

(ii) Put the leaf in hot ethanol for 10 minutes to remove the green chlorophyll.

(iii) Wash the leaf in hot water and add brown iodine solution.

(iv) The parts of the leaf which have been photosynthesising and contain starch will turn dark blue.

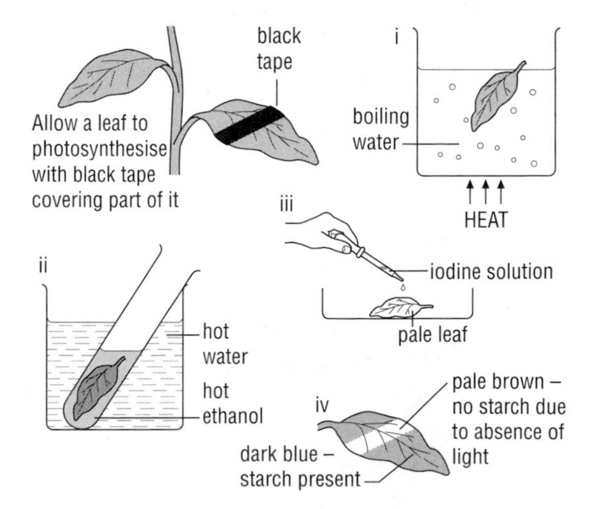
black tape
Allow a leaf to photosynthesise with black tape covering part of it
i
boiling water
HEAT
ii
hot water
hot ethanol
iii
iodine solution
pale leaf
iv
pale brown – no starch due to absence of light
dark blue – starch present

4 Factors affecting photosynthesis

- The **frequency** (colour) of light – chlorophyll absorbs red, orange, blue, indigo and violet light most effectively and therefore photosynthesis occurs faster in the light with these colours.
- The **intensity** (brightness) of light – the brighter the light, the faster the rate of photosynthesis.
- The **supply of water** – in a plentiful supply of water, the rate of photosynthesis increases. In drought conditions, the rate of photosynthesis is very slow.
- The **concentration of carbon dioxide** – the rate of photosynthesis increases if the concentration of carbon dioxide increases. This explains why commercial growers boost carbon dioxide levels in greenhouses.
- The **temperature** – as the temperature rises from 0°C, the rate of photosynthesis increases. Above 35°C, photosynthesis decreases as enzymes necessary for the process are denatured.

5 Using the products of photosynthesis

The diagram shows the four main uses of glucose and oxygen from photosynthesis.

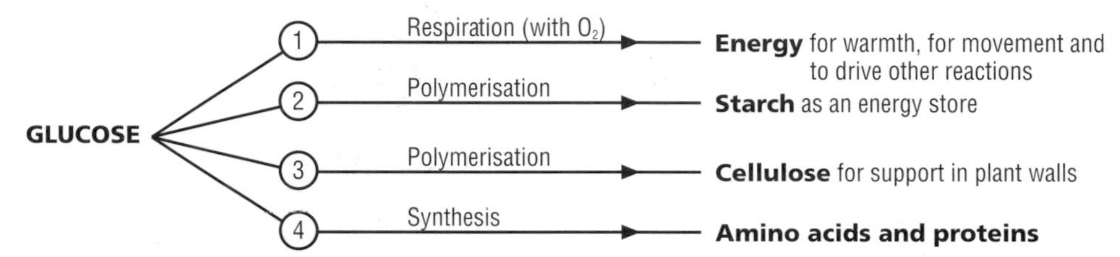

STUDY QUESTIONS

Objective questions

Questions 1 to 4

The diagram shows a vertical section through a thin green leaf. Use the letters A, B, C, D and E to identify

1 a cell in the spongy layer of the leaf.

2 a palisade cell.

3 a guard cell at the side of a stoma.

4 a cell next to the cuticle.

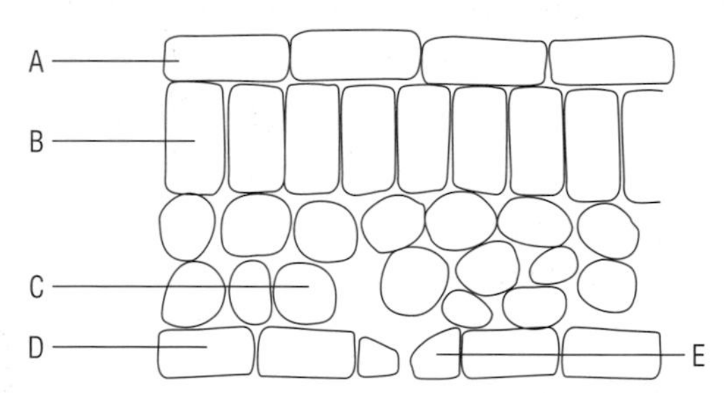

Questions 5 to 10

The diagram shows the movement of substances into and out of a green leaf during photosynthesis.

5 What form of energy enters the plant?

6 What gases enter from the atmosphere?

7 What gas is lost to the atmosphere?

8 What substance is stored in the leaf?

9 What substance enters the leaf from the roots?

10 What substance(s) move(s) down the stem to the rest of the plant?

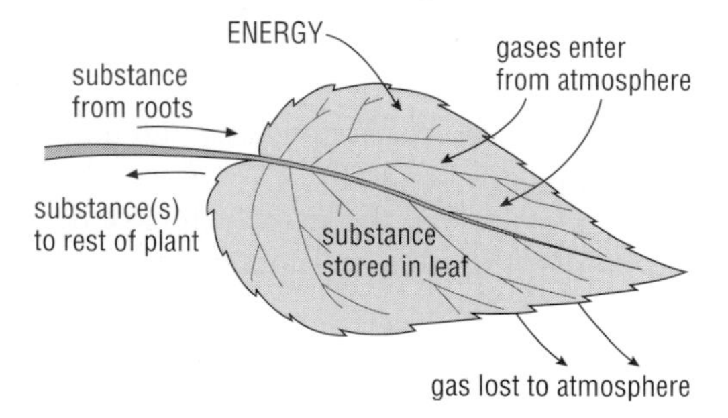

Short questions

11 The graph shows the average light intensity in a woodland area throughout the year.

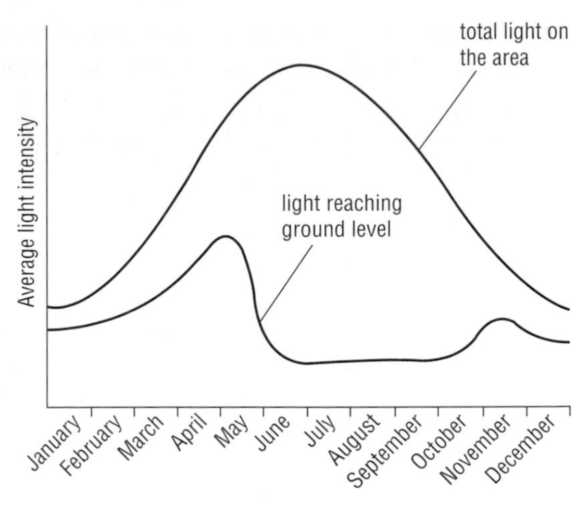

a) (i) In which month will the rate of photosynthesis in the trees in the woodland be the greatest? *1 mark*
(ii) Explain your answer. *1 mark*

b) (i) In which month would you expect the rate of growth of plants living on the ground under the trees to be the greatest? *1 mark*
(ii) Explain your answer. *1 mark*

12 The graph shows how the rate of photosynthesis is affected by the concentration of carbon dioxide.

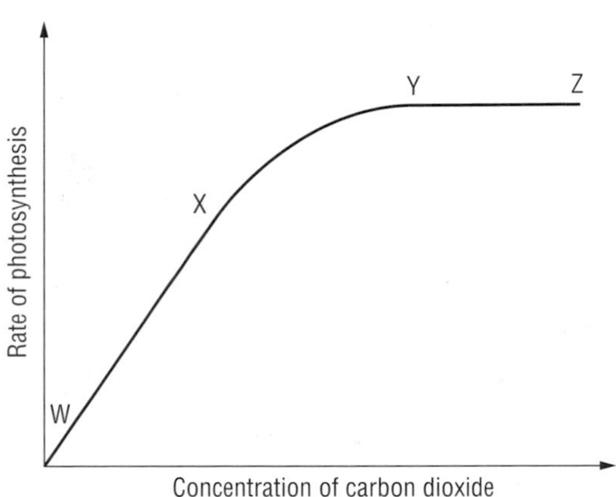

a) How and why does the rate of photosynthesis change between W and X? *3 marks*

b) Why is there no change in the rate of photosynthesis between Y and Z? *2 marks*

c) Some gardeners provide extra carbon dioxide to their greenhouse plants. Why is a graph like this of use to them? *2 marks*

13 Plants are often grown in greenhouses to extend the growing season. Plants make their own food and grow as a result of photosynthesis.

$$6CO_2 + 6H_2O \xrightarrow{\textit{sunlight}} C_6H_{12}O_6 + 6O_2$$

In winter, heated greenhouses ensure that photosynthesis continues at an optimum rate. Suggest *three* other steps that could be taken to ensure that the plants continue to photosynthesise as they would in the summer months. *3 marks*

14 A biologist was investigating the production of sugar by sugar cane plants. He collected leaves from the plants at different times during a 24 hour period and analysed them to find the percentage of sugar. The results are shown in the table.

Time from start/hrs	Percentage of sugar
0	1.0
3	1.5
6	1.9
9	2.0
12	1.5
15	0.8
18	0.5
21	0.5
24	0.5

a) Plot a graph to show how the percentage of sugar varies during the 24 hour period. *4 marks*

b) Draw an arrow on your graph to show the time of dusk. *1 mark*

15 The presence of starch in a leaf is often taken as evidence that photosynthesis has occurred. The flow diagram shows the main stages in testing a leaf for starch.

a) What is the purpose of stage A? *1 mark*
b) What is the purpose of stage B? *1 mark*
c) What is the purpose of stage C? *1 mark*
d) What happens in stage D if the leaf contains starch? *1 mark*

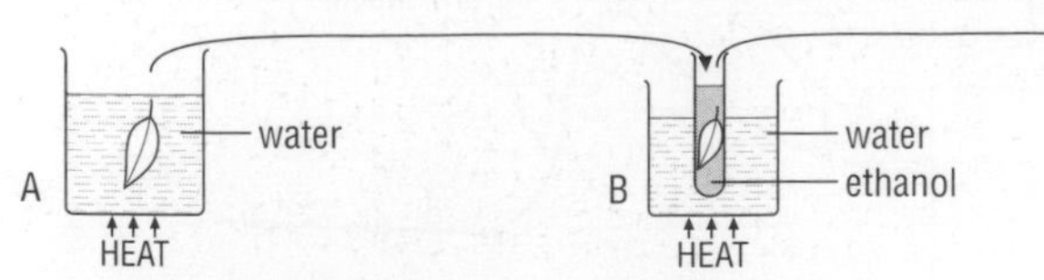

Further examination questions

16 A farmer planted corn seeds in three different positions on ridges and on flat soil to find out which position was the best. He planted the same number of seeds in each position in early spring. The diagram shows how well the corn plants grew in the three different positions.

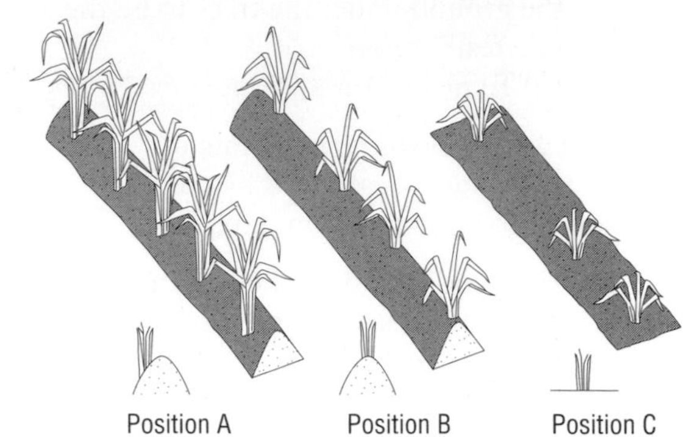

He took the temperature of the soil around the corn in three places. The graph shows the temperatures at different times of the day.

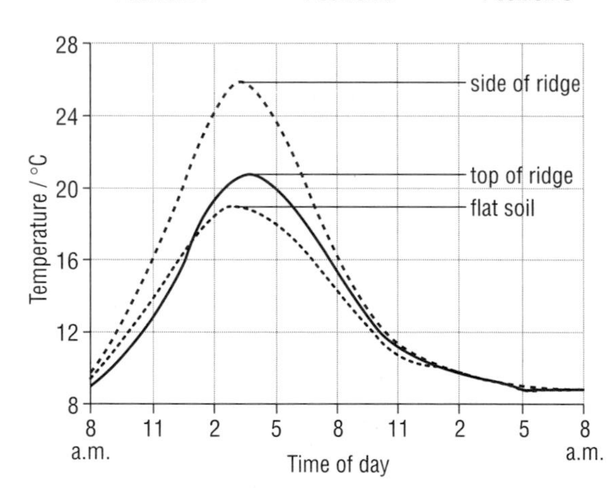

a) (i) Which position warmed up the fastest between 8 a.m. and 11 a.m.? *1 mark*
(ii) Which position stayed above 16°C for the longest time? *1 mark*
b) Using the information in the graph, explain why the corn grew better in some positions compared with others. *2 marks*
c) Write down *two* factors which the farmer would try to keep the same in the three positions. *2 marks*
d) The farmer normally planted corn in early spring and harvested it in early autumn. Faster growing varieties of corn are available. Suggest *two* reasons why the farmer might plant some of these early varieties. *2 marks*
e) Which of the alternatives – A, B, C or D – shows the products of photosynthesis in corn leaves? *1 mark*

A	carbon dioxide and oxygen
B	carbon dioxide and water
C	oxygen and sugar
D	oxygen and sunlight

f) The same amount of leaf tissue was taken from two types of corn plant and tested for the amount of chlorophyll. The results are shown in the table.

Amount of chlorophyll/mg	
Type 1 corn plant	**Type 2 corn plant**
2.15	0.97

Using the information in the table explain why Type 2 plants might grow slower than Type 1 corn plants.
2 marks

OCR

17 A green plant was placed in a dark cupboard. After 24 hours, some of the leaves were tested for starch. No starch was found in any leaf. The same plant was then placed in sunlight. One leaf, Q, was treated as shown in the diagram. After a further 24 hours, leaves P and Q on the plant were tested for starch.

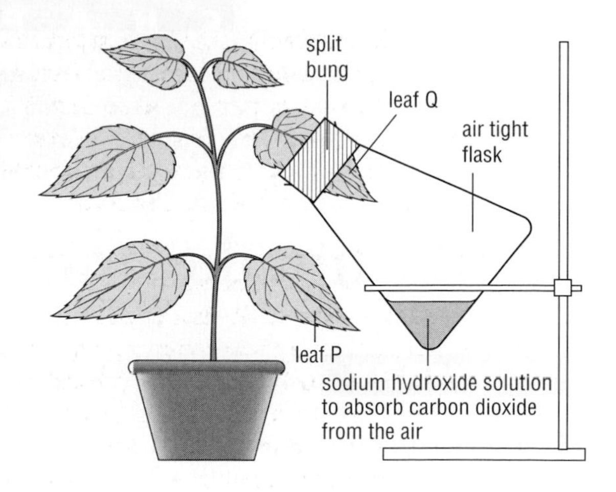

a) Why was there no starch in the leaves after the plant had been kept in the dark cupboard for 24 hours? *2 marks*

b) (i) Name the substance in green leaves which helps plants to make glucose. *1 mark*
(ii) Name the gas produced by green leaves when they make glucose for starch production. *1 mark*

c) The diagrams show the results of starch tests on discs taken from leaves P and Q. The diagrams also show the parts of the leaves from which the discs were taken.
Explain the reasons for the differences between discs P_3 and Q_3 and between discs P_2 and Q_2. *3 marks*

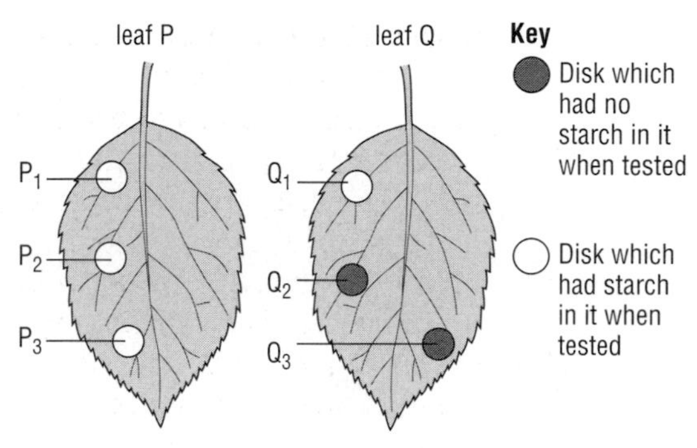

EDEXCEL

18 The diagram suggests three limiting factors which are affecting photosynthesis in the aquarium tank.

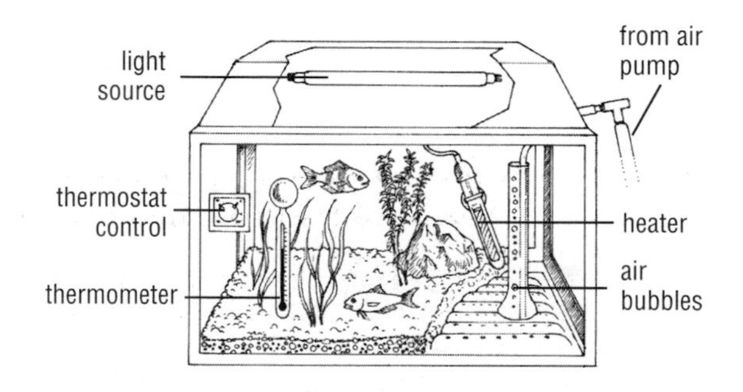

a) What is meant by a limiting factor? *1 mark*

b) Name *two* limiting factors that are suggested in the diagram. *2 marks*

c) The graph shows the changes in concentration of two gases in an outdoor pond over a 4-day period.
(i) Identify the gases represented on the graphs by A and B. *2 marks*
(ii) Explain the shapes of the graphs for gases A and B. *4 marks*
(iii) Suggest a reason for the lowest peak at point X. *1 mark*
(iv) What shape would you expect the graphs to be if the same measurements were taken in the balanced aquarium shown in part a)? *1 mark*

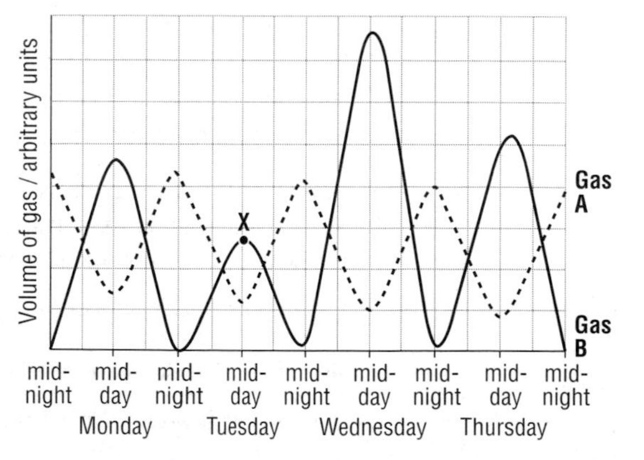

WJEC

6

Water uptake and transport in plants

SUMMARY

1 Water and mineral uptake by plants

In order to grow, plants must take in water and essential minerals from the soil.

- Water and minerals are taken up by plants through the **root hair cells**.
- Millions of root hair cells provide a large surface area with thin cell walls to help the uptake of water and minerals.
- The water absorbed through the roots is required for:
  - photosynthesis (see chapter 5)
  - support of the plant
  - transport of solutions up and down the plant.

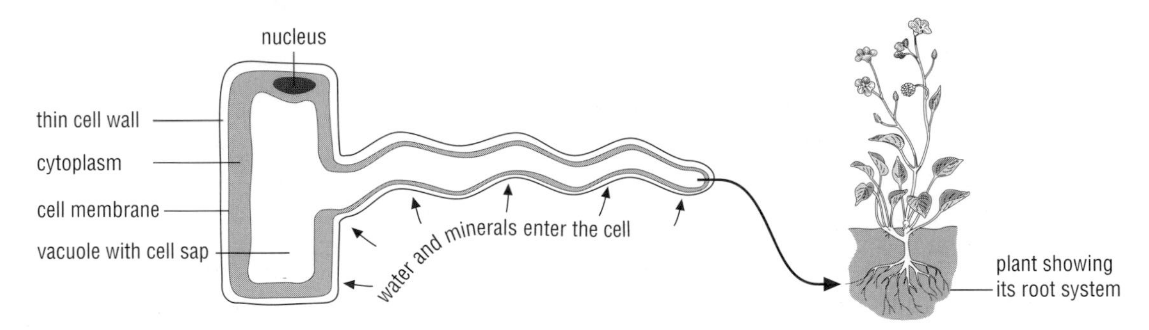

Most of the water absorbed through the roots eventually evaporates from cells in the leaves. The evaporation of water from the leaves is called **transpiration**. The continuous flow of water from the roots, through the plant to the leaves is called the **transpiration stream**.

2 Water and support in plants

Unlike animals, plants have strong, rigid cell walls. The cell wall is completely permeable to water and solutes, but just below the cell wall is the cell membrane which is selectively permeable. Water and dissolved minerals can pass from one cell to another from the roots of a plant to the leaves. When the cells have a good supply of water, they fill up and pack tightly. The cells press on one another and make the leaves and stem firm yet flexible. The swollen plant cells are described as **turgid** and the support they give to the plant is called **turgor**.

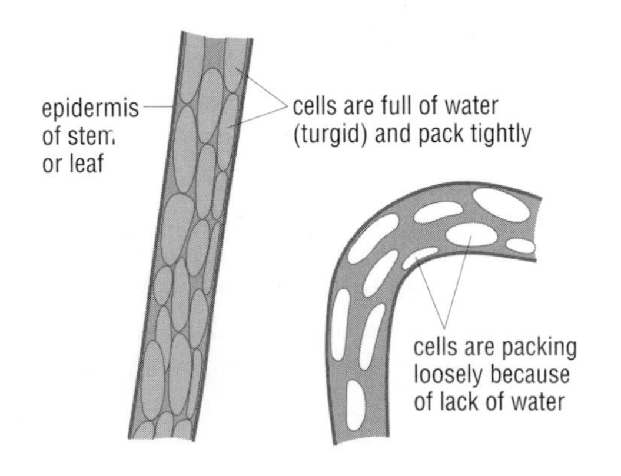

A well-watered plant stem A wilting plant stem

3 Water and transport in plants

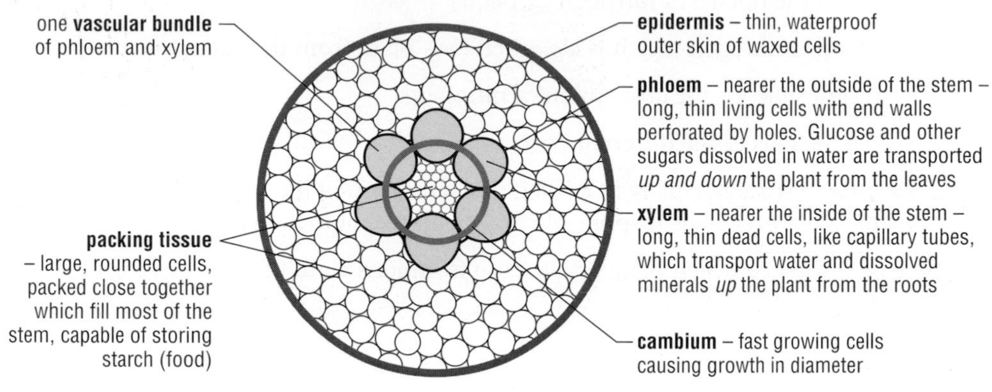
one **vascular bundle** of phloem and xylem
epidermis – thin, waterproof outer skin of waxed cells
phloem – nearer the outside of the stem – long, thin living cells with end walls perforated by holes. Glucose and other sugars dissolved in water are transported *up and down* the plant from the leaves
xylem – nearer the inside of the stem – long, thin dead cells, like capillary tubes, which transport water and dissolved minerals *up* the plant from the roots
packing tissue – large, rounded cells, packed close together which fill most of the stem, capable of storing starch (food)
cambium – fast growing cells causing growth in diameter

Cross-section of a young plant stem

4 Mineral requirements of plants

Plants require certain **essential elements** to grow and flourish. The most important essential elements are:

- **carbon** from carbon dioxide in the air.
- **hydrogen** from water in the soil.
- **oxygen** from the air.

The next three major elements for plants are:

- **nitrogen** from nitrates and ammonium salts in the soil to synthesise proteins and nucleic acids. Lack of nitrogen results in small plants with yellow leaves.
- **phosphorus** from phosphates in the soil to synthesise ATP and nucleic acids. Lack of phosphorus results in a poor root system and purple coloration of leaves.
- **potassium** from potassium salts in the soil to help in the transfer of materials across cell membranes. Lack of potassium results in yellow leaves which die back.

STUDY QUESTIONS

Objective questions

Questions 1 to 6

The diagram shows a section through a thin side root of a young plant.

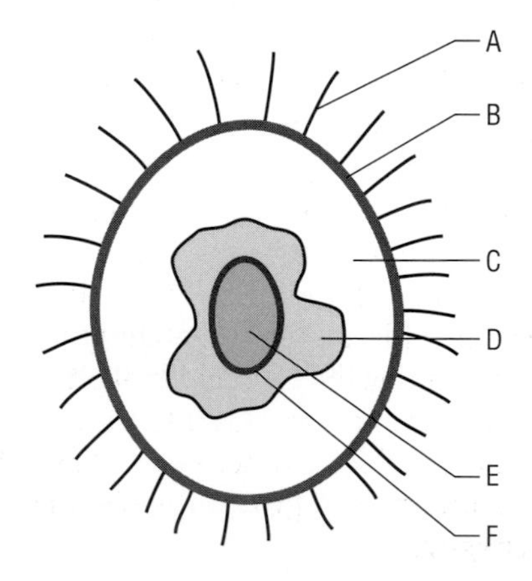
A
B
C
D
E
F

Use the letters, A, B, C, D, E and F to identify

1 cambium

2 epidermis

3 packing tissue

4 phloem

5 a root hair

6 xylem

Questions 7 to 11

Questions 7 to 11 concern the uptake of nitrogen by plants.

7 Name an ion containing nitrogen which is absorbed by a plant from the soil.

8 Which cells are responsible for the absorption of nitrogen?

9 What tissue carries the ions to the different parts of a plant?

10 Nitrogen-containing ions are normally in a lower concentration in the soil than in the cells of a plant. What process is responsible for the uptake of the ions?

11 Name *one* type of substance which plants synthesise from ions containing nitrogen.

Short questions

12 a) Name the *two* types of transport tissue found in tree trunks. *2 marks*

b) When trenches are being dug by mechanical diggers, the roots of nearby trees are sometimes cut. What effect will this have on the rate of water transport? Explain your answer. *3 marks*

c) When trenches are dug, the roots may be exposed to the air.
(i) What can now happen to the water in the root cells? *1 mark*
(ii) How can this be prevented? *1 mark*

d) When the soil is returned to the trench, it must be packed around the roots without leaving large air spaces. Why is this? *1 mark*

13 The diagram shows the flow of water through a plant. This is caused by transpiration from the leaves.

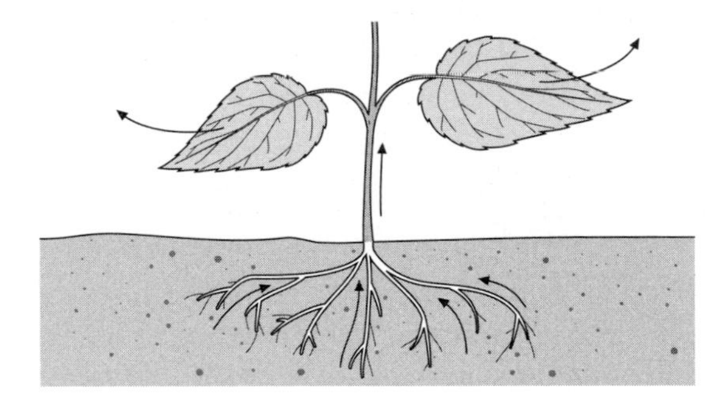

a) State *two* reasons why the transpiration stream is important. *2 marks*

b) Water enters the root hairs by osmosis. How does this occur? *4 marks*

14 The graph shows the change in mass of two similar detached leaves under the same experimental conditions. Y and Z are oak leaves with either the upper or lower surface covered with vaseline.

a) (i) Which oak leaf had vaseline on its lower surface? *1 mark*
(ii) Explain your answer. *2 marks*

b) Why did the decrease in mass of leaf Z cease after about 100 minutes? *1 mark*

c) What would you expect the loss in mass of leaf Y to be after 30 minutes if its surface is not treated in any way? *2 marks*

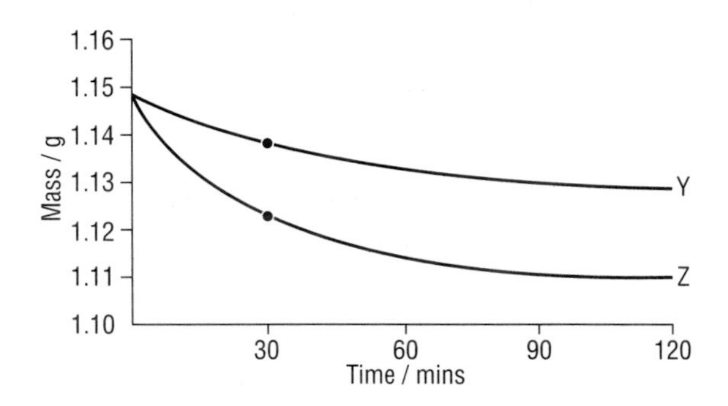

Further examination questions

15 Lettuce plants can grow with their roots dangling in a nutrient solution. No soil is needed. The plants grow in plastic tubes through which the nutrient solution flows. The solution is pumped from a tank through all the tubes. When it returns to the tank, air is bubbled into it and nutrients are replaced. Then the solution is pumped through the tubes again. The diagram shows how the tubes are arranged.

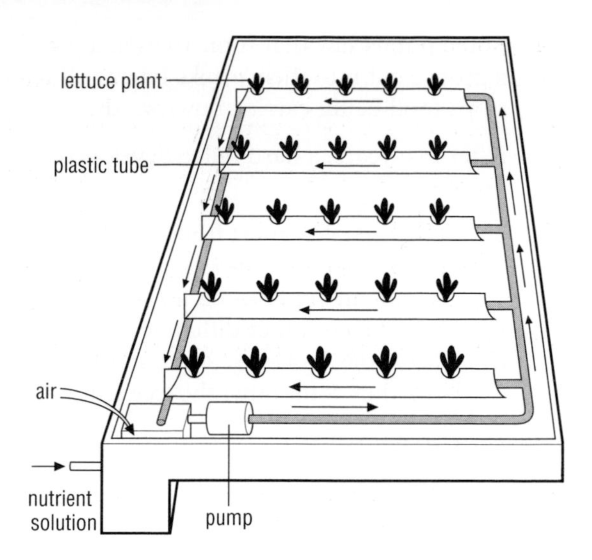

a) Why would the nutrients need replacing after the solution had been circulated? *1 mark*

b) The grower changed the concentration of oxygen in the air which was bubbled into the solution. The graph shows what happened to the amount of nutrients absorbed by the lettuce roots.

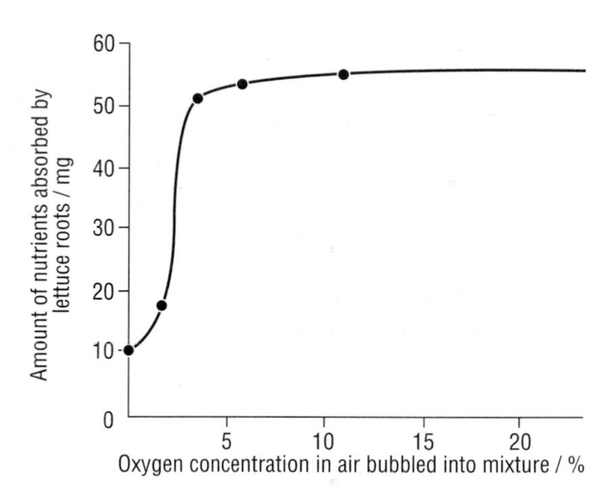

(i) Use this information to write a statement which links the amount of nutrients absorbed to the concentration of oxygen supplied to the plants. *2 marks*
The grower tried the same investigation on rice plants. The graph below shows the results.
(ii) State *two* ways in which the results for rice roots differ from the results for lettuce roots. *2 marks*
(iii) Explain why different oxygen concentrations affect the amount of nutrients absorbed. *2 marks*
(iv) The solution supplies oxygen directly to the roots. Describe another way in which oxygen might reach the roots. *1 mark*
(v) In winter, the grower heats the nutrient solution. Explain why the uptake of nutrients is affected by temperature. *2 marks*

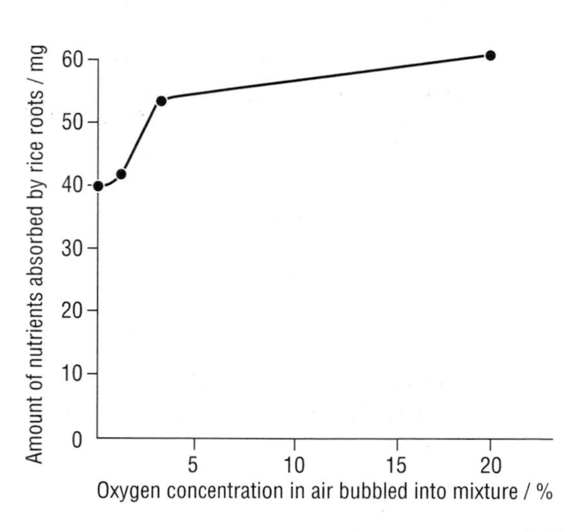

OCR

16 Some pupils saw that some trees in a park had a groove cut into their trunks (stems). This stops food being carried downwards.

After five years, a hard bulge developed above the groove.

a) State *two* jobs, apart from carrying food, of a tree trunk. *2 marks*
b) Suggest an explanation for the development of the bulge in the tree trunk. *2 marks*
c) Describe *one* other difference in the tree trunk after five years. *1 mark*
d) The pupils looked at a tree stump and saw different markings. From a science book they found that the tree stump had different cells.

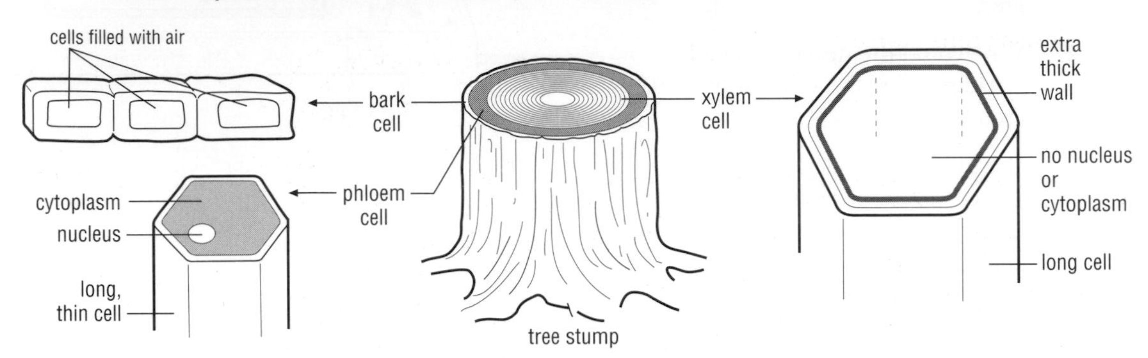

(i) Which type of cell will protect the tree? *1 mark*
(ii) Which type of cell will carry water and help to support the tree? Give *two* reasons for your choice. *3 marks*
(iii) Which type of cell is alive? Give *one* reason for your choice. *2 marks*
(iv) Only one type of cell is strong enough to make wooden furniture. Which type of cell is it? *1 mark*

OCR

17 The diagram shows a plant used in an experiment.

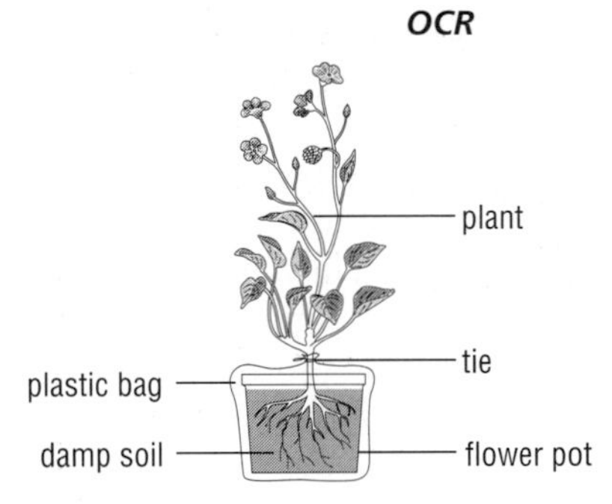

The mass of water lost from the plant by evaporation was measured over 14 hours. The whole apparatus was left outside on a warm, sunny day and weighed every two hours. Here are the results.

Time	Mass of plant/g
8 a.m.	365
10 a.m.	363
12 noon	358
2 p.m.	353
4 p.m.	347
6 p.m.	344
8 p.m.	342
10 p.m.	341

a) (i) Plot the data as a line graph. Credit will be given for accurate plotting and for labelling the axes correctly. Join only the points you have plotted. *4 marks*
(ii) What was the mass of the plant at 1 p.m.? *1 mark*
(iii) What was the average rate of water loss in g per hour between 12 noon and 4 p.m.? Show your working. *2 marks*

b) (i) Suggest why a plant may die when the rate of water loss is faster than the rate at which it is replaced. *1 mark*
(ii) State how a plant may be able to reduce this water loss. *1 mark*

WJEC

7 Communication and nerves

SUMMARY

1 Senses and stimuli

Living things depend on their **senses** and **sense organs** for survival. We have five sense organs.

Sense organ	Sense
eyes	sight
ears	hearing, balance
nose	smell
tongue	taste
skin	temperature, touch

Our sense organs have specialised cells called **receptors** which are sensitive to **stimuli** (changes around us).

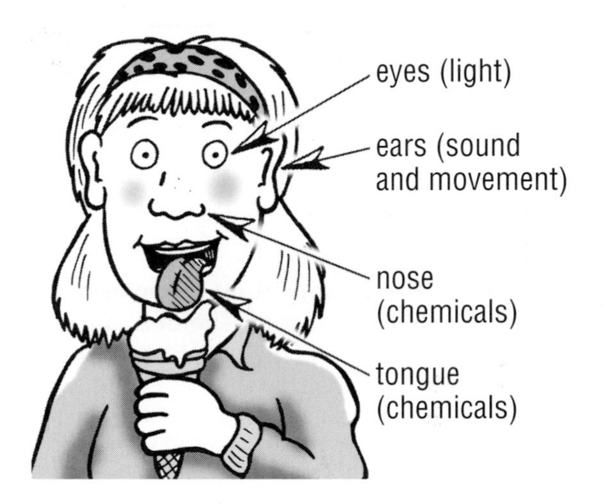

Senses and stimuli

2 The nervous system

The nervous system carries messages (impulses) from one part of the body to another. It co-ordinates and controls the whole body.

Receptors in our sense organs can turn (transduce) the energy from stimuli into **impulses**. These tiny electric impulses pass along nerve cells called **sensory neurones** to the **central nervous system**. The central nervous system is the **brain** and the **spinal cord**.

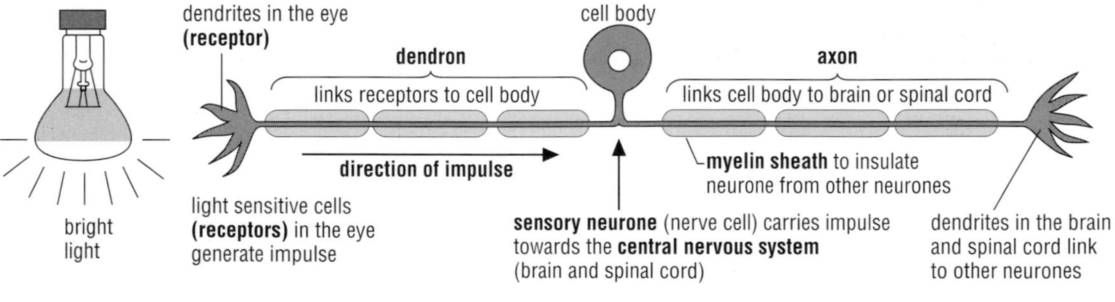

When an impulse reaches the spinal cord or the brain, it is transferred to a **relay neurone** to pass on (relay) the message.

In the brain, relay neurones connect to **pyramidal neurones** which have thousands of branches to other neurones like the junctions in a computer. The junctions between neurones are called **synapses**.

3 Responses and reflex actions

Interconnections between neurones in the spinal cord and the brain lead to **responses**. **Motor neurones** carry the response impulses from the central nervous system to muscles and glands (**effectors**).

- Rapid and automatic responses to stimuli are described as **reflex actions**. In reflex actions, the stimulus and response usually travel by the shortest route which may not include the brain. These shortest routes are called **reflex arcs**.

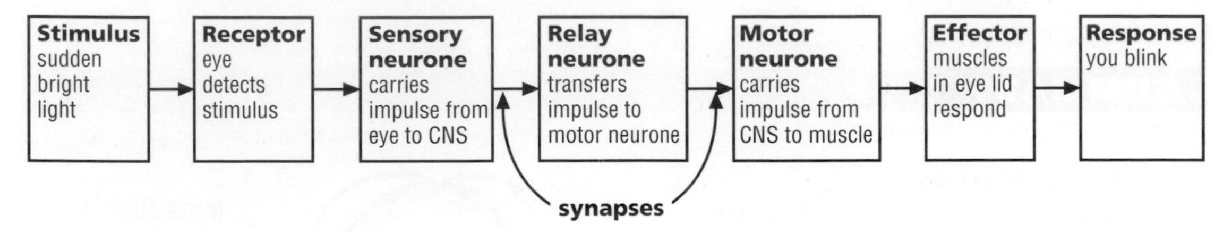

Stimulus
sudden bright light
Receptor
eye detects stimulus
Sensory neurone
carries impulse from eye to CNS
Relay neurone
transfers impulse to motor neurone
Motor neurone
carries impulse from CNS to muscle
Effector
muscles in eye lid respond
Response
you blink
synapses

A reflex arc showing how we respond to a sudden bright light

- Thoughtful and considered responses are described as **conscious actions**. In conscious actions, impulses must travel to the brain which co-ordinates a response.

4 The eye – an important sense organ

Our eyes respond to the stimulus of light. The structure of the eye and the parts crucial to our response to light are demonstrated in the diagram.

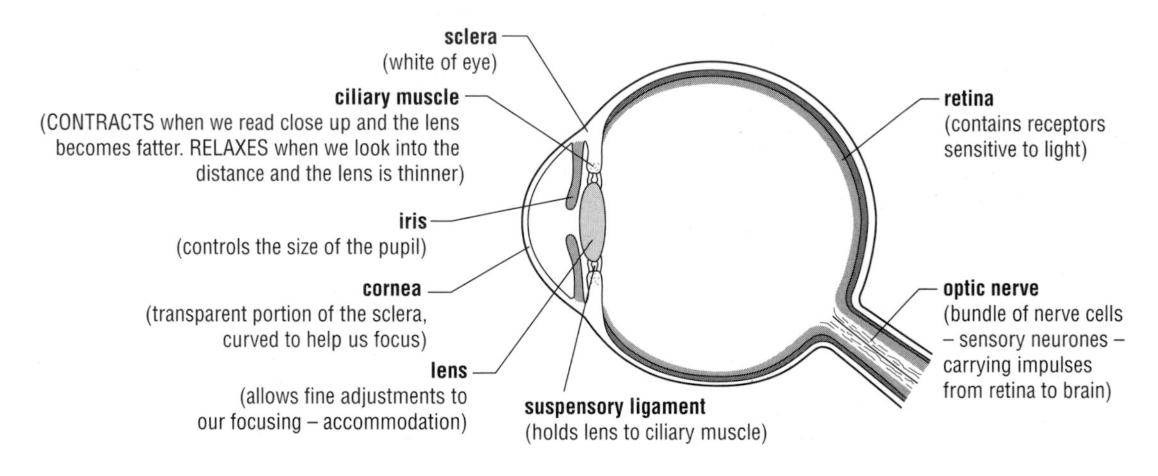

sclera
(white of eye)
ciliary muscle
(CONTRACTS when we read close up and the lens becomes fatter. RELAXES when we look into the distance and the lens is thinner)
iris
(controls the size of the pupil)
cornea
(transparent portion of the sclera, curved to help us focus)
lens
(allows fine adjustments to our focusing – accommodation)
suspensory ligament
(holds lens to ciliary muscle)
retina
(contains receptors sensitive to light)
optic nerve
(bundle of nerve cells – sensory neurones – carrying impulses from retina to brain)

STUDY QUESTIONS

Objective questions

Questions 1 to 5

In questions 1 to 5, choose from A, B, C, D, E or F which is the correct answer. The diagram shows a reflex arc.

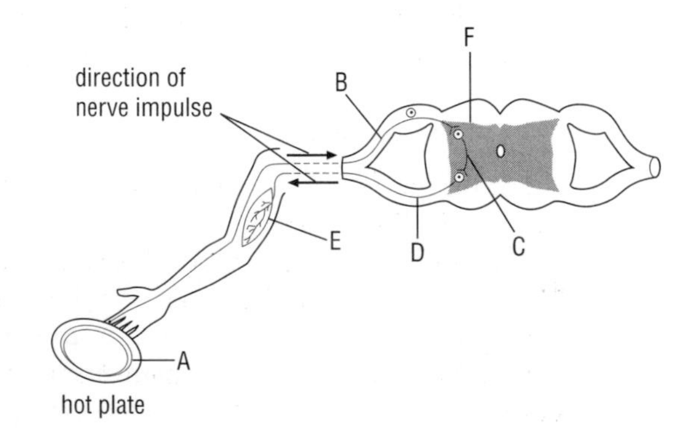

direction of nerve impulse
F
B
E
D
C
A
hot plate

1 Which is a motor neurone?
2 Which is a relay neurone?
3 What is the stimulus?
4 Which is the effector?
5 Which is the spinal cord?

Questions 6 to 13

Questions 6 to 13 relate to the diagram, showing a section through a human eye.

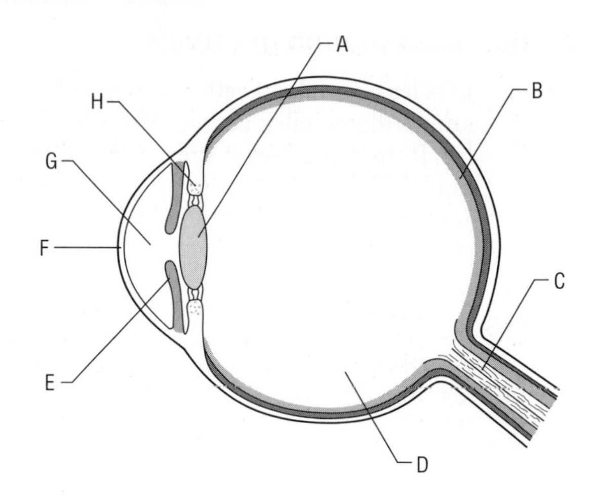

6 Name the part of the diagram labelled A.

7 Name the part of the diagram labelled E.

8 & 9 Which *two* labelled parts help us to focus light on B?

10 Which labelled part is sensitive to light?

11 Which labelled part contains muscles for accommodation?

12 & 13 Which *two* labelled parts control the amount of light entering the eye?

Short questions

14 Look at the diagram of a human cell. This type of cell can be very long.

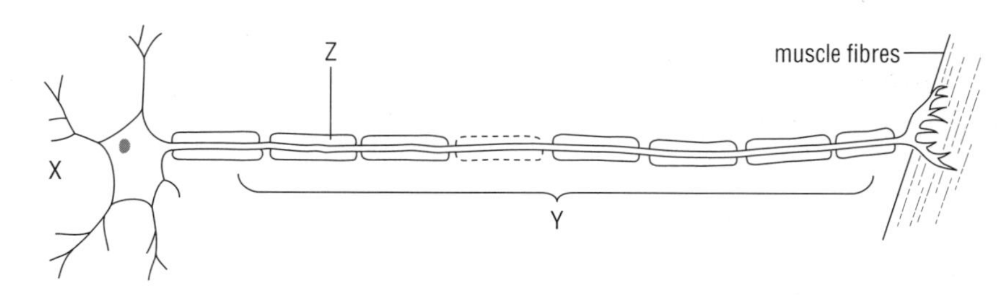

a) What is the name for this type of cell? *1 mark*
b) In which part of the body would you find the end labelled X? *1 mark*
c) Why does the part labelled Y need to be so long? *1 mark*
d) Z is part of the myelin sheath. What is its function? *1 mark*

15 A dog runs across the road in front of a car. The driver slams her foot on the brakes. Explain how the nervous system brings about this response. Use the following words in your answer: effector, impulse, motor neurone, receptor, response, sensory neurone, stimulus. *5 marks*

16 a) The skin is a sense organ. The skin senses two types of stimuli. What are they? *2 marks*
b) What is the name of the action when you put your foot into icy, cold water and withdraw it immediately? *1 mark*
c) To survive, animals need to avoid accidents and their consequences. What is the advantage of actions such as that in part b)? *1 mark*
d) Why are there more nerve endings at the tips of our fingers than on the balls of our feet? *2 marks*

17 a) New born babies need to communicate. What is their main form of communication? *1 mark*
b) What is often the stimulus for their main form of communication? *1 mark*
c) When a baby is about six weeks old, how can it respond to someone talking to it? *1 mark*
d) Babies thrive on being stimulated. Suggest *two* ways to stimulate a baby. *2 marks*

Further examination questions

18 a) Jo is learning about reflex actions. She knows that a reflex action will stop her from being hurt if she touches a hot object. Jo's teacher writes these six sentences on the board for the class to put into the right order.

A The receptor in the skin detects the high temperature.
B A message passes along the motor nerve to the effectors.
C The finger touches a hot object.
D The effector muscles contract and pull the finger away.
E The message passes to a motor nerve in the spinal cord.
F A message is sent along a sensory nerve to the spinal cord.

Write down A, B, C, D, E and F in the correct order. *6 marks*

b) The human body contains many different types of cell. The nervous system contains special cells called neurones. The diagram shows a sensory neurone.
Explain how the structure of the neurone allows it to do its job. *3 marks*

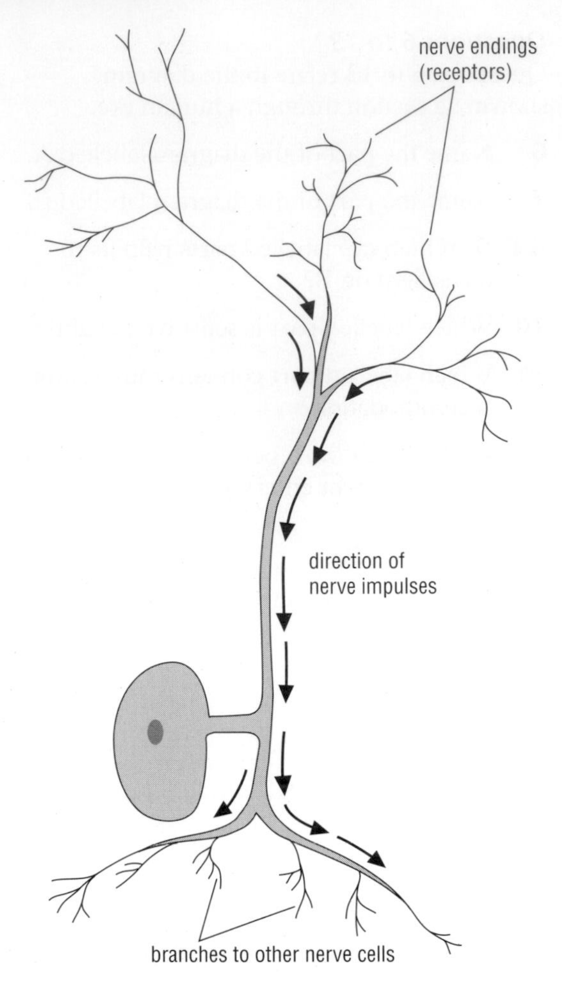

OCR

19 The doctor is testing the child's nervous system by tapping the tendon just below the knee. This pulls cells which are sensitive to stretching.

a) What are cells which are sensitive to stimuli called? *1 mark*
b) These cells send information to the spinal cord. In what form is this information sent? *2 marks*
c) The healthy response to the stimulus is the straightening of the leg. What is the effector in this response? *1 mark*
d) This response is one example of a reflex action. Describe *one* other example of a reflex action in terms of

stimulus → receptor → co-ordinator → effector → response
5 marks

AQA

20 The diagram shows a section through an eye.

a) Name the parts A and B. *2 marks*
b) Which part focuses light by changing shape? *1 mark*
c) If a bright light shines on the eye, structure X will change shape due to a reflex action. Describe this change. *4 marks*
d) Tissue Z contains two types of light sensitive cell. Name the two types of cell and state the function of each type. *2 marks*
e) The diagram shows some of the details of the region between a pair of neurones in a reflex arc.
(i) Name the microscopic gap shown in the diagram. *1 mark*
(ii) Describe how nerve impulses are transmitted across this gap. *3 marks*

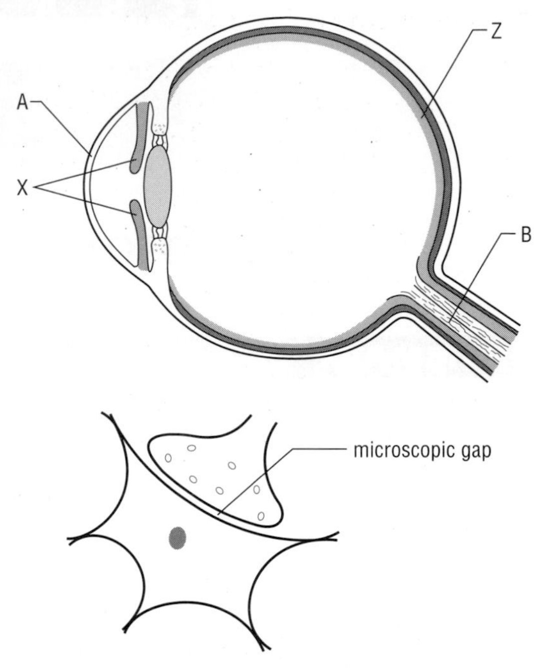

EDEXCEL

21 The diagram represents a reflex arc.

a) Use the information in the diagram to help describe the path taken by a nerve impulse in a named reflex action. *6 marks*
b) State *two* features which are characteristic of all reflexes. *2 marks*

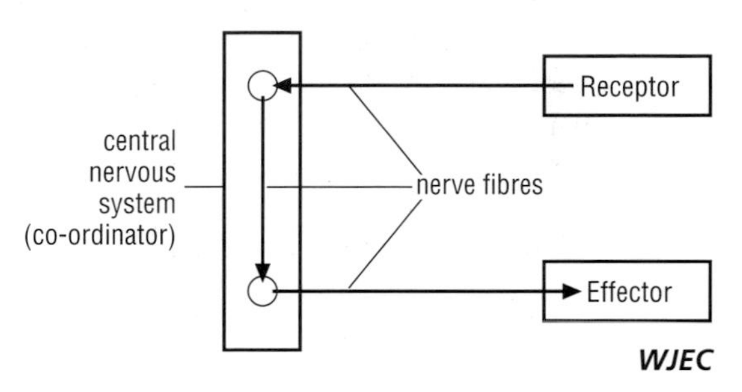

WJEC

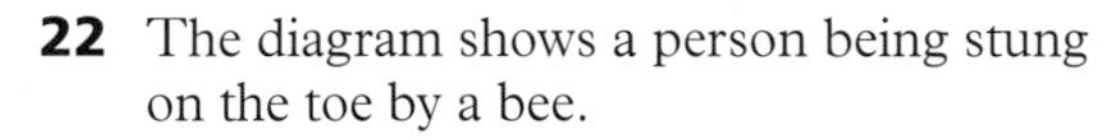

22 The diagram shows a person being stung on the toe by a bee.

a) (i) What is the name given to this type of action? *1 mark*
(ii) What use is this type of action to you? *1 mark*
b) Write an account of the action shown in the diagram. *4 marks*
c) What is protecting the central nervous system in this example? *1 mark*
d) Explain why the type of action shown is different from riding a bicycle. *1 mark*
e) Besides using nerves, what other system in your body helps to co-ordinate your behaviour and development? *1 mark*

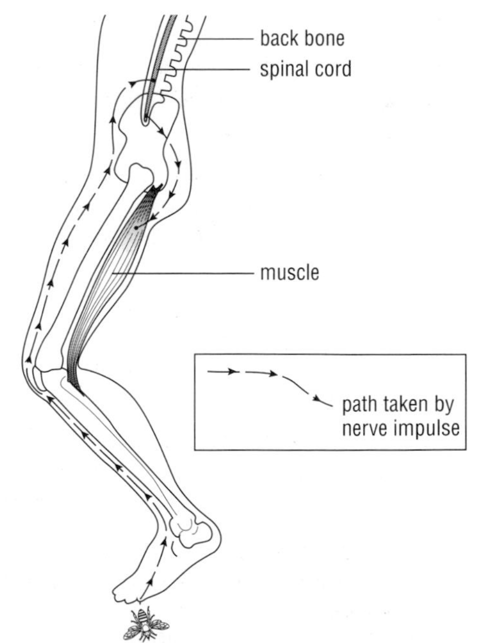

WJEC

8

Hormones and control

SUMMARY

1 Hormones

Hormones are produced in our **glands** and then released (secreted) into the bloodstream in tiny amounts. They control our **growth**, our **sexual characteristics** and our **body metabolism**. Like the central nervous system, they carry 'messages' from one part of the body to another.

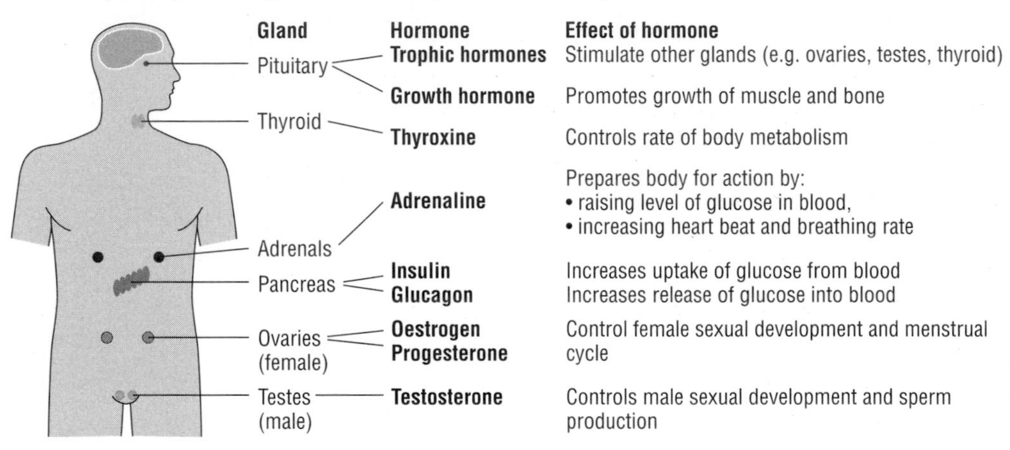

Gland
Hormone
Effect of hormone
Pituitary
Trophic hormones
Stimulate other glands (e.g. ovaries, testes, thyroid)
Growth hormone
Promotes growth of muscle and bone
Thyroid
Thyroxine
Controls rate of body metabolism
Adrenals
Adrenaline
Prepares body for action by:
• raising level of glucose in blood,
• increasing heart beat and breathing rate
Pancreas
Insulin
Increases uptake of glucose from blood
Glucagon
Increases release of glucose into blood
Ovaries
(female)
Oestrogen
Progesterone
Control female sexual development and menstrual cycle
Testes
(male)
Testosterone
Controls male sexual development and sperm production

The positions of our glands, the hormones they release and the effects of these hormones

2 Controlling glucose levels in the blood

Cells will only metabolise effectively if the glucose concentration stays fairly steady. Our bodies control the level of glucose in our blood by two hormones released by the **pancreas**.

- If the glucose concentration in the blood is too high, the pancreas releases **insulin** into the bloodstream. The insulin causes **liver cells** to convert glucose into glycogen which is stored in the liver.
- If the glucose concentration in the blood is too low, the pancreas stops releasing insulin and releases **glucagon**. The glucagon causes the **liver** to convert glycogen into glucose and release it into the blood.

This control of insulin and glucagon output and hence glucose levels by feeding back information to the pancreas is called **feedback control**. It is an example of **homeostasis**.

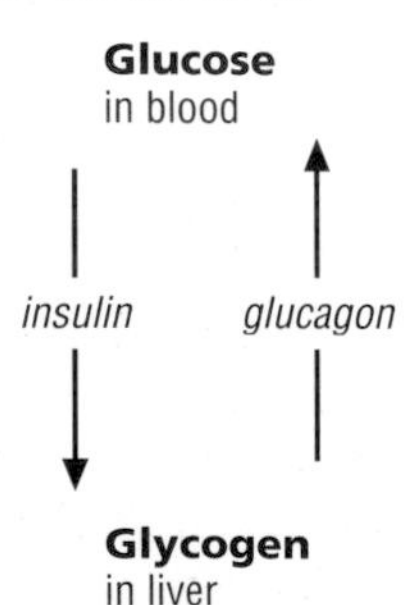

Some people cannot produce enough insulin to keep their blood glucose at a steady level. This condition is known as **diabetes**. Without treatment, their glucose level becomes too high, leading to a coma.

3 Controlling menstruation

Three hormones control menstruation and the menstrual cycle:

- The **follicle stimulating hormone** (**FSH**) is secreted by the pituitary gland.
- **Oestrogens** and **progesterone** are secreted by the ovaries.

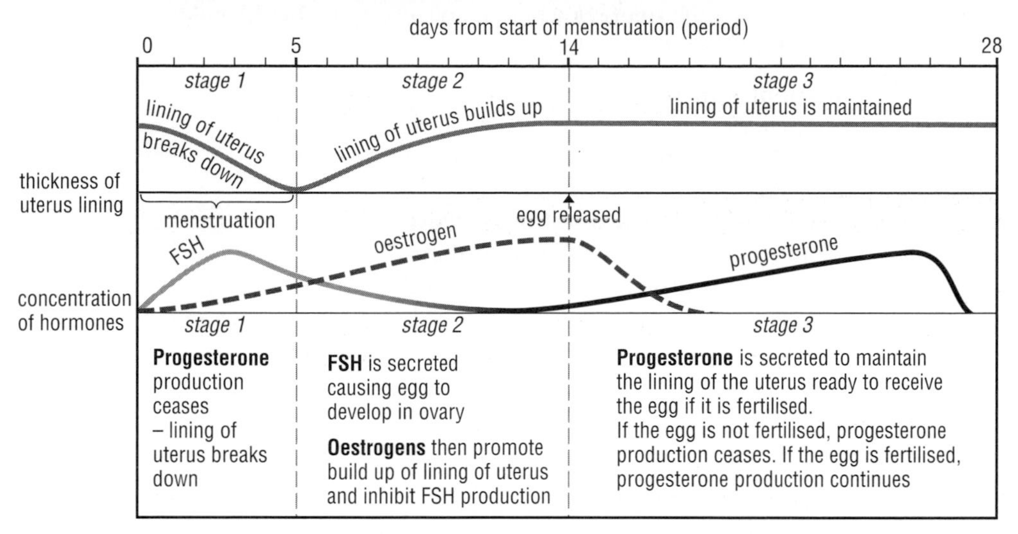

The effects of FSH, oestrogens and progesterone in the key stages of the menstrual cycle

4 Controlling body temperature

Our body temperature must stay close to 37°C for our cells and overall metabolism to work effectively.

Our **skin** plays an important part in maintaining a constant body temperature.

Temperature receptors in the skin send impulses to the brain which then controls the responses of the skin. The skin can respond and control body temperature in four ways.

- Variable blood flow – by controlling the cross section of blood vessels and capillaries and hence heat loss.
- Shivering – involves muscle contraction which stimulates respiration and hence heat production.
- Sweating – as sweat evaporates, it takes heat from the body and causes cooling.
- Hairs and muscle activity – tiny erector muscles can raise and lower hairs. In cold weather, the muscles contract, raise hairs and trap more air. The trapped air acts as an insulator.

5 Controlling the water and salt content of our bodies – the kidneys

After our cells have used all the nutrients from the blood, it is necessary to remove waste and unwanted materials. The kidneys play a major role in removing urea, salts and excess water from the blood.

- **Urea** is a poisonous waste product which is produced when proteins are metabolised.
- **Salts**, particularly sodium chloride from our diet, collect in the blood.
- **Water** is an essential constituent of blood which needs control. This is done via **osmoregulators** in the brain which direct the release of **antidiuretic hormone** (**ADH**) from the pituitary to control water transfer into and out of our capillaries.

The kidneys receive an excellent supply of blood via the renal arteries. As branches of the renal arteries pass into the kidneys, they divide into a network of capillaries. These capillaries are wrapped around thousands of looped **tubules**. As blood flows in the capillaries over the tubules, water, salt and urea are filtered out as **urine**. Urine flows down the **ureters** where it collects in the bladder. At intervals, urine is excreted through the urethra.

6 Controlling the growth of plants

Plants respond to three major stimuli – light, gravity and moisture. These responses are caused by **plant hormones** which control cell growth and development in a similar way to hormones in animals.

- The response of plants to light is called **phototropism**.
- The response of plants to gravity is called **geotropism**.

The hormones which control the growth of plant roots and shoots are called **auxins**. Some auxins which increase growth are produced in those parts of a plant which receive less light so they:

- promote the growth of roots,
- promote the growth of tips rather than side shoots,
- promote growth on the shaded side of a plant so that it bends towards the light.

STUDY QUESTIONS

Objective questions

Questions 1 to 6

The diagram shows some of the glands in the female body.

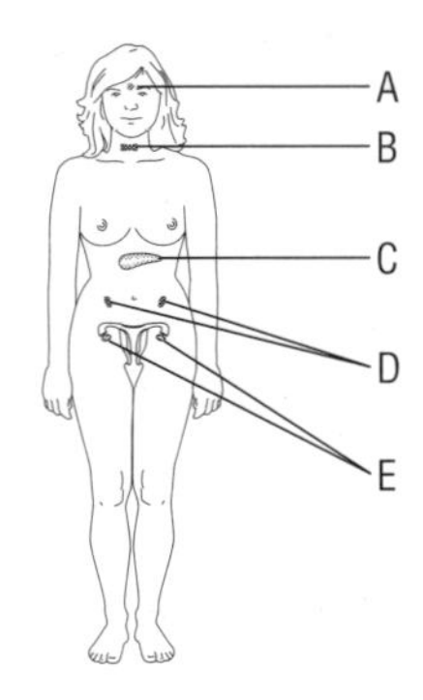

1 Name the body system to which these glands belong.

2 Write down the letter which labels the pancreas.

3 Write down the letter which labels the adrenal glands.

4 Write down the letter which labels the thyroid gland.

5 Name the gland which is sometimes called the master gland.

6 Name the two hormones produced by E.

Questions 7 to 12

The diagram shows organs in human skin.

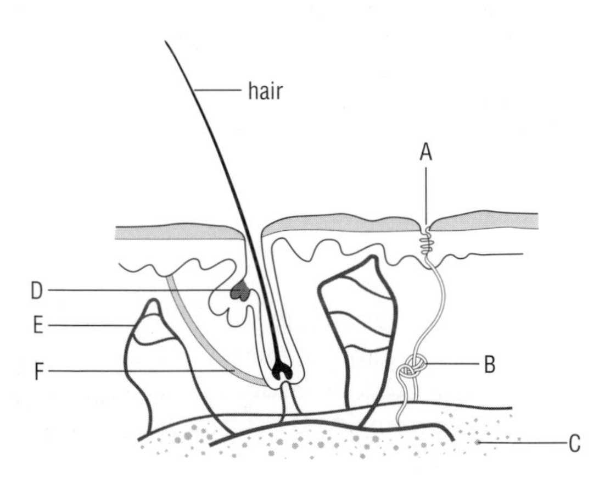

7 Which labelled part is a sweat pore?

8 Which labelled part is an erector muscle?

9 Which labelled part is an oil (sebaceous) gland?

10 What happens to B when you get hot?

11 What liquid is carried in E?

12 What happens to E when you get cold?

Short questions

13 The graph shows the relative concentrations of oestrogen and hormone X during the menstrual cycle.

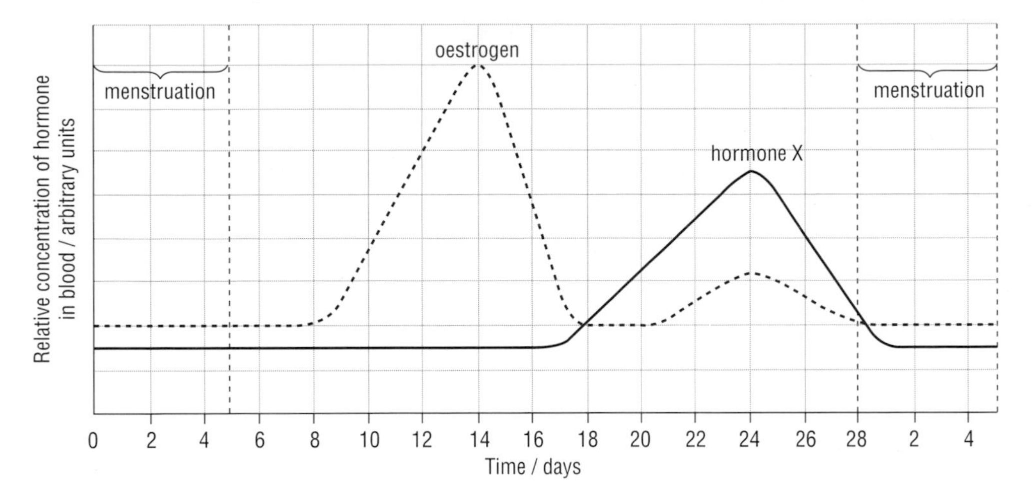

a) When is the concentration of oestrogen equal to that of X? *2 marks*
b) Name hormone X. *1 mark*
c) Where in the body is X produced? *1 mark*
d) What happens when the concentration of X falls? *1 mark*
e) What has happened if the concentration of X remains high after day 24? *1 mark*

14 Chemicals called auxins can affect the growth of plants. If the growing stem of a young plant is put into auxin solution, its effect can be investigated by measuring the increase in length of the stem. The diagram shows how this can be done.

Short pieces of young stem were measured and floated on auxin solutions of different concentration. After 48 hours the pieces were re-measured. The results are shown in the table.

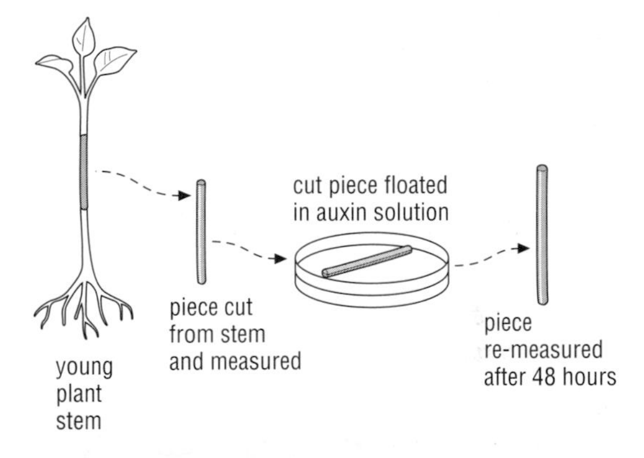

Auxin concentration /parts per million	Initial length of stem piece/mm	Final length of stem piece/mm	Change in length of stem piece/mm	Percentage change in length/%
0	25	26	1	4
1	25	27	2	8
2	25	28	3	12
5	25	30	5	20
10	25	31	6	24

a) Plot a graph of the percentage change in length of stem piece (vertical) against the auxin concentration (horizontal). *4 marks*
b) How is the percentage change in length related to the auxin concentration? *2 marks*

15 Our body temperature stays fairly steady. The graph shows the small changes in body temperature of a person over 24 hours.

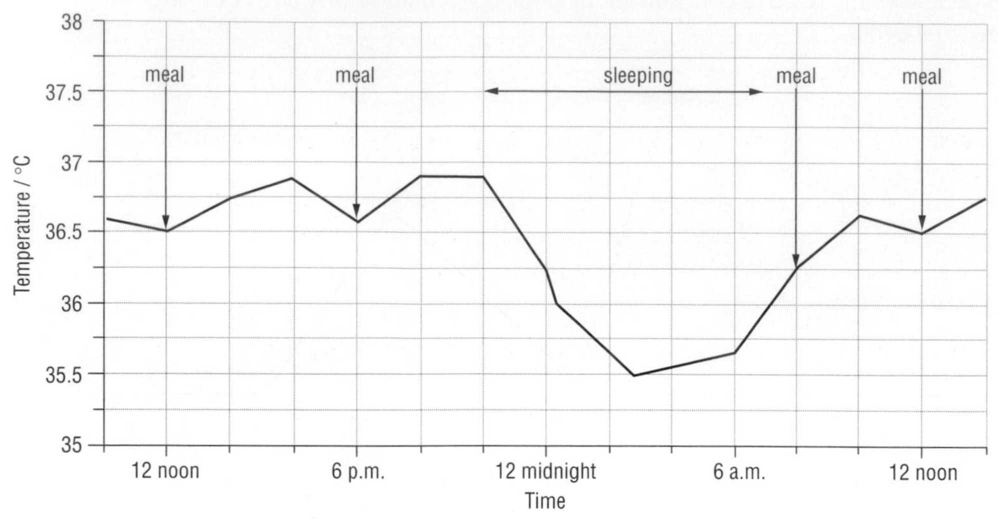

a) State one way in which the person's body generated heat. *1 mark*

b) (i) What happens to the person's body temperature in the first part of sleep from 10.00 p.m. to 2.00 a.m.? *1 mark*
(ii) Explain this change. *1 mark*

c) (i) What happens to the person's body temperature in the latter part of sleep from 3.00 a.m. to 6.00 a.m.? *1 mark*
(ii) Explain this change. *1 mark*

16 a) Sweating helps to keep our bodies at a constant temperature. How does it do this? *3 marks*

b) Shivering helps to keep our bodies at a constant temperature. How does it do this? *3 marks*

17 The graph shows how the concentration of glucose in a normal person's blood changes over a few hours.

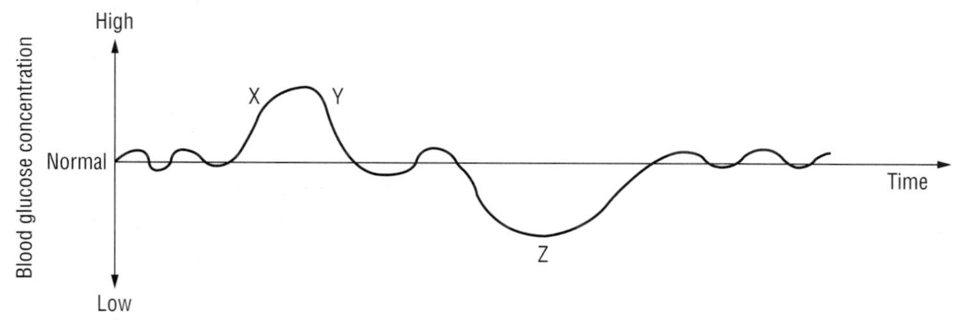

a) Why do you think the concentration of blood glucose increased at X? *1 mark*

b) Name the hormone which helps the concentration of glucose to return to normal at Y. *1 mark*

c) Explain how this hormone reduces the concentration of glucose in the blood. *3 marks*

d) The concentration of blood glucose becomes very low at Z. Which hormone helps to return the glucose concentration to normal? *1 mark*

e) During exercise, our muscles need extra glucose. How does the hormone, adrenaline, help our muscles to obtain extra glucose? *2 marks*

18 Read these details about hormone patches which men can stick onto their skin.

- 100 men aged 65 and over were randomly given a testosterone patch or a look-alike dummy patch.
- Scientists were trying to determine whether raising the men's testosterone levels to that of a 40-year old would result in improved levels of bone calcium, muscle size, strength and sex drive.
- At the start of the study, both groups of men had average testosterone levels of 0.10 mg per man.
- Those being treated with the testosterone patch reached average levels of about 0.50 mg per man. This is the normal level of a 40-year old man.
- The men's blood was regularly checked for changes in the number of red blood cells and level of cholesterol.

a) Explain why the two kinds of patches were given to the men randomly. *1 mark*

b) Suggest why giving the testosterone by patches is better than injecting it directly into the blood. *1 mark*

c) (i) Suggest one possible side effect of using testosterone patches. *1 mark*
(ii) Explain why this side effect might be harmful. *1 mark*

OCR

Further examination questions

19 The diagram shows part of the urinary system of a human.

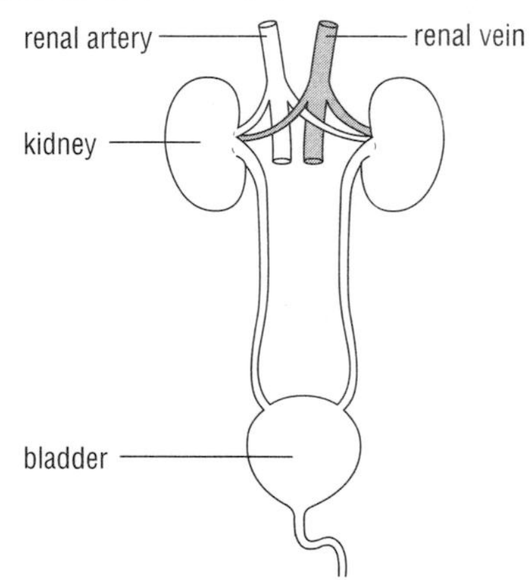

a) Write down the job of the kidney and the job of the bladder. *2 marks*

b) The diagram below shows a kidney cut in half vertically.

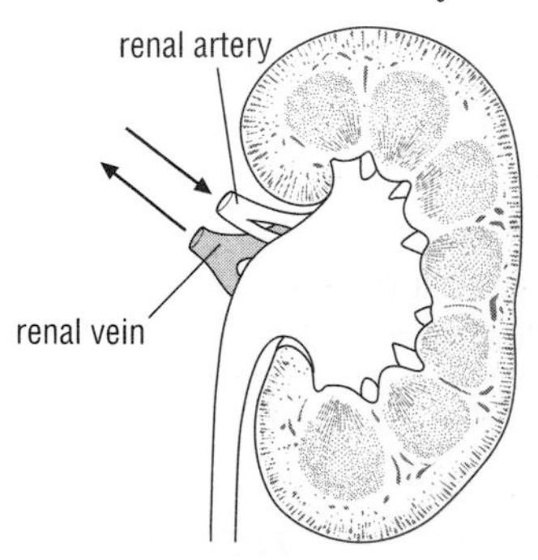

Blood enters the kidney through the renal artery and leaves through the renal vein. The artery has thick walls of elastic tissue. The vein is wider than the artery and has thinner walls.
(i) Explain why the artery and vein have different structures. *2 marks*
(ii) Describe *three* changes that take place in the blood as it passes through the kidney. *3 marks*
(iii) The urine leaving the kidney contains water, urea and other waste substances. Describe how urine is formed. *3 marks*

c) People whose kidneys do not function are treated by a process called dialysis. Blood is taken from a vein and passed between sheets of a membrane. This membrane is surrounded by a solution of salts in the same concentration as they usually occur in blood. The blood is then returned to the vein. The process of dialysis is shown in the diagram over the page.
(i) Explain the purpose of the membrane in dialysis. *3 marks*
(ii) Explain how, by having salts in the dialysis fluid, the concentration of salts in the blood is regulated. *2 marks*

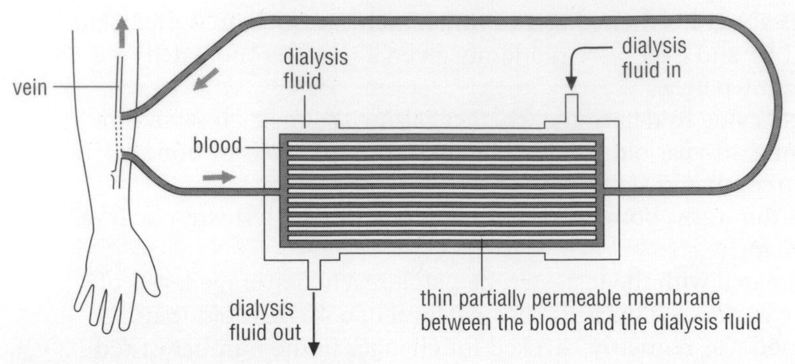

d) The diagram below shows how the water content of blood is controlled when a large volume of water is drunk.

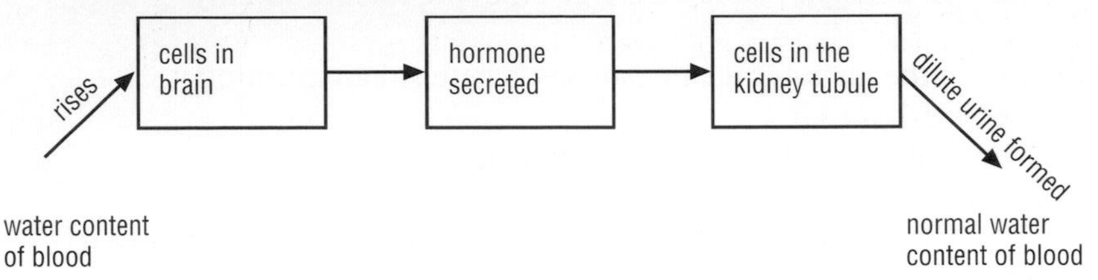

Describe the stimulus, receptor and effector in this system. *3 marks*

OCR

20 The graph shows the levels of the hormones oestrogen and progesterone in a woman's blood during a month when she becomes pregnant.

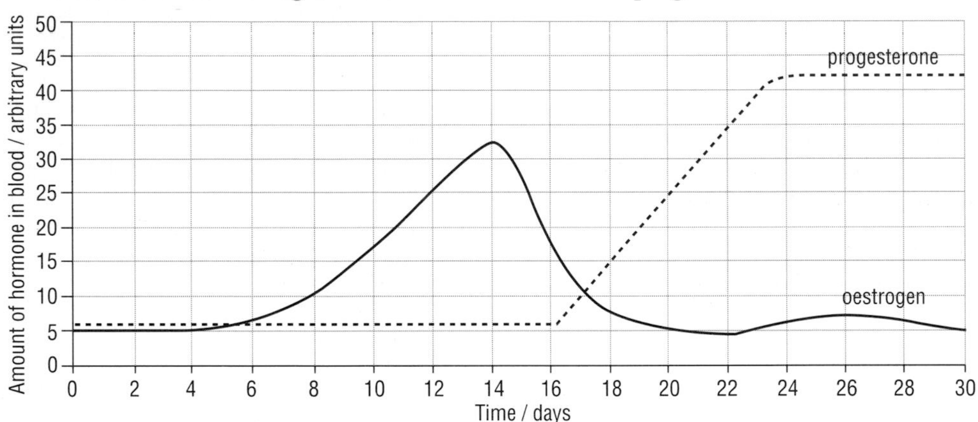

Use the information in the graph and your own knowledge to answer the questions below.

a) When are the levels of oestrogen and progesterone equal? *1 mark*
b) Which process occurred between day 0 and day 5? *1 mark*
c) Give *one* function of each of these hormones.
   (i) Oestrogen *1 mark*
   (ii) Progesterone *1 mark*
d) What evidence from the graph shows that an ovum was fertilised? *1 mark*
e) On the graph draw a line between day 16 and day 30 to show a probable level of progesterone which would be found in the woman's blood if she had not become pregnant. *2 marks*
f) How does the lining of the uterus help in the development of a fertilised ovum? *4 marks*

EDEXCEL

21 Read the following passage which is from an advice book for diabetics.

Insulin Reactions

Hypoglycaemia, or 'hypo' for short, occurs when there is too little sugar in the blood. It is important always to carry some form of sugar with you and take it immediately you feel a 'hypo' start. A 'hypo' may start because:

- you have taken too much insulin, or
- you are late for a meal, have missed a meal altogether, have eaten too little at a meal or
- you have taken a lot more exercise than usual.

The remedy is to take some sugar.
An insulin reaction usually happens quickly and the symptoms vary – sweating, trembling, tingling of the lips, palpitations, hunger, pallor, blurring of vision, slurring of speech, irritability, difficulty in concentration.
Do not wait to see if it will pass, as an untreated 'hypo' could lead to unconsciousness.

a) Many diabetics need to take insulin.
(i) Explain why. *2 marks*
(ii) Explain why there is too little sugar in the blood if too much insulin is taken. *3 marks*
(iii) Explain why there is too little sugar in the blood if the person exercises more than usual. *3 marks*

b) Suggest why sugar is recommended for a 'hypo' rather than starchy food. *3 marks*

c) Explain how the body of a healthy person restores blood sugar level if the level drops too low. *3 marks*

AQA

22 In an experiment, half a field of parsnips was sprayed with growth hormone. The other half of the field was left untreated.

The table below gives the average masses of treated and untreated parsnips.

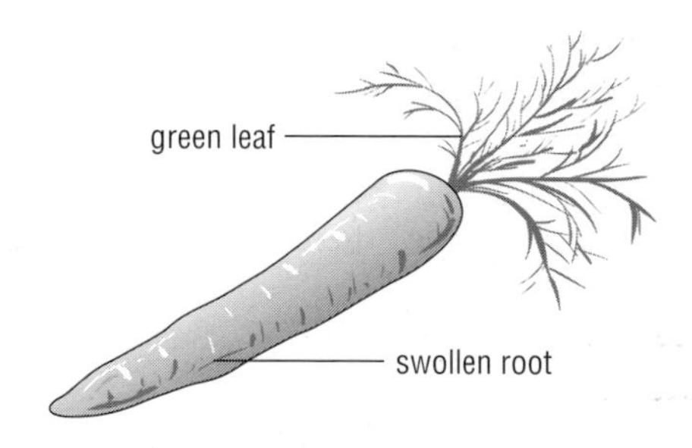

	SHOOT		ROOT	
	Mass when fresh/g	Mass after drying/g	Mass when fresh/g	Mass after drying/g
Hormone treated	300	66	100	25
Untreated (control)	60	24	175	25

a) Calculate the percentage of water in the untreated shoot. *2 marks*

b) What can you conclude about the effect of the hormone on the growth of the parsnips? *4 marks*

c) Parsnip roots are eaten as a vegetable. Is it a good idea to treat the parsnips with growth hormone? Explain your answer as fully as you can. *3 marks*

AQA

9

Chains and cycles

SUMMARY

1 Communities, habitats and ecosystems

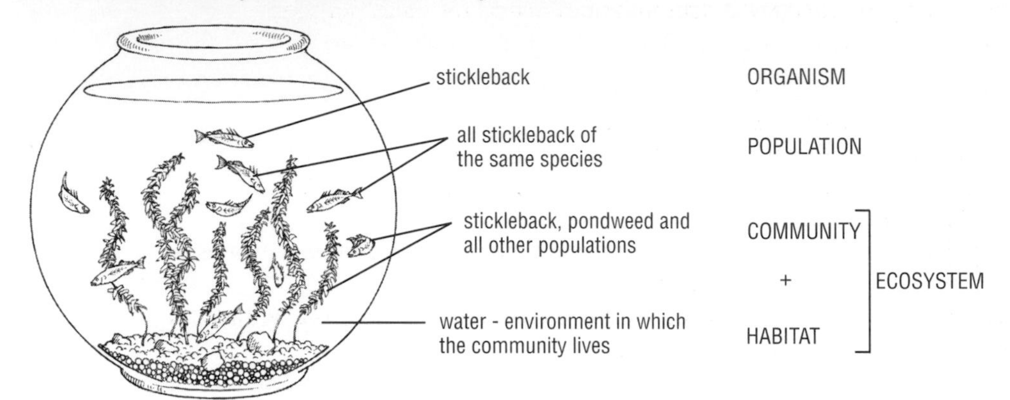

An ecosystem inside a fish bowl

2 Food chains, food webs and pyramids

- Food chains and webs involve the transfer of energy from one organism to another.
- The ultimate source of energy for most ecosystems is the Sun.
- Plants can produce their own food. They are described as **producers** in an ecosystem.
- Animals consume plants or other animals for their food and energy. They are called **consumers**.
- Sequences of feeding interactions from producers to consumers such as seaweed → crab → seagull are called **food chains**.
- Interlinked food chains form a **food web**.
- The number of organisms in the populations at each step in a food chain can be presented as a **pyramid of numbers**.

Organisms	Pyramid of numbers	Trophic level	Pyramid of biomass
heron	1	4	1 kg
fish	1000	3	10 kg
water insects	1 000 000	2	100 kg
leaves of pond weed	1 000 000 000	1	1000 kg

Pyramids of numbers and biomass for the trophic levels in a short stream

- A more accurate description of the food and energy transfer in a food chain is shown by a **pyramid of biomass** based on the total dry mass (biomass) at each **trophic level**.
- Pyramids of biomass, unlike pyramids of numbers, are almost always neat pyramids with the biomass of organisms decreasing from one trophic level to the next. This indicates a loss of energy (food) along the chain due to:
  - respiration using energy for movement and warmth;
  - energy (heat) lost to the surroundings;
  - energy and materials lost in faeces and urine.
- Food production can be made more efficient by:
  - reducing the number of stages in our food chains (eating cereals rather than meat from animals fed on cereals);
  - restricting the energy loss from animals reared for meat (limiting their movements).

3 Energy chains and important cycles

- The source of energy for every food chain is the Sun.
- Energy from sunlight allows plants (producers) to photosynthesise. This converts light energy into chemical energy in carbohydrates, fats and proteins in plants.
- The chemical energy in these substances is then passed along the food chain to animals (consumers).
- Eventually, all the energy is lost from the chain as the animals die and decay, releasing carbon dioxide, water and nitrogen.

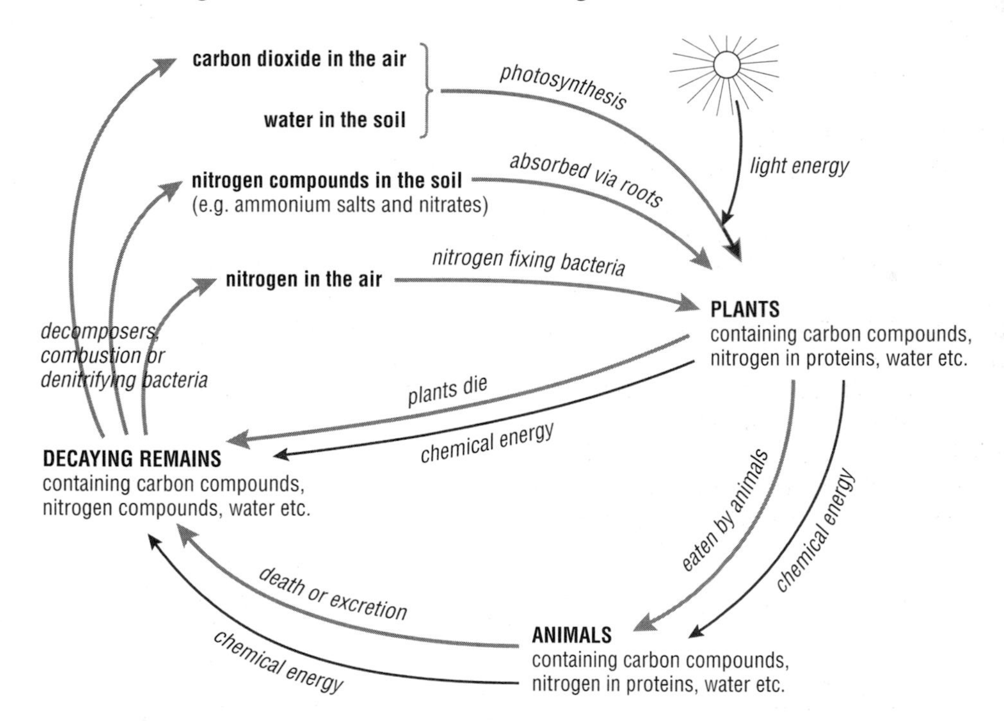

The energy chain and cycles for carbon, nitrogen and water. (Where would you put arrows for respiration? Two arrows have been omitted)

- Superimposed on this energy chain there is a continual cycling of the elements and compounds in plants and animals, including carbon, nitrogen, water and minerals.

STUDY QUESTIONS

Objective questions

Questions 1 to 6
The diagram below shows a food chain.

Oak tree → aphids → lacewing larvae → blue tits → sparrow hawk

1 Name the producer in this food chain.

2 Name the secondary consumer.

3 At which trophic level do the blue tits feed?

4 What is the source of energy for the oak tree?

5 What do the arrows indicate?

6 What is the prey for the blue tits?

Questions 7 to 10
The diagram shows part of the nitrogen cycle.

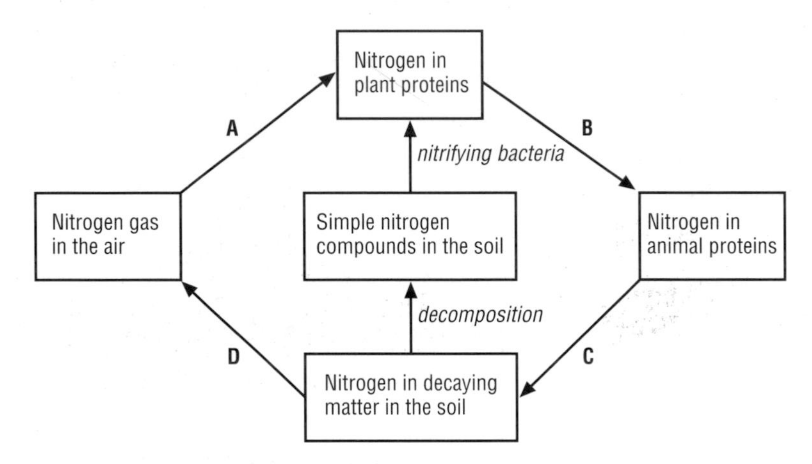

7 Name the process along arrow A.

8 Name the process along arrow B.

9 Name the process along arrow C.

10 Name the process along arrow D.

Questions 11 to 13
Here is a newspaper report about carbon dioxide in the atmosphere.

> The amount of carbon dioxide in the atmosphere has increased in the last 20 years. This is because we burn so much fossil fuel. Scientists think that the extra carbon dioxide is changing the Earth's climate. This will cause sea levels to rise and could lead to many other problems.

11 How do scientists expect the climate to change?

12 What is the name for this change?

13 Why will rising sea levels cause problems?

Short questions

14 In a hedgerow ecosystem, sparrowhawks eat sparrows and thrushes; sparrows eat grass seeds; thrushes eat blackberries and snails; snails eat grass and leaves.

a) Use the information to draw a food web of this ecosystem. *4 marks*
b) (i) Microbes are decomposers. Add decomposers to the food web to show how they link in with the other organisms. *1 mark*
(ii) Decomposers are important in an ecosystem. Explain why they are important. *2 marks*

OCR

15 The diagram shows energy transfer in a food chain.

a) Which is the primary consumer? *1 mark*
b) What is happening to the energy at X? *1 mark*
c) Calculate the percentage of energy which is ultimately transferred from grass to humans. *2 marks*
d) Give *two* reasons why cattle fed on silage in sheds lose less energy than cattle grazing in an open field. *2 marks*

16 The energy content of each population in one hectare of a heathland foodchain is shown below.

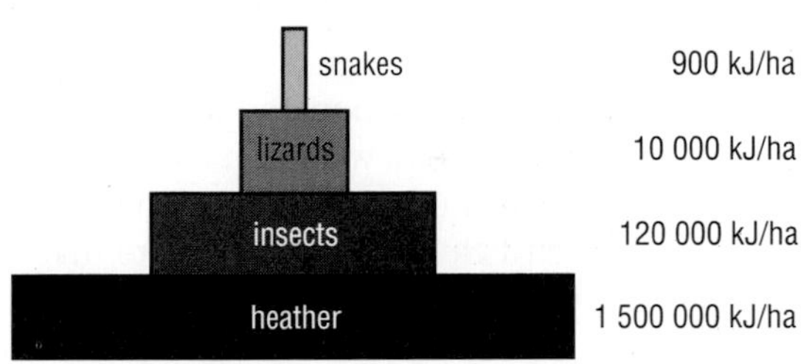

a) What percentage of energy in the heather plants reaches the snakes? *2 marks*
b) Some energy is not transferred from one level to the next. Suggest *three* ways in which this energy is lost. *3 marks*
c) If a dead lizard is not eaten by another animal, what happens to the usable energy in it? *2 marks*

17 Cow pats in a field are broken down by decay organisms such as bacteria and fungi.

a) Maggots and dung beetles tunnel through the cow pats and this increases the rate of decay. Why do the tunnels increase the rate of decay? *2 marks*
b) The grass at the edge of cow pats is often longer and greener than that in other parts of the field. Why is this? *3 marks*

18 The sequence below shows part of the nitrogen cycle affecting dead leaves and some of the organisms involved.

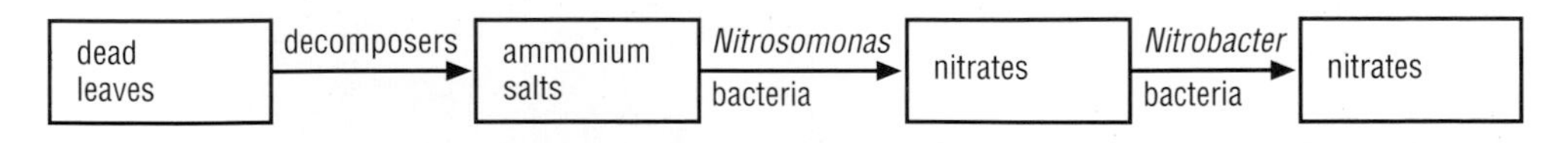

a) Name *two* groups of organisms which decompose leaves. *2 marks*
b) Give *four* effects on the sequence above if *Nitrosomonas* bacteria died out. *2 marks*
c) Suggest *two* factors which would speed up the decomposition of leaves. *2 marks*

Further examination questions

19 Barn owls nest in barns and other old buildings. They often hunt along the sides of roads. Their main prey are voles. Voles eat seeds and plant shoots.

a) Use this information to draw a food chain. Write the names of the organisms with arrows. *2 marks*

b) In recent years, the number of barn owls has gone down rapidly. One reason might be that more owls are being hit by cars. Suggest one other reason for the decline in the barn owl population. Explain your answer as fully as you can. *2 marks*

Every five years a bird club carries out a survey of barn owls in their county. The results are shown on the graph.

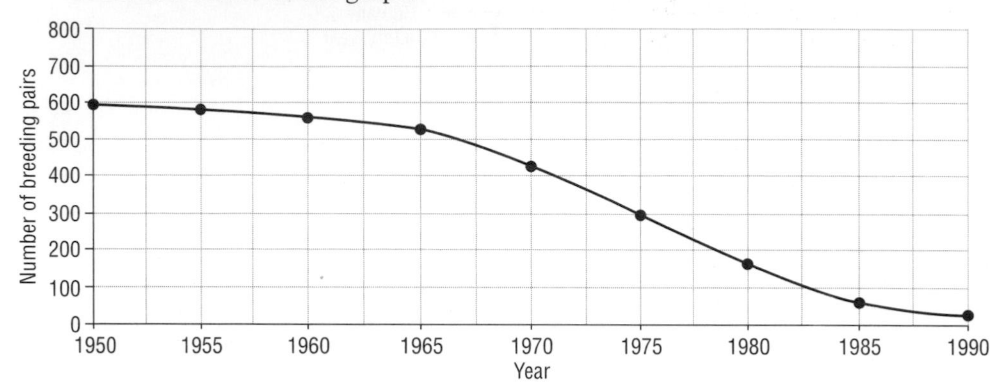

c) A survey of vole numbers was also carried out over the same time. At first, the number of voles went up rapidly. Then the numbers levelled off. Suggest why the number of voles changed in this way. Explain as fully as you can. *2 marks*

d) The bird club members want to make sure that their surveys are as accurate as possible. Suggest *two* ways they could do this. *2 marks*

OCR

20 This question is about a food web in an oak woodland. Foxes and tawny owls eat woodmice. The tawny owls also eat blue tits. The blue tits feed on moths and greenfly. Woodmice, moths and greenfly feed on the oak trees.

a) Use the information to complete the food web found in an oak tree. *3 marks*

One year there are not as many greenfly as usual. This could make the moth population increase or it could make it decrease.

b) (i) Explain why this could make the moth population increase. *1 mark*
(ii) Explain why this could make the moth population decrease. *1 mark*

c) Draw a pyramid of biomass for the oak trees, woodmice and tawny owls. Label the parts of your diagram. *1 mark*

d) Look at the drawing of a tawny owl. Owls hunt mice and other small animals at night. How is the owl adapted to this way of life? Write down *three* features shown in the drawing. Explain your answers. *3 marks*

OCR

21 This question is about decay. Look at the list of things Eileen finds in her local park.

aluminium can   animal droppings   broken bottle   dead worm   fallen branch
leather shoe

a) (i) Write down *two* things from the list that would decay quickly. *2 marks*
(ii) Write down *two* things from the list that would not decay. *2 marks*

Eileen knows that bread can decay (go mouldy). She wants to see if she can make bread decay more quickly. She puts pieces of bread in sealed plastic bags. She then leaves the bags in these three places – in a fridge, in a cupboard and by a radiator.

b) (i) Where would the bread probably decay most quickly? *1 mark*
(ii) Explain your choice as fully as you can. *3 marks*

Eileen wants to make bread decay even more quickly.

(iii) Write down one thing she can add to the bread to help it decay more quickly. *1 mark*

OCR

22 The diagram shows the energy flow through an ecosystem. The width of the arrows represents the relative amount of energy transferred at each stage.

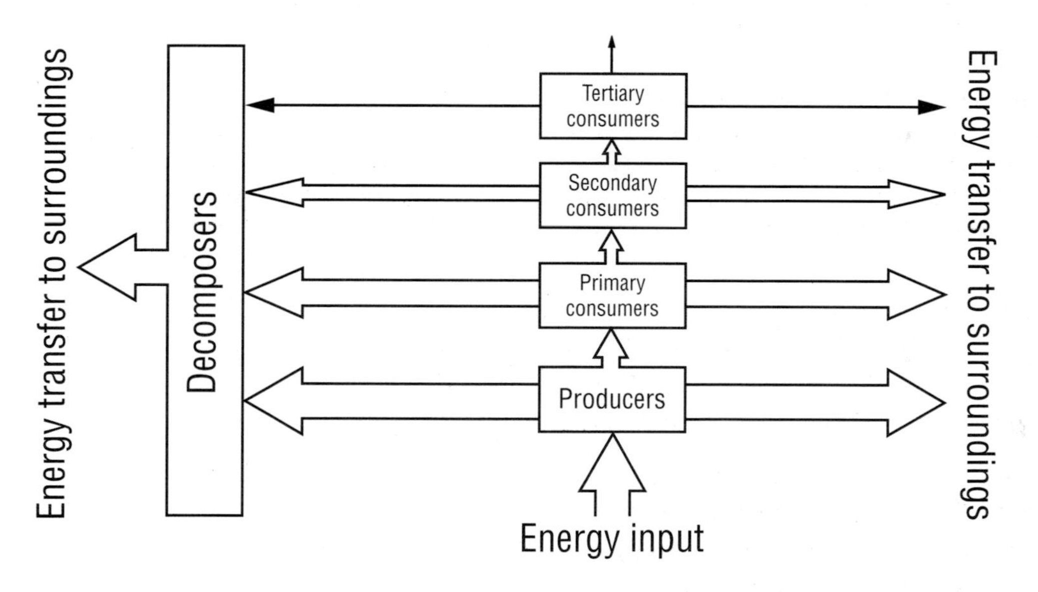

a) Give *two* life processes which transfer energy from the consumers to the surroundings. *2 marks*
b) What is likely to happen to the biomass of primary consumers when the rate of photosynthesis increases? Give *one* reason for your answer. *2 marks*
c) A farmer wanted to grow a good crop of barley without using a fertiliser. He grew pea plants in the field the previous year to encourage a good crop of barley. Explain how growing the pea plants could improve the barley crop the following year. *3 marks*
d) Describe the chemical changes involved when nitrifying bacteria decompose dead plant material. *3 marks*

AQA

CHAPTER

10

Reproduction and genetics

SUMMARY

1 The nucleus, chromosomes, genes and DNA

- The **nucleus** of every cell contains long threadlike structures which are visible after staining through a light microscope.
- The threadlike structures are **chromosomes** composed of molecules of the polymer **DNA** (**deoxyribonucleic acid**).
- Sections of the DNA polymer in each chromosome make up a **gene** and each gene controls the synthesis of a particular **enzyme**.

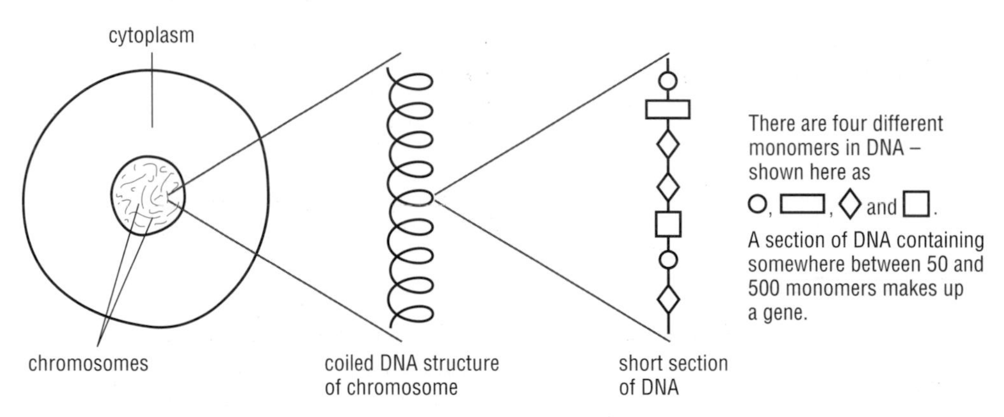
cytoplasm
chromosomes
coiled DNA structure of chromosome
short section of DNA
There are four different monomers in DNA – shown here as ○, ▭, ◇ and □.
A section of DNA containing somewhere between 50 and 500 monomers makes up a gene.

The monomers in DNA, genes and chromosomes

- Humans have 46 chromosomes which are present in all cells (except sex cells – **gametes**) as 23 **homologous pairs**. One chromosome of the pair comes from the male parent and the other from the female parent. Gametes have half the number of unpaired chromosomes (i.e. 23 in humans).
- During fertilisation, a **sperm cell** (male gamete) fuses with an **egg cell** (female gamete) to form a fertilised egg called a **zygote**. The zygote grows by dividing into two cells, then again into four cells and so on, eventually forming an **embryo**.

2 Comparing mitosis and meiosis

In the diagram, we have assumed that the cell has only four chromosomes (two homologous pairs) for simplicity.

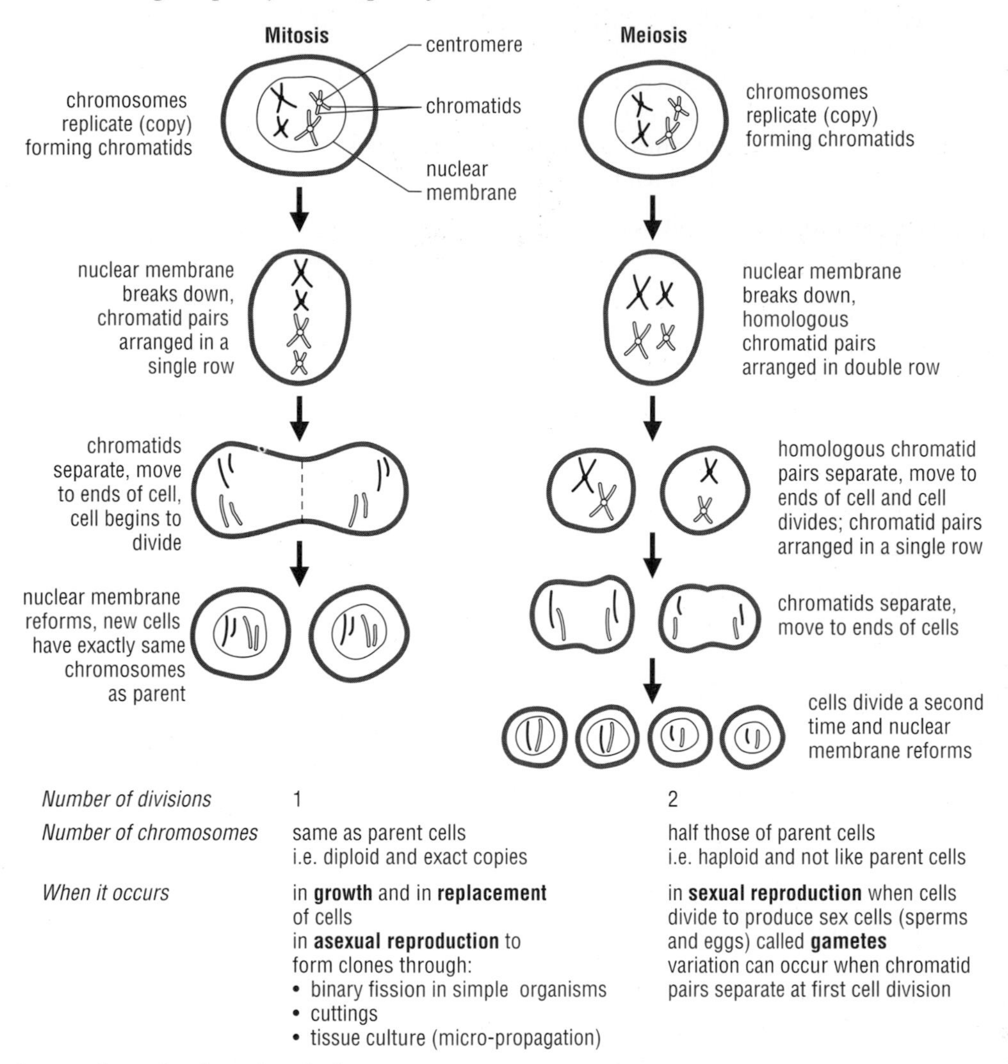

	Mitosis	Meiosis
Number of divisions	1	2
Number of chromosomes	same as parent cells i.e. diploid and exact copies	half those of parent cells i.e. haploid and not like parent cells
When it occurs	in **growth** and in **replacement** of cells in **asexual reproduction** to form clones through: • binary fission in simple organisms • cuttings • tissue culture (micro-propagation)	in **sexual reproduction** when cells divide to produce sex cells (sperms and eggs) called **gametes** variation can occur when chromatid pairs separate at first cell division

Comparing mitosis and meiosis

STUDY QUESTIONS

Objective questions

Questions 1 to 3

Humans reproduce by internal fertilisation and internal development of the embryo in the uterus. Look at the list of animals below.

A thrush B frog C trout D cat E grass snake F dolphin

Use the letters A to F to identify:

1 two animals which reproduce by internal fertilisation and internal development.

2 two animals which reproduce by internal fertilisation and external development.

3 two animals which reproduce by external fertilisation and external development.

Questions 4 to 6

Sperm cells may have an X or a Y chromosome. Egg cells have only X chromosomes. During fertilisation the following may fuse:

A one X sperm cell with one egg cell.
B one Y sperm cell with one egg cell.
C two X sperm cells separately with two egg cells.
D two Y sperm cells separately with two egg cells.
E one X sperm cell with one egg cell which then form two embryos.
F one Y sperm cell with one egg cell which then form two embryos.

Which of the processes A to E will result in

4 a boy?

5 identical twin boys?

6 non-identical twin girls?

Questions 7 and 8

The diagram shows a human sperm cell.

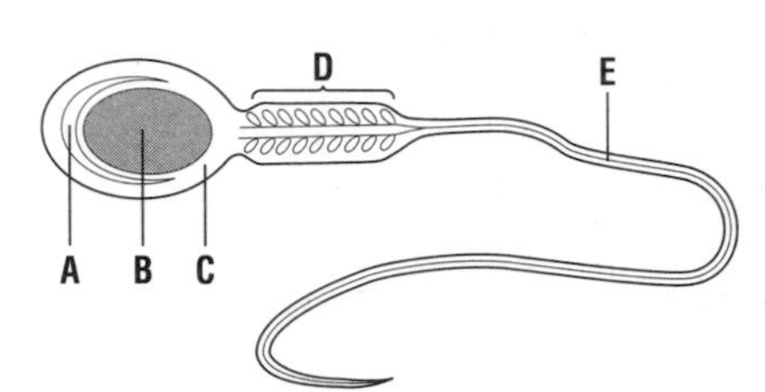

7 Which part produces the energy necessary for movement?

8 Which part is the nucleus?

Questions 9 to 11

Consider the following arrangements of sex chromosomes.

A X only B Y only C XX D YY E XY

9 Which arrangement represents a male zygote?

10 Which arrangement represents an ovum?

11 Which arrangement represents a female zygote?

12 Which of the following techniques can be used to breed new types of disease-resistant barley?

A cloning from cuttings
B cross pollination and selection
C treating plants with pesticide
D growing barley in harsh conditions

Short questions

13 The diagram shows two ways in which a hen cell can divide. The number of chromosomes in each cell is shown.

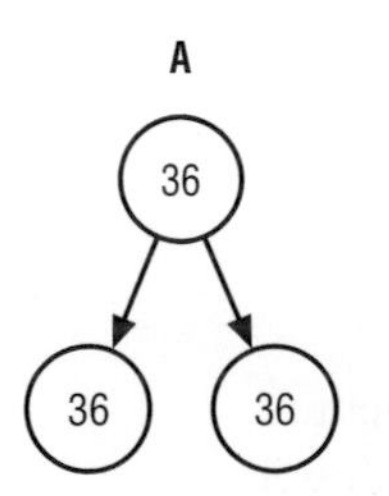

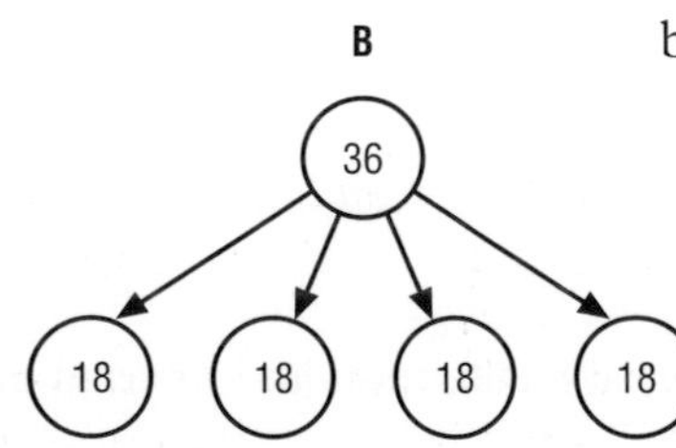

a) (i) What type of cell division is A? *1 mark*

(ii) What type of cell division is B? *1 mark*

b) Most hen cells have 36 chromosomes. However, egg cells only have 18 chromosomes. Why is it important for egg cells to have only 18 chromosomes? *2 marks*

14 The diagram shows a sperm cell swimming towards an egg cell.

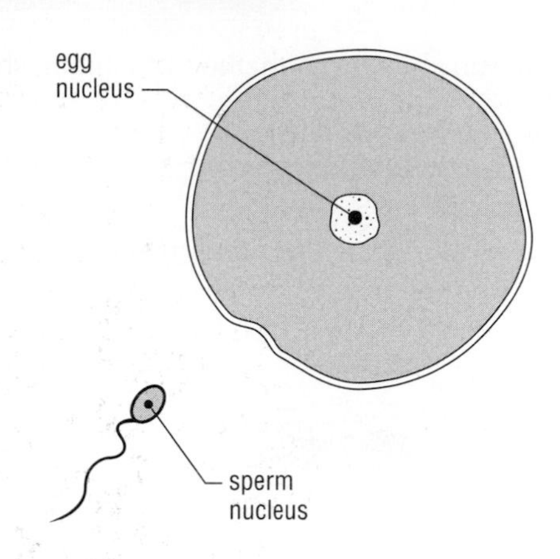

Complete the following sentences.

When the sperm cell enters the egg cell, the two nuclei fuse together. This process is called ________ . The new cell formed is described as a ________ . The nucleus of the new cell contains chromosomes from each parent made up of a complex polymer called ________ . This complex polymer is divided into sections called ________ .
4 marks

15 Variation in organisms occurs when gametes combine.

a) What type of cell division takes place when gametes form? *1 mark*
b) Explain the other source of genetic variation which occurs during this type of cell division. *1 mark*
c) How many different gametes can form from a simple organism with only three homologous pairs of chromosomes? Explain your answer. *2 marks*

Further examination questions

16 Newts are amphibians. This question is about the way they breed. The male newt does a special tail-waving dance and then produces a small sac of sperm. The female newt pulls the sperm sac into her body. After a few days, she lays the fertilised eggs one by one onto pondweed. Later, the eggs hatch as newt tadpoles.

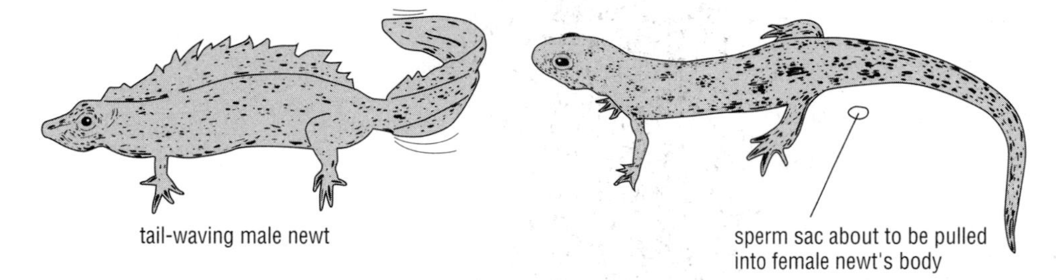

a) Suggest why the male newt does the tail-waving dance. *2 marks*
b) The sperm fertilise the eggs inside the female's body. This is called internal fertilisation.
(i) Write down *one* advantage of internal fertilisation. *1 mark*
(ii) Write down the name of *one* vertebrate animal which has external fertilisation. *1 mark*
c) The 30 to 60 eggs laid by the female newt develop outside her body. This is called external development.
(i) Write down *one* advantage of external development. *1 mark*
(ii) Write down *one* disadvantage of external development. *1 mark*
(iii) Write down the name of *one* vertebrate animal which has internal development. *1 mark*
d) The fertilised egg cells from the female newt develop into tadpoles. What changes must happen to these cells to produce this development? Write down *two* changes. *2 marks*

OCR

17 The diagram shows one method of cloning sheep.

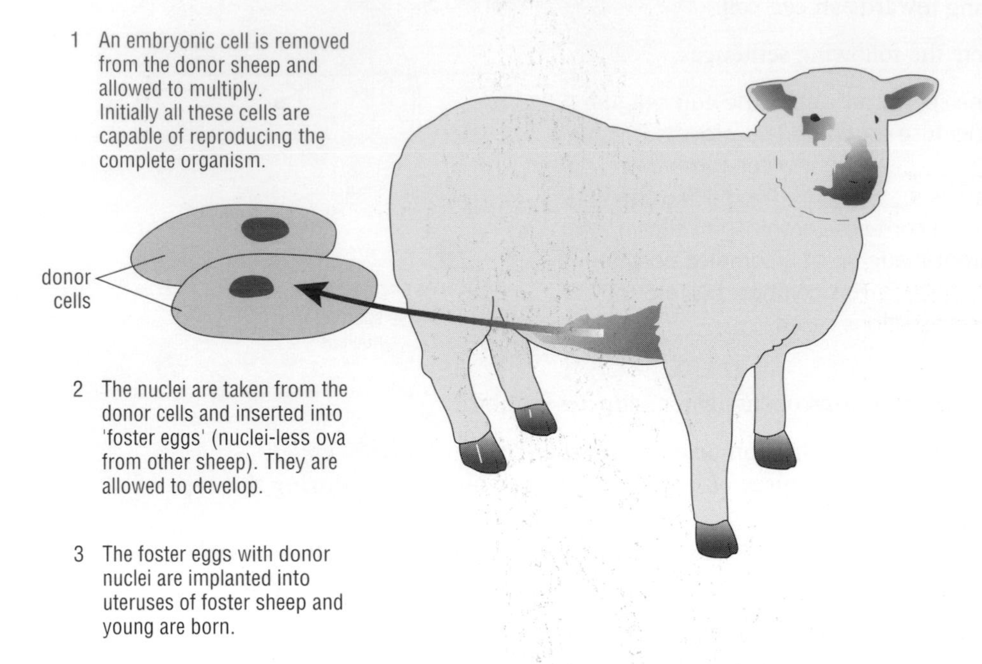

1 An embryonic cell is removed from the donor sheep and allowed to multiply. Initially all these cells are capable of reproducing the complete organism.
donor cells
2 The nuclei are taken from the donor cells and inserted into 'foster eggs' (nuclei-less ova from other sheep). They are allowed to develop.
3 The foster eggs with donor nuclei are implanted into uteruses of foster sheep and young are born.

a) Explain why the lambs produced by this technique are identical to each other. *2 marks*

b) Explain why the lambs are not genetically identical to the sheep which produced the 'foster' eggs. *2 marks*

c) Explain the drawback of widespread use of just a few clones of sheep. *3 marks*

AQA

18 The diagram shows three types of cells in a life history of a simple animal.

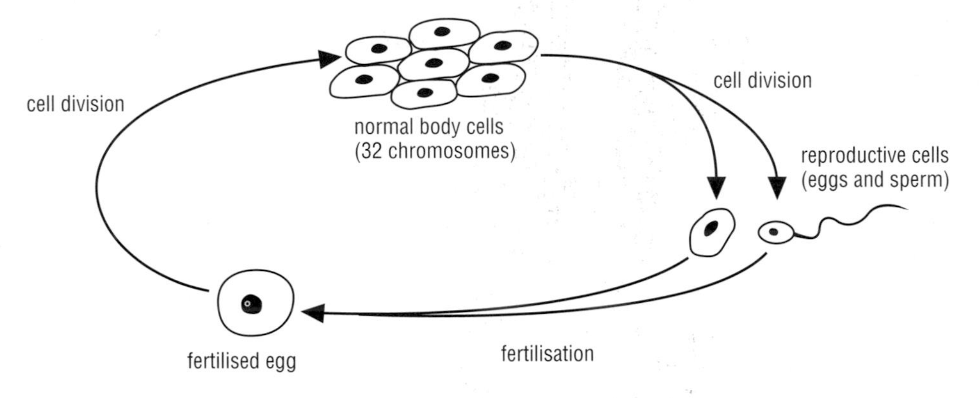

a) How do the chromosomes of the body cells compare with the chromosomes in the fertilised egg from which they came? *1 mark*

b) Describe what happens to chromosomes in the nucleus of a body cell when it forms reproductive cells. *4 marks*

AQA

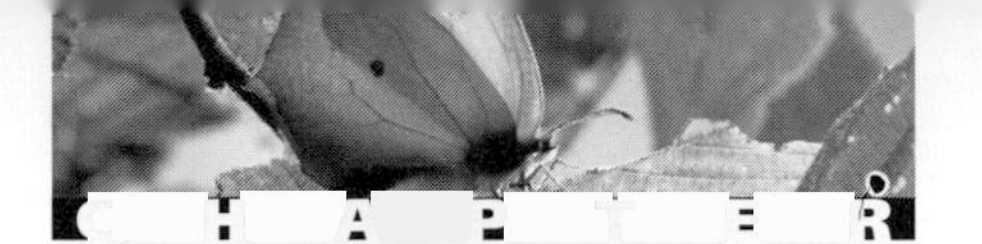

11

Genetics and inheritance

SUMMARY

1 Genetics and genes

- **Genetics** is the study of genes and inheritance.
- Chromosomes contain long threads of DNA and sections of this DNA are divided up into genes.
- Genes control the making of proteins, including enzymes which can in turn determine a characteristic such as eye colour.
- Some characteristics are controlled by a single pair of genes. One gene in the pair comes from the mother, the other comes from the father. This is called **monohybrid inheritance**.
- Genes may be **dominant** (brown eyes, B) or **recessive** (blue eyes, b). Dominant genes always express themselves. Recessive genes only express themselves if *both* of the genes in the pair are recessive. The alternative genes which can make up a pair, e.g. B and b, are called **alleles**.
- The possible pairing of genes (e.g. BB, Bb or bb) are called **genotypes** (genetic types). These genotypes determine the characteristics (**phenotype**) of an organism.
- Genotypes with two identical genes (e.g. BB or bb) are described as **homozygous**. Genotypes with two different genes (e.g. Bb) are described as **heterozygous**.

Eye colour phenotypes	brown eyes	blue eyes
Alleles	B – dominant	b – recessive
Possible genotypes	Bb heterozygous	BB bb homozygous

2 Inherited characteristics

- The phenotypes of offspring can be predicted by drawing a Punnett square if the genotypes of the parents are known.

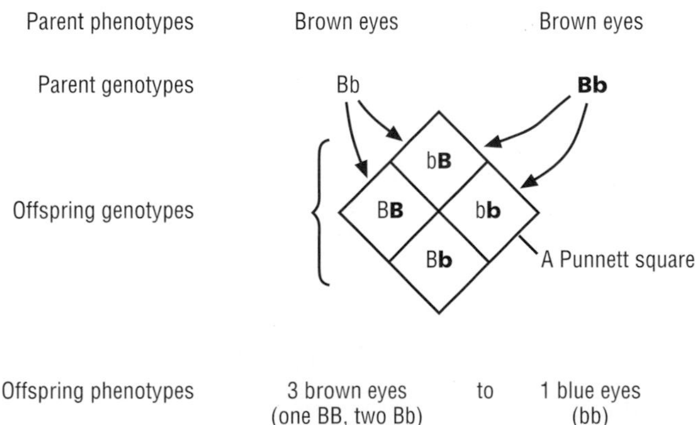

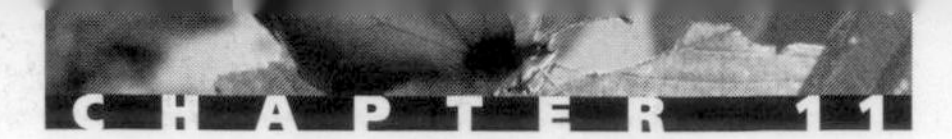

- Some diseases (e.g. cystic fibrosis) are genetic in origin and can be inherited from parents who are simply **carriers**.
- Some other diseases (e.g. muscular dystrophy, haemophilia) are much more common in males than females. This is because they are linked to recessive alleles on the X chromosome and males with an XY chromosome pair are more likely to have the disease than females with an XX pair.

3 Variation

Variations within a species can arise genetically or environmentally. Genetically caused variations result during:

- meiosis (chromosome pairs separate randomly),
- fertilisation (chromosomes from both parents),
- mutation (chromosomes may not copy accurately or they may be changed by chemicals or radiation).

Environmental causes of variation include differences in diet, climate and fashion.

4 Adaptation, selection and evolution

- Variations within a species may result in some individuals being better **adapted** to their environment than others. The animals and plants that are better adapted are usually the fittest (i.e. avoiding predators, free of disease, surviving the climate, strong competitors for food, etc.)
- This survival of the fittest and the slow change within a species over many generations is known as **natural selection**.
- The changes which occur in a species over millions of years result in new species (e.g. humans from apes). This is called **evolution**.
- In some cases, the organisms within a species are unable to adapt to changing conditions and this leads to their **extinction**.
- The best evidence for evolution comes from fossils.

STUDY QUESTIONS

Objective questions

Questions 1 to 4

A biologist wrote the following statements after observing a colony of rabbits in a cold climate.

A The rabbits had ears of different lengths.
B Large numbers of rabbits were born.
C Short ears are an advantage in a cold climate as they lose heat less readily.
D Long-eared rabbits tend to produce young with long ears.

Charles Darwin recorded some important observations about animals as evidence for natural selection and evolution. Which of the biologist's statements, A to D, about rabbits match Darwin's observations in questions 1 to 4?

1 Some characteristics are inherited.

2 All populations of organisms show variations.

3 Organisms which are best adapted to their surroundings will survive.

4 Most populations over produce.

Questions 5 to 10

Sickle cell anaemia is a genetic disease. People with the disease have two alleles (SS) which cause the production of sickle-shaped red blood cells compared with the normal allele, A. People with the allele A are normally healthy although they may be carriers. Look at the family tree. The genotypes of Q and R are already shown.

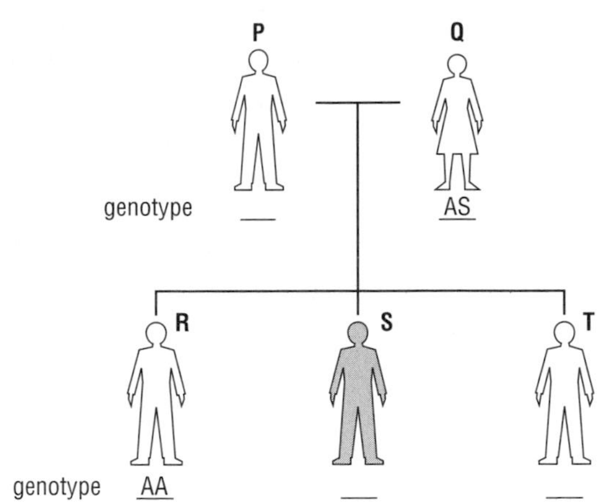

5 What is the genotype of P?

6 What is the genotype of S?

7 What are the possible genotypes of T?

8 Which two people are definitely homozygous?

9 What is the phenotype of S?

10 What are the possible phenotypes of T?

Questions 11 to 14

In questions 11 to 14 choose from A, B, C or D which is the correct answer.

11 All male (sperm) gametes contain
A one X chromosome
B one Y chromosome
C either one X or one Y chromosome
D both an X and a Y chromosome

12 A healthy couple produce a child with cystic fibrosis. The gene (allele) for cystic fibrosis is recessive. This means that
A one parent is homozygous and the other is heterozygous
B both parents are heterozygous
C both parents are homozygous dominant
D both parents are homozygous recessive

13 When organisms are well adapted to their surroundings, they usually breed more successfully than those species which are poorly adapted. Over many generations the well-adapted species may change and this change is called
A evolution
B mutation
C selection
D variation

14 New types of disease-resistant wheat may be produced by
A cloning from diseased species
B growing diseased species
C selection after cross-fertilisation
D spraying with pesticide

Short questions

15 Cells from the eye of a fruit fly contain eight chromosomes.

a) How many chromosomes are there in the cells of the fruit fly from the following body parts
(i) sperm, (ii) leg, (iii) fertilised egg? *3 marks*

b) Why do all cells of the fruit fly not contain the same number of chromosomes? *2 marks*

c) During cell division things can go wrong with chromosomes.
(i) What is this called? *1 mark*
(ii) Suggest one way in which the chromosomes can go wrong. *1 mark*

16 Some adult rabbits normally have black ears, tail, nose and feet.

Scientists wanted to find out whether the black colour was controlled by a gene or by the cold conditions in which the rabbit grew up. A cold pad was placed on the back of a young rabbit. A warm pad was placed on the same rabbit's ear. When the pads were removed, the rabbit looked like the diagram below.

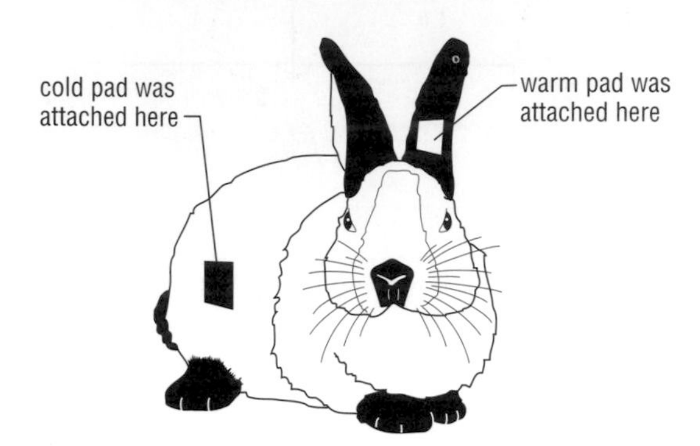

a) How did this treatment affect the colour of the rabbit's fur? *1 mark*

b) Use these results to suggest why the adult rabbits normally have white bodies and black ears, feet, nose and tail. *2 marks*

c) Two rabbits that had warm pads on their ears and body were mated together. All their offspring had black ears. Suggest an explanation for this. *2 marks*

OCR

17 A pale blue male butterfly and a dark blue female butterfly mated producing a range of offspring. A biologist caught five of the offspring – four were pale blue and one was dark blue.

a) Why might the biologist suspect that the pale blue colour is dominant? *1 mark*

b) Why can the biologist only 'suspect' that pale blue is dominant? *2 marks*

c) Using the symbol P for the pale allele (assumed to be dominant) and p for the recessive allele, show the gene arrangement (genotype) for
(i) the male parent,
(ii) the female parent,
(iii) the possible offspring. *3 marks*

18 a) What is meant by the term selective breeding? *2 marks*

b) Give *two* examples of selective breeding. *2 marks*

c) What name is given to the differences between individuals in the same species? *1 mark*

d) Suggest *two* ways by which differences in the same species arise. *4 marks*

Further examination questions

19 Ricky breeds rabbits. He had one black male (genotype BB) and one white female (genotype bb). He allowed them to mate. All of the babies in this first litter were black. Ricky kept two of these baby black rabbits. When they were adult, he allowed them to mate and produce several lots of babies. A quarter of the babies were white and the rest were black.

a) Explain these results using a genetic diagram. *4 marks*

b) Ricky wants to produce a family with equal numbers of black and white rabbits.

(i) What must the genotypes of the parents be in order to get this result? *1 mark*

(ii) Suggest which rabbits he could use to be sure of this result. *1 mark*

c) The genetic information is carried in the chromosomes of each cell. During nuclear division, the genetic information is passed on to the next generation. The diagram shows the type of nuclear division that produces sex cells (gametes).

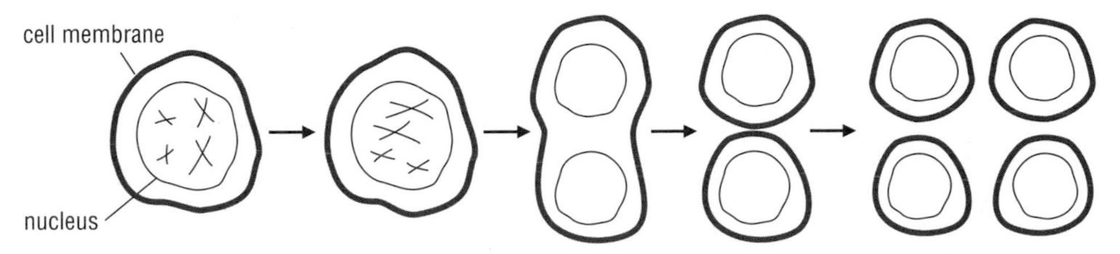

(i) Redraw the diagram and complete it by drawing in the chromosomes. The first two have been done for you. *3 marks*

(ii) Name the type of nuclear division that produces sex cells (gametes). *1 mark*

(iii) Explain how this type of nuclear division followed by fertilisation leads to genetic variation. *2 marks*

d) Mutation is the cause of further variation. What is meant by mutation? *1 mark*

OCR

20 Darwin found finches on the Galapagos Islands. The diagram shows the heads of some of these finches and the foods they eat.

a) Which feature shows the most variation? *1 mark*

b) What is the advantage to the finches of this variation? *1 mark*

c) Explain how this variation has been brought about by evolution. In your answer you should write about genetic variation, natural selection and reproductive success. *4 marks*

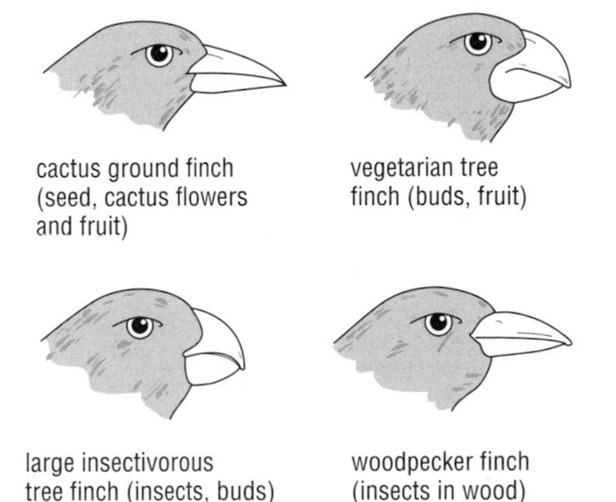

EDEXCEL

21 a) Haemophilia is a sex linked inherited disease. The table gives some data about the five possible combinations of genotype and phenotype. Redraw and complete the table by

(i) drawing the correct chromosomes in boxes 2, 3 and 5;

(ii) writing the alleles for normal blood clotting (N) and haemophiliac trait (n) in the correct positions on the chromosomes;

(iii) showing the sex of the person having these chromosomes in their nuclei. *6 marks*

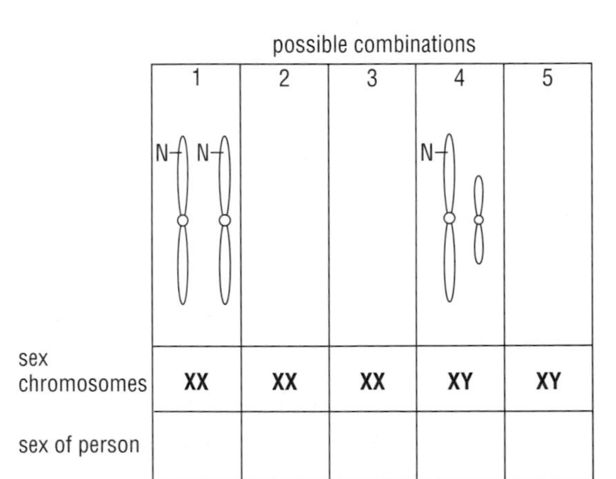

	possible combinations				
	1	2	3	4	5
	N– N–			N–	
sex chromosomes	**XX**	**XX**	**XX**	**XY**	**XY**
sex of person					

b) Sex is determined by the pair of sex chromosomes. How many other pairs of chromosomes are there in the nucleus of a human cheek cell? *1 mark*

c) The family tree shows the inheritance of normal clotting and haemophiliac trait alleles.
What is the genotype of person 1? Give reasons for your answer. *3 marks*

d) Why is it very rare to find a haemophiliac female? *2 marks*

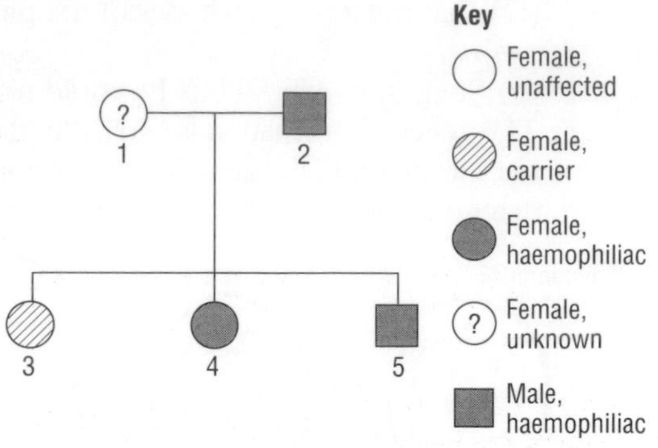

EDEXCEL

22 a) In Koala bears, brown or white hair is controlled by two alleles. B for brown hair is dominant to b for white hair.
(i) Name the part of the nucleus of a cell which contains the alleles. *1 mark*
(ii) Redraw and complete the table. Homozygous means 'having the same alleles from both parents'. *3 marks*

Genotype (alleles present)	Phenotype (outward appearance)	Homozygous or Heterozygous
	white	
BB		homozygous
Bb		

b) A zoo keeper in Australia had a pair of brown Koala bears. These were mated together and the offspring was white.
(i) What must have been the genotypes of the parents? *1 mark*
(ii) Complete the boxes to show how matings with a white Koala bear would help to decide the genotype of a brown one. *4 marks*

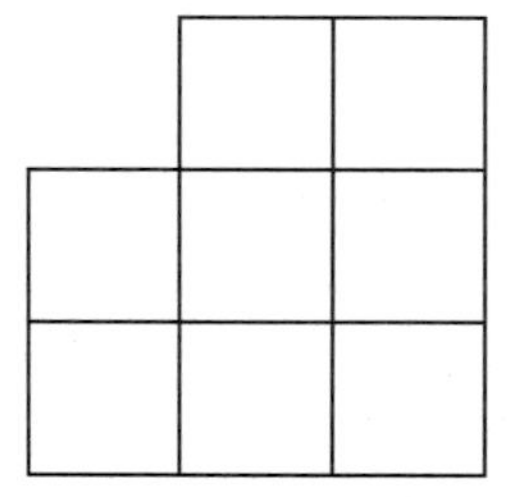

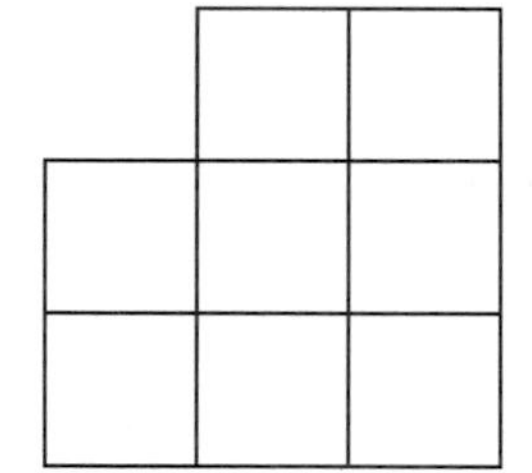

WJEC

23 a) In sweet pea plants, the allele for red flowers is dominant over the allele for white flowers. The allele for red flowers can be represented by R, and r can represent the allele for white flowers.
(i) What is the genotype of a plant with white flowers? *1 mark*
(ii) What are the possible genotypes of a plant with red flowers? *2 marks*

b) A pure breeding (homozygous) red flowered plant is crossed with a pure breeding white flowered plant.
(i) What is the phenotype of all the offspring? *1 mark*
(ii) What is the genotype of all the offspring? *1 mark*

c) Two plants, both with the genotype Rr, are crossed. This produces 100 seeds. Use a genetic diagram to show the possible genotypes of the gametes and offspring and predict the number of red offspring and white offspring. *4 marks*

AQA

CHAPTER

12

Living things in their environment

SUMMARY

1 **Classifying living things**

- Living things can be divided into (classified as) five major groups known as **kingdoms** – animals, plants, fungi, protoctists (e.g. amoeba) and bacteria.
- Kingdoms are divided into **phyla** and then phyla into **classes**. For example, the animal kingdom has two phyla – vertebrates and invertebrates. The vertebrate phylum has five classes – mammals, birds, reptiles, amphibians and fish.
- Further divisions of classes occur, eventually reaching **species**. All the organisms in a species are alike and they can breed together.
- Each species has a common name (e.g. cat) and a systematic name derived from Latin. Systematic names give the **genus** (in italic with a capital first letter) followed by the species (in italic). For example, *Panthera catus* is the systematic name for a cat.

2 **Populations, competition and adaptation**

Living things can only survive as populations if the species is well adapted to its habitat. Various factors can affect organisms and the size of populations.

- The supply of food and water
- Climate related to availability of water and air temperature
- Space or lack of it
- Waste products which accumulate can cause pollution and disease
- Predators.

Life on earth is a competition. Both plants and animals experience competition for food (nutrients), water and space. Many organisms have developed features which enable them to compete more favourably and survive in their normal environment. This is called **adaptation**.

3 Our effect on the environment – pollution

Pollutant	Source	Effect	Method of control
Soot and smoke	Burning fuels	Deposit soot on buildings, clothes, etc.	Use of smokeless fuels. Improve supply of air to burning fuels.
Carbon dioxide	Burning fuels	Increased concentrations of CO_2 in atmosphere cause more heat to be retained by the Earth – 'greenhouse effect'.	Burn less fossil fuels. Develop non-fossil fuel sources of energy
Carbon monoxide	Incomplete burning of fuels e.g. car exhaust fumes	Poisonous to humans and other animals. Prevents haemoglobin from carrying oxygen.	More efficient motor vehicle engines. Fit catalytic converters to vehicles ($CO \rightarrow CO_2$)
Sulphur dioxide and nitrogen oxides	Burning fossil fuels	Acidic non-metal oxides which react with water to form acid rain which harms plants, kills fish and attacks stonework.	Burn less fossil fuels. Remove SO_2 (using basic oxides), NO and NO_2 (catalytic converters $\rightarrow N_2$) from waste gases.
Lead compounds	Vehicle exhaust gases	Poisonous to humans and other animals – harms nervous system.	Reduce lead additives in petrol. Use unleaded petrol.

Air pollution

Pollutant	Source	Effect	Method of control
Sewage	Humans	Bacteria grow on nutrients in sewage, using up O_2 dissolved in the water. Lack of oxygen causes bacteria, aquatic plants and animals to die.	Sewage treatment. Separation of sewage from water supplies.
Artificial fertilisers	Excessive use of fertilisers	Rain water washes fertilisers into rivers and lakes.	Bacteria and water plants grow rapidly, use up all the oxygen and then die. The decaying matter makes the water smelly – eutrophication.

Water pollution

STUDY QUESTIONS

Objective questions

Questions 1 to 4

In certain parts of the world, large areas where plants once grew have become deserts. This process is called **desertification**. Six causes of desertification are:

A fewer plant roots B compacted ground C low rainfall
D minerals in soil used up E over-irrigation F windbreaks removed.

1 Which cause of desertification is not directly caused by humans?

2 Which two causes of desertification are most likely if all trees are cut down?

3 Which cause of desertification is most likely when the same crops are grown year after year?

4 Which cause of desertification is most likely in areas where there are large herds of cattle?

Questions 5 to 7

Look at the three animals.

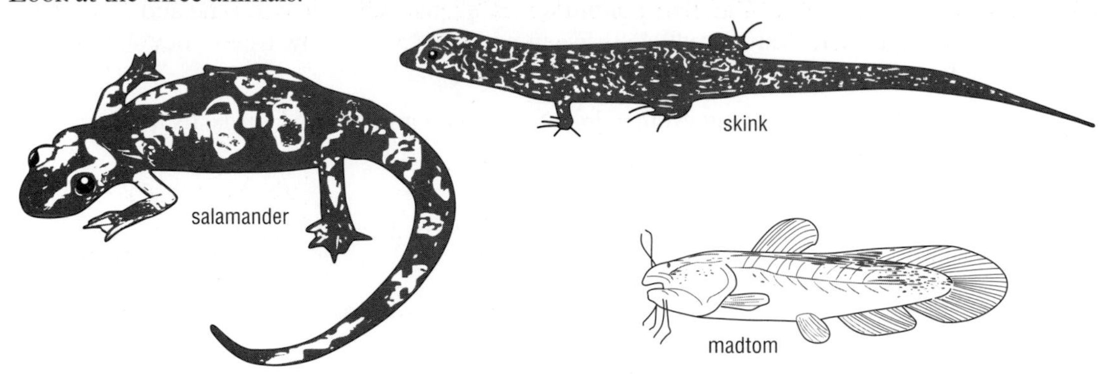

Fish have fins and a tail. Amphibians have webbed toes and smooth skin. Reptiles have open toes and scaly skin.

5 Which of the three animals is a fish?

6 Which of the three animals is an amphibian?

7 Which of the three animals is a reptile?

8 Which *one* of the following fuels does not add to the greenhouse effect?
A coke B natural gas C petrol D uranium

Questions 9 to 12

In questions 9 to 12, choose from A, B, C or D which is the correct answer.

9 The worst two pollutants in car exhaust gases are nitrogen oxides and carbon monoxide. Many new cars are now fitted with catalytic converters which convert these two pollutants to
A carbon and nitrogen
B carbon dioxide and nitrogen
C carbon and nitrogen dioxide
D carbon dioxide and nitrogen dioxide.

10 Lead-free petrol has been developed to reduce
A global warming
B the cost of motoring
C atmospheric pollution
D the use of fossil fuels.

11 A fallen tree is left to decay in a woodland area. What conditions will allow it to decay fastest?
A dry and cold
B dry and warm
C wet and cold
D wet and warm.

12 Which *one* of the following will reduce the efficiency of beef production, but produce healthier cattle?
A Giving the cattle more food.
B Keeping the cattle out of doors throughout the year.
C Leaving calves with their mothers.
D Putting the cattle in small indoor pens.

Short questions

13 Bats are furry, intelligent, social, flying animals. Female bats give birth in June and July. The young bats are fed on milk. Bats live in buildings and hollow trees. They hibernate from October to April.

a) (i) To which of the following groups do bats belong – amphibians, birds, fish, mammals, reptiles? *1 mark*
(ii) Give one reason for your answer. *1 mark*

b) Suggest why it is an advantage for the young bats to be born in June and July. *1 mark*

c) From June to September, adult bats eat large quantities of insects and put on extra weight. Explain why they need this extra weight. *1 mark*

OCR

14 Battery-farmed hens are kept in a controlled environment to maintain a high yield of eggs. Two of the factors that are controlled are water supply and light. Give *four* other factors that can be controlled for battery-farmed hens. *4 marks*

15 A pupil used a 50 cm quadrat to investigate the distribution of daisies in part of a lawn. The number of daisies in each quadrat is shown in the diagram.

(Each quadrat: 50 cm × 50 cm)

1	0	2	1	0
0	3	7	6	1
1	2	8	3	1
1	1	3	2	0
0	0	6	0	1

a) What is the area of each quadrat in cm^2? *1 mark*

b) What is the area of each quadrat in m^2? *1 mark*

c) Calculate the mean number of daisies per square metre in the area of lawn studied by the pupil. *2 marks*

d) Suggest *two* reasons for the uneven distribution of daisies in the area studied. *2 marks*

16 An investigation was carried out to study how different plants are affected by road traffic. The table shows the data collected for three different plants growing by a main road.

Plant type	Average number of plants per square metre at the distance shown from the road				
	0.5 m	1.0 m	1.5 m	2.0 m	2.5 m
dandelion	15.2	12.5	9.9	7.5	5.0
nettle	10.3	8.4	6.2	4.6	3.7
thistle	0.1	0.9	1.6	2.3	0.3

a) Which of the plants grows best closest to the road? Explain your choice. *2 marks*

b) Suggest *two* factors which might affect the growth of plants in such an environment. Say whether the factor will make plants grow better or worse. *4 marks*

c) Suggest a reason for the increase in thistles up to 2 m from the road and then a decrease. *1 mark*

Further examination questions

17 Flamingos are birds which live in warm, shallow lakes in tropical Africa. They feed by sieving single-celled green algae and bacteria from the lake water.

a) Lakes in tropical Africa support large numbers of flamingos. There would not be enough algae in a British lake to feed large numbers of these birds. Suggest one environmental factor which might account for the difference in the number of algae in African and British lakes. *1 mark*
Explain how this factor causes a difference in the number of algae in the lakes. *2 marks*

b) In Kenya, the large numbers of flamingos in one of the lakes, Lake Nakuru, is a big tourist attraction. The Kenyan government want to expand a nearby industrial town. One effect will be an increase in the amount of sewage which flows into Lake Nakuru.
(i) Suggest and explain how increasing amounts of sewage are likely to affect the organisms in the lake. *4 marks*
(ii) Suggest and explain why the Kenyan government wants to increase the size of the industrial town even though this might affect tourism in the area. *3 marks*

AQA

18 a) The diagram shows the transmission of infra-red radiation from the Sun to the Earth's atmosphere.

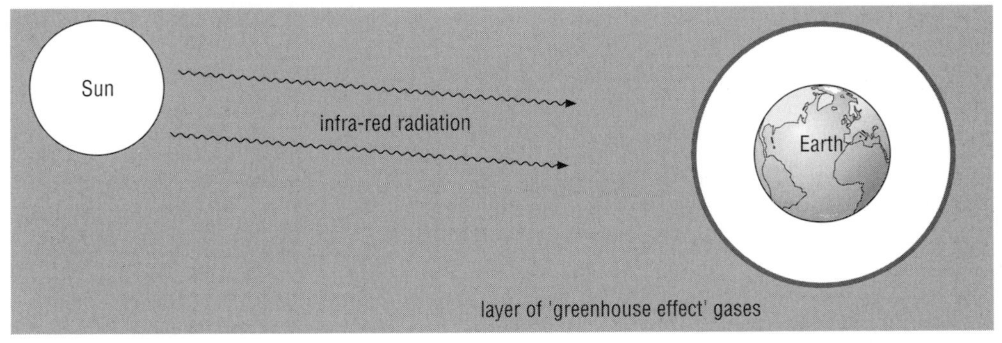

Using the information in the diagram explain how the 'greenhouse effect' may be caused. *3 marks*

b) The table shows a number of 'greenhouse effect' gases, their abundance in the troposphere and their 'greenhouse effect' factors. The troposphere is the bottom 15 kilometres of the atmosphere around the Earth. The 'greenhouse effect' factor is a measure of the 'greenhouse effect' of the gas relative to carbon dioxide.

'Greenhouse effect' gas	Percentage abundance in the troposphere	'Greenhouse effect' factor
carbon dioxide	0.035	1.0
nitrous oxide	0.000 03	160.0
methane	0.000 17	30.0
chlorofluorocarbons	0.000 000 04	23 000.0
water vapour	1.0	0.1
ozone	0.000 004	2000.0
oxygen	20.0	negligible

Use the data to work out which *two* gases have the most powerful 'greenhouse effect' in the troposphere. Show your working. *4 marks*

c) Describe *one* possible sequence of events which could be caused by an increased 'greenhouse effect'. *3 marks*

d) Suggest *two* ways in which 'greenhouse effect' gases could be reduced. *2 marks*

EDEXCEL

19 The pie chart shows the proportions of four greenhouse gases produced by human activities in the 1980s.

a) Calculate the percentage contribution to the greenhouse gases of methane. Show your working. *2 marks*

b) Give *two* ways, other than respiration, by which human activities increase the proportion of carbon dioxide in the atmosphere. *2 marks*

c) What is the principal source of the 'human-made' methane in the atmosphere? *1 mark*

d) Explain how increases in the proportion of greenhouse gases in the atmosphere lead to global warming. *3 marks*

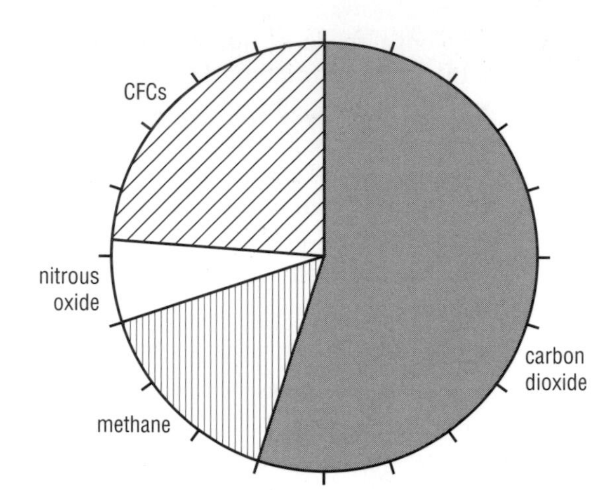

AQA

20 On farm land in the United Kingdom, many fields are bordered by hedgerows. The graph shows how the total length of hedgerows has changed over the past 50 years.

a) Explain *one* advantage to a farmer of removing hedgerows from his fields. *2 marks*

b) Give *two* disadvantages of removing hedgerows. *2 marks*

c) Explain each of the following claims.
(i) The over-use of artificial fertilisers is bad for the soil. *2 marks*
(ii) The over-use of artificial fertilisers can lead to the pollution of rivers. *3 marks*

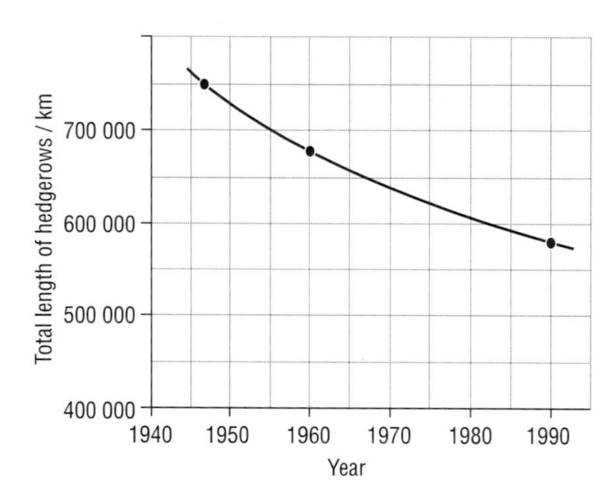

AQA

13

Mixtures, elements and compounds

SUMMARY

1 A **material** is a form of matter which can be used in some way.

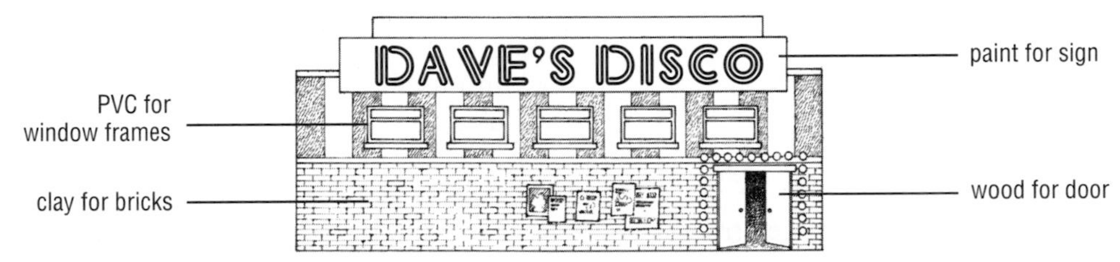

2 **Substances** are different materials which can be identified and named such as oxygen, steel, petrol or grass.

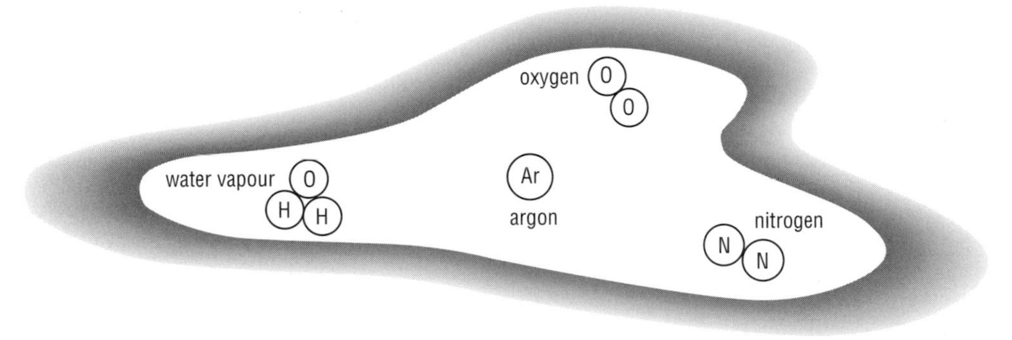

There are four different substances in air

3 Materials can be classified as **naturally-occurring**, such as salt or crude oil, or **man-made**, such as chlorine or polythene. Materials can also be classified and divided into five groups according to their properties.

- metals (e.g. steel, aluminium, copper)
- plastics (e.g. polythene, PVC)
- ceramics or pottery (e.g. china, concrete, bricks)
- glasses (e.g. bottle glass, window glass)
- fibres (e.g. paper, wood, wool)

4 In choosing materials for different uses, two criteria are important:

- the **properties** of the material
- the **cost** of the material.

5 **Composite materials** are made of two different materials which work together to produce a more suitable product than either of the separate materials such as reinforced concrete and bone. Usually, one of the materials provides a matrix and the other provides fibres or rods.

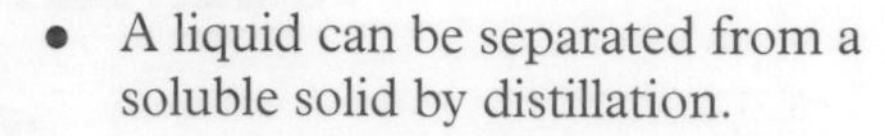

6 Methods of separation

The separation of a mixture depends on a difference in a property of the substances being separated.

- An insoluble solid can be separated from a liquid by filtering, decanting or centrifuging.

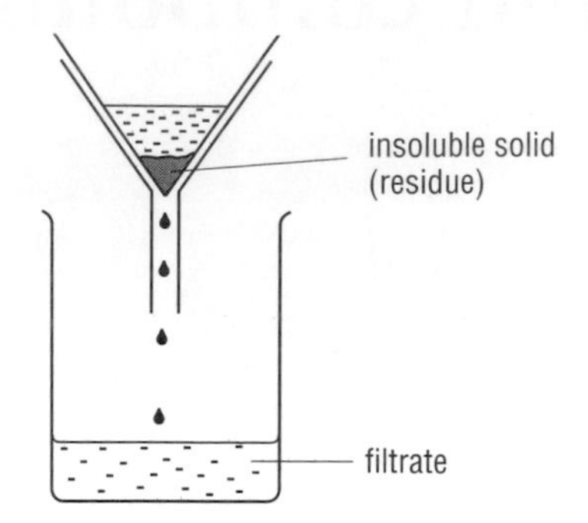

Filtering

- A soluble solid can be separated from a liquid by evaporating off some of the liquid and allowing the solid to crystallise.

- A liquid can be separated from a soluble solid by distillation.

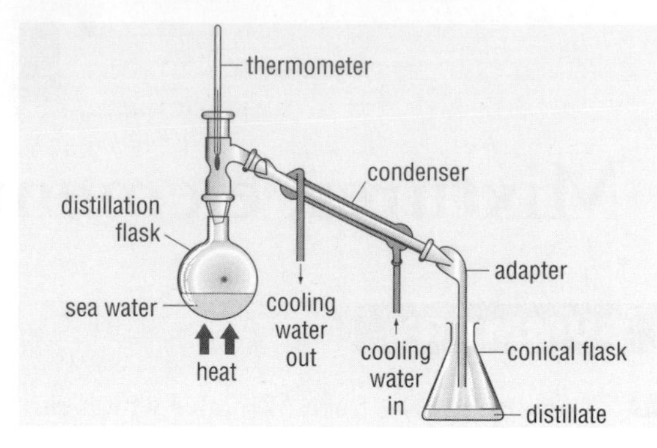

Distilling

- Immiscible liquids can be separated using a separating funnel.
- Miscible liquids can be separated by fractional distillation.
- Very similar substances can sometimes be separated by fractional distillation, fractional crystallisation or chromatography.

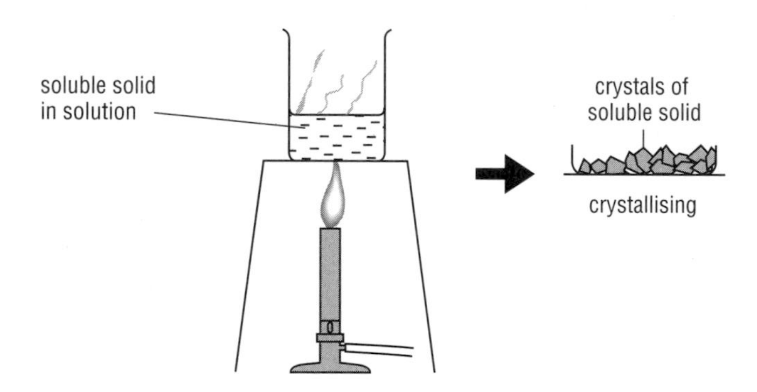

Evaporating

7 **Elements** are the simplest substances. They cannot be broken down into simpler substances by chemical reactions. Elements are most conveniently classified as **metals** and **non-metals**. The easiest way to check whether an element is a metal or non-metal is to see if it conducts electricity.

Property	Metals	Non-metals
melting point	usually HIGH	usually LOW
boiling point	usually HIGH	usually LOW
density	HIGH	LOW
lustre	SHINY	DULL
conduction of heat and electricity	GOOD	POOR

The properties of metals and non-metals

8 **Compounds** are substances containing two or more elements combined together (e.g. salt contains sodium and chlorine).

9 **Mixtures** contain two or more substances which are not combined together. They may be mixtures of elements, mixtures of compounds or mixtures of elements and compounds.

10 When elements combine to form compounds, we say they have reacted and the process is called a **chemical reaction**. The initial substances are called **reactants** and the substance(s) produced are called **products**.

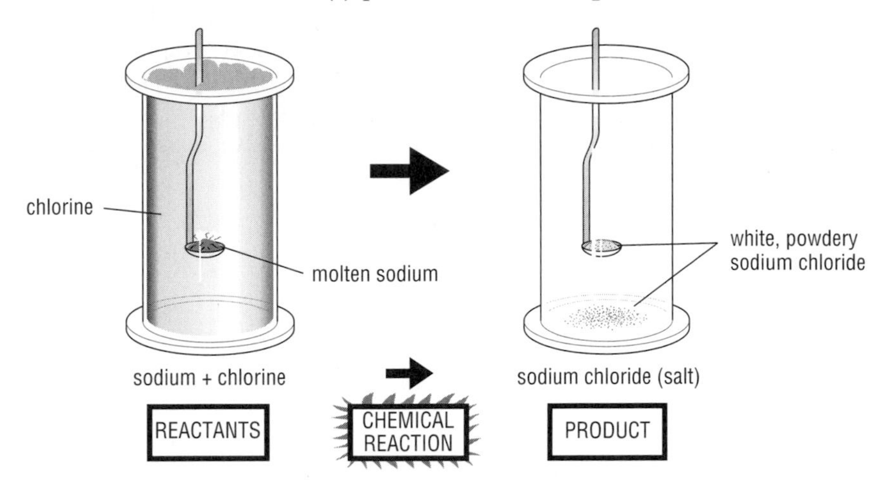

11 Decomposition is the splitting of one substance into two or more simpler substances. Synthesis is the joining together of two or more substances. For example:

$$\text{hydrogen} + \text{oxygen} \underset{\textit{decomposition}}{\overset{\textit{synthesis}}{\rightleftarrows}} \text{water}$$

12 **The Law of Conservation of Mass** is one of the most important scientific concepts. It summarises the fact that matter and materials cannot be created or destroyed. When substances react, the total mass of the products equals the total mass of the reactants.

STUDY QUESTIONS

Objective questions

Questions 1 to 6

In questions 1 to 6, choose from A, B, C or D which is the correct answer.

1 Which one of the following properties is correct for any metal?
A It has a high melting point.
B It is solid at room temperature.
C It sinks in water.
D It conducts electricity.

2 Iron is an element because it
A is formed when iron ore is heated at high temperatures.
B has been known for many centuries.
C has never been decomposed.
D combines with oxygen to form a compound.

3 Salt (sodium chloride) is a compound because it
A is not decomposed by heat.
B all melts at the same temperature.
C forms pure white crystals.
D is formed when sodium and chlorine react.

4 Water can be purified by boiling or chlorination. Water authorities do not use boiling because it
A does not kill bacteria.
B cannot be done on a large scale.
C cannot remove all impurities.
D is too expensive.

5 Pure water can be made from river water by
A boiling B chlorination
C distillation D filtration

6 Which of the following is the correct order in purifying salt from impure rock salt, using water as a solvent?
A dissolve, filter, evaporate, crystallise
B dissolve, evaporate, filter, crystallise
C filter, dissolve, crystallise, evaporate
D filter, dissolve, evaporate, crystallise

Questions 7 to 10

A condensation B crystallisation C distillation D evaporation E filtration

Select from the list A to E, the process which

7 occurs when solid sugar forms from liquid honey.

8 causes dew to form on a cold morning.

9 is used to separate nitrogen and oxygen from liquid air.

10 is used to make dried milk.

Short questions

11 A gas was produced when a solid was heated. How would you test to see if the gas is:

a) carbon dioxide *2 marks*
b) water vapour *3 marks*
c) oxygen? *2 marks*

12 At one time, gutters and drainpipes were made of iron. Today, they are made of plastics like PVC.

a) What properties of iron made it useful for gutters and drainpipes? *2 marks*
b) What were the disadvantages of using iron? *2 marks*
c) Why has iron been replaced by plastics? *2 marks*

13 A colourless gas A relights a glowing splint.

a) Identify the substances A, B, C, D and E.
(i) A reacts with copper to give a black solid B.
(ii) A reacts with a hot, dark grey solid C to give a colourless gas D.
(iii) D turns lime water milky.
(iv) B and D are produced when the green powder E decomposes on heating.
5 marks
b) Write word equations for the reactions in (i), (ii) and (iv). *4 marks*

14 Sea water contains dissolved solids including salt. Describe how you would find the mass of solid dissolved in 10 cm^3 of sea water. *6 marks*

Further examination questions

15 Some properties of six elements, A to F, are given in the table.

Element	Density /g cm^{-3}	Boiling point/°C	Electrical conductivity
A	3.12	58	poor
B	8.65	765	good
C	19.30	2970	good
D	3.4×10^{-3}	−152	poor
E	0.53	1330	good
F	2.07	445	poor

a) Which of the six elements are metals? *3 marks*
b) Which element will float on water as a solid at 20°C? *1 mark*
c) Which element is a gas at room temperature? *1 mark*
d) Which non-metal may be a solid at 60°C? *1 mark*
e) Which metal is probably a liquid at 2000°C? *1 mark*
f) What is the boiling point of the least dense metal? *1 mark*
g) The elements in the list are cadmium, gold, lithium, bromine, krypton and sulphur. Which element is which? *6 marks*

16 A sample of crude oil was distilled in the apparatus shown in the diagram.

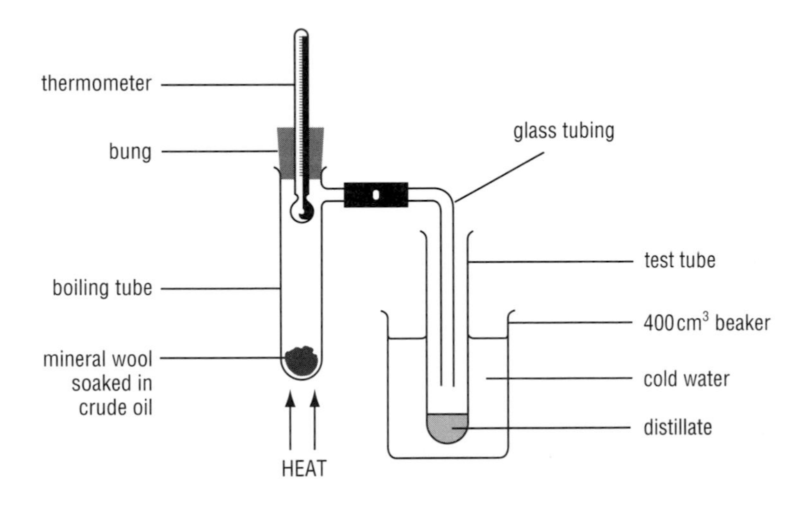

Four fractions of distillate were collected in the temperature ranges:

20–70°C, 70–120°C, 120–170°C, 170–220°C

a) The crude oil is not heated alone but with mineral wool. Why is mineral wool used? *1 mark*
b) Why is the receiving tube in a beaker of cold water? *1 mark*
c) Why would it be better to use a condenser to condense the vapours? *2 marks*
d) Which fraction will have condensed least efficiently? *1 mark*
e) Which fraction of the distillate will collect first? *1 mark*
f) Which fraction of distillate could be used as petrol? *1 mark*
g) Which fraction of distillate is most flammable? *1 mark*
h) Which fraction burns with the smokiest flame? *1 mark*
i) Which fraction would be darkest in colour? *1 mark*
j) Why is the mineral wool black at the end of the experiment? *1 mark*

17 At different stages in the development of aircraft and cars, various materials have been used in their construction. These include materials which can be classified as fibres, metals, glasses, plastics, ceramics and composite materials.

a) One of the materials which has been used for light aircraft wings and bodies is glass reinforced plastic. This is a composite material. Explain what is meant by a composite material. *2 marks*

b) The table gives some information about the materials which could be used for the wings of aircraft.

Material	Density/g/cm^3	Strength/GPa	Stiffness/GPa	Relative cost
aluminium alloy	2.8	0.6	90	low
steel	7.8	1.0	210	low
wood	0.6	0.1	20	low
glass reinforced plastic	1.9	1.5	21	medium
Kevlar plastic	1.5	3.0	190	high

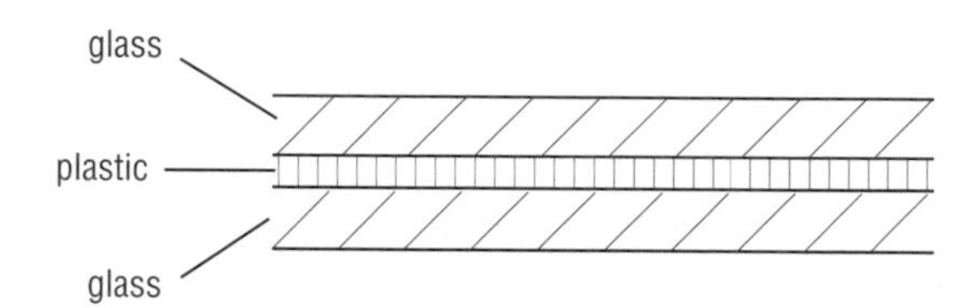

(i) Give *two* reasons why glass reinforced plastic is a suitable material for the wings of a small aircraft. *2 marks*

(ii) Which material would you choose for the wings of a large jet aircraft? Explain your choice. *2 marks*

c) The diagram above shows a section through a laminated car windscreen. Laminated windscreens crack when struck hard but remain in one piece. This givcs greater protection in the event of an accident. Use your knowledge of the properties of plastics and glass to help you to suggest:

(i) a reason, other than transparency, why car windscreens are made of glass. *1 mark*

(ii) two reasons why laminated glass windscreens crack rather than shatter when struck hard. *2 marks*

AQA

18 Three elements are burned in oxygen to form oxides.

Element	pH of oxide	Reaction of oxide with dilute hydrochloric acid	Reaction of oxide with sodium hydroxide solution
sulphur	2	no reaction	reaction
zinc	7	reaction	reaction
sodium	13	reaction	no reaction

a) (i) Which of the following chemicals is used to find the pH of the oxide?

hydrogencarbonate indicator   limewater   litmus   Universal Indicator. *1 mark*

(ii) Which of the elements shown in the table produces an acidic oxide? *1 mark*

(iii) Zinc oxide is an amphoteric oxide. What is meant by the term amphoteric oxide? *1 mark*

b) When a pure substance, X, is burnt in oxygen it forms a mixture of two oxides. One oxide is carbon dioxide and the other is water. What can you conclude about X? *2 marks*

OCR

CHAPTER

14

Particles, reactions and equations

SUMMARY

1 Classifying materials

Materials and substances can be classified in different ways:

- as naturally-occurring and man-made
- as elements, compounds and mixtures
- as solids, liquids and gases.

Solids, **liquids** and **gases** are called the **three states of matter**.

2 The kinetic theory of moving particles

- All materials and substances are made up of incredibly small moving particles. This is called the **kinetic theory of matter**.
- In different substances the particles are either **atoms**, **molecules** or **ions**.

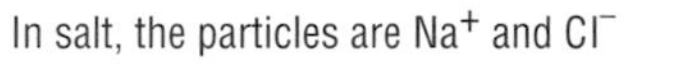

In the metal copper, the particles are Cu	**atoms**	
In water, the particles are H_2O	**molecules**	
In salt, the particles are Na^+ and Cl^-	**ions**	

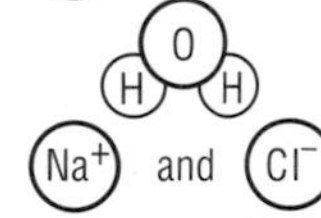

- Small particles move faster than larger particles at the same temperature.
- As the temperature rises, the particles have more energy and move around faster.

3 Properties of solids, liquids and gases

- In a **solid**
  - *the particles are very close together* with *strong forces* between them ∴ solids have a high density and they cannot be compressed.
  - *the particles can only vibrate* about fixed points ∴ solids have a fixed volume and a fixed shape and they cannot flow.
- In a **liquid**
  - *the particles are a little further apart* and the *forces between them are not so strong* as solids ∴ liquids have medium density and they can be compressed slightly.
  - *the particles can roll around each other* ∴ liquids flow easily, they can change their shape, but keep a fixed volume.
- In a **gas**
  - *the particles are far apart* with *no forces between them* ∴ gases have low densities and they can be compressed a lot.
  - *the particles move very fast* in all the space available ∴ gases flow easily and fill the whole of their container.

4 Atoms, ions and molecules

Atoms, ions and molecules can be represented using the **symbols** for different elements (e.g. the symbol for oxygen is O).

An	**atom**	is the smallest particle of an element	an *atom* of chlorine Cl
A	**molecule**	contains two or more atoms chemically joined together	a *molecule* of chlorine Cl_2
An	**ion**	is formed from an atom by the loss or gain of one or more electrons	a chloride *ion* Cl^-

5 Compounds

- When non-metals react they form compounds containing molecules e.g. H_2O, CO_2, HCl.
- When metals react with non-metals, they form compounds containing ions e.g. salt (sodium chloride) is NaCl or more precisely Na^+Cl^-.
- All metal ions are positive and all non-metal ions (except H^+) are negative.
- Most metal ions have a charge of 2^+.
- The only common metal ions with a charge of 1^+ are Ag^+, Na^+ and K^+ (AgNaK).
- The only common metal ions with a charge of 3^+ are Cr^{3+}, Al^{3+} and Fe^{3+} (CrAlFe).

6 Relative atomic masses

- Atoms of carbon are given a relative mass of exactly 12. The masses of other atoms are obtained by comparison with carbon.
  e.g. The relative atomic mass of copper atoms is 63.5 and that of oxygen atoms is 16.0.
  That is: $A_r(Cu) = 63.5, A_r(O) = 16.0$
  This is sometimes just written as Cu = 63.5, O = 16.
- The relative atomic mass in grams of every element contains 6×10^{23} atoms. This is called **Avogadro's Constant**.
  That is: 12 g of carbon and 63.5 g of copper each contain 6×10^{23} atoms.
- The amount of an element which contains 6×10^{23} atoms is known as **one mole**.
  So, 12 g of carbon is one mole of carbon.
- The mole is the chemist's counting unit for the amounts of substances because this indicates the number of particles.

$$\text{Number of moles} = \frac{\text{mass in grams}}{\text{mass of 1 mole}}$$

- Using relative atomic masses, it is possible to calculate relative formula masses.
  e.g. relative formula mass of carbon dioxide (CO_2) = relative formula mass of carbon (C) + 2 × relative formula mass of oxygen (O)
  $= 12 + 2 \times 16 = 44$

7 Equations

- A chemical equation is a summary of the reactants and products in a reaction.

  e.g. hydrogen + chlorine → hydrogen chloride

- Balanced chemical equations can be developed from word equations by
  - writing symbols and formulae

    e.g. $H_2 + Cl_2 \rightarrow HCl$

  - and then balancing the number of atoms of each element on both sides of the equation.

    e.g. $H_2 + Cl_2 \rightarrow 2HCl$

- Using equations with relative atomic masses, it is possible to calculate the masses of reactants and products.

e.g.	H_2	+	Cl_2	$\longrightarrow$	2HCl	$A_r(H) = 1.0$
	1 mole H_2 = 2 g		1 mole Cl_2 = 71 g		2 moles HCl = 73 g	$A_r(Cl) = 35.5$

STUDY QUESTIONS

Objective questions

Questions 1 to 5

In the boxes the different shadings represent different atoms.

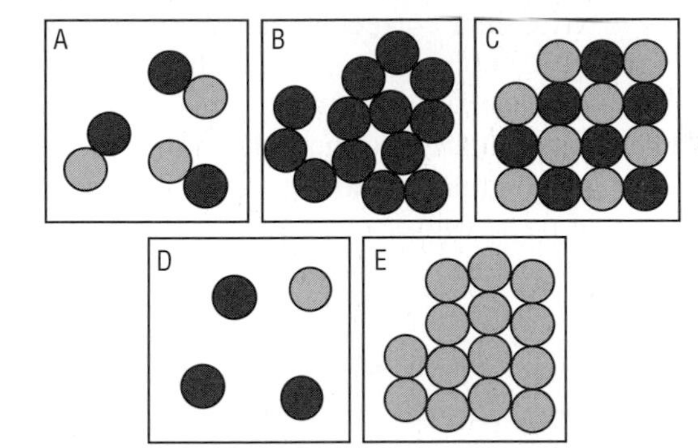

Which box contains

1 a close packed metal?

2 a solid compound?

3 a liquid?

4 a mixture?

5 a diatomic molecule?

Questions 6 to 10

In questions 6 to 10, choose from A, B, C and D which is the correct answer.

6 Gases diffuse faster than liquids because gas molecules are
A more compressible than liquid molecules.
B lighter than liquid molecules.
C freer to move than liquid molecules.
D more elastic than liquid molecules.

7 As a liquid freezes, its particles
A slow down, but move around each other.
B stop moving and form a regular arrangement.
C slow down, but only move around fixed points.
D stop moving as they get closer to each other.

8 A formula shows the relative number of
A atoms of each element in a compound.
B different elements in a compound.
C different atoms in a compound.
D atoms of different elements.

9 What is the relative mass of ammonium nitrate, NH_4NO_3?
($A_r(N) = 14, A_r(H) = 1, A_r(O) = 16$)
A 9 B 45 C 66 D 80

10 Titanium oxide is used as a white pigment for paint. A sample of pure titanium oxide contained 96 g of titanium combined with 64 g of oxygen. What is the formula of titanium oxide? ($A_r(Ti) = 48, A_r(O) = 16$)
A TiO B TiO_2 C Ti_2O D Ti_2O_4

Short questions

11 Look carefully at John Dalton's symbols for the elements in the photograph.

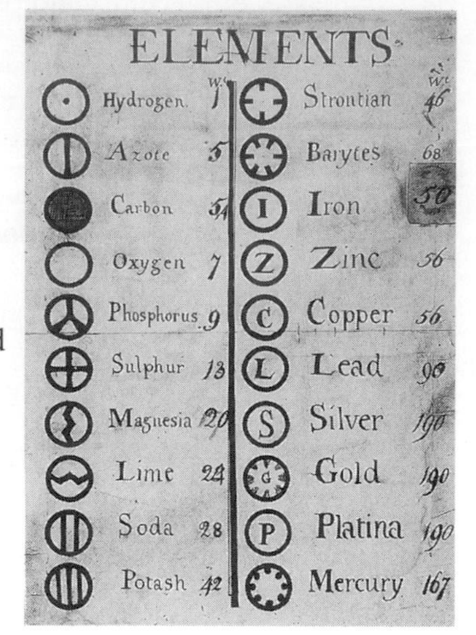

a) From Dalton's list pick out
(i) two gases at room temperature
(ii) a liquid at room temperature
(iii) two solid non-metals at room temperature
(iv) four solid metals at room temperature. *9 marks*

b) Six of the substances in Dalton's list are compounds and not elements. Pick out two of these compounds and write their correct chemical names. *2 marks*

c) The second element on Dalton's list is named 'Azote'. What do we call 'Azote' now? *1 mark*

d) The numbers to the right of the names in Dalton's list are the relative atomic masses which he used. Only one of these values agrees with those we use today. Which one is this? *1 mark*

12 The diagram shows two gases, of equal volume, separated by a barrier.

barrier

a) (i) What happens if the barrier is removed? *1 mark*
(ii) Use the particle theory of matter to explain what happens if the barrier is removed. *2 marks*
(iii) What is the scientific name for this process? *1 mark*

b) With the barrier removed, the container is warmed. Explain, in terms of the particles present, what happens to the pressure inside the container. *4 marks*

13 Particles in oxygen gas and nitrogen gas are represented in these diagrams.

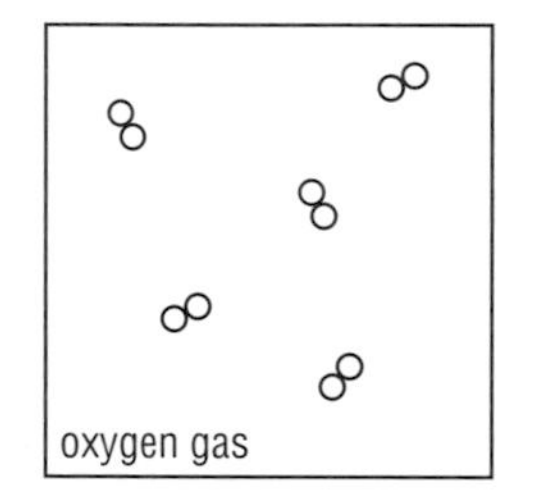

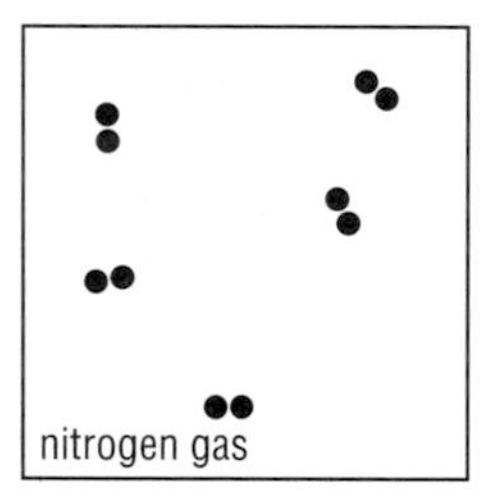

a) (i) What does ○ represent?
(ii) What does ● represent?
2 marks

b) Why are the particles in these diagrams best described as molecules? *2 marks*

c) Describe the movement of the particles in these gases. *2 marks*

d) What does ○● represent? *2 marks*

14 The symbol equation with state symbols summarises a chemical reaction between solutions of lead nitrate and potassium chloride.

$$Pb(NO_3)_2(aq) + 2KCl(aq) \rightarrow 2KNO_3(aq) + PbCl_2(s)$$

The equation gives the formulae of the two products of the reaction.

a) What are the names of the two products? *2 marks*

b) What other information does the equation give you about these products? *2 marks*

15 Write balanced equations for the following reactions.

a) magnesium + chlorine → magnesium chloride *1 mark*
b) aluminium + oxygen → aluminium oxide *2 marks*
c) copper(II) carbonate → copper(II) oxide + carbon dioxide *1 mark*
d) calcium oxide + water → calcium hydroxide *1 mark*
e) iron + hydrochloric acid → iron(II) chloride + hydrogen *2 marks*

16 Rewrite and balance the following equations which already have the correct formulae for the substances involved.

a) $Fe + Cl_2 \rightarrow FeCl_3$ *1 mark*
b) $NH_3 + H_2SO_4 \rightarrow (NH_4)_2SO_4$ *1 mark*
c) $CH_4 + O_2 \rightarrow CO_2 + H_2O$ *1 mark*
d) $K + H_2O \rightarrow KOH + H_2$ *1 mark*
e) $FeS + O_2 \rightarrow Fe_2O_3 + SO_2$ (difficult) *1 mark*

Further examination questions

17 When molten (liquid) sodium chloride is electrolysed, changes occur at the electrodes.

Negative electrode (cathode) $2Na^+ + 2e^- \rightarrow 2Na$
Positive electrode (anode) $2Cl^- - 2e^- \rightarrow Cl_2$

The overall reaction is $2NaCl \rightarrow 2Na + Cl_2$

Relative atomic mass of sodium, $A_r(Na) = 23.0$
Relative atomic mass of chlorine, $A_r(Cl) = 35.5$
1 mole of a gas occupies 24 dm^3 at 25°C and 1 atmosphere.

a) Calculate the relative formula mass (M_r) of sodium chloride. *1 mark*
b) How many grams of sodium chloride are needed to produce 92 g of sodium at the negative electrode? Show your working. *3 marks*
c) How many grams of chlorine could be produced at the positive electrode, whilst 92 g of sodium are formed at the negative electrode? Show your working. *3 marks*
d) What is the volume of chlorine, at 25°C and 1 atmosphere pressure, which could be produced at the positive electrode whilst 92 g of sodium are formed at the negative electrode? Show your working. *3 marks*

WJEC

18 Some lead oxide was heated in the apparatus shown.

The lead oxide was reduced to lead.

a) Write a word equation for the reaction of lead oxide and hydrogen. *1 mark*
b) The mass of the combustion tube, the mass of the combustion tube and lead oxide before heating, and the mass of the combustion tube and lead after heating were measured.

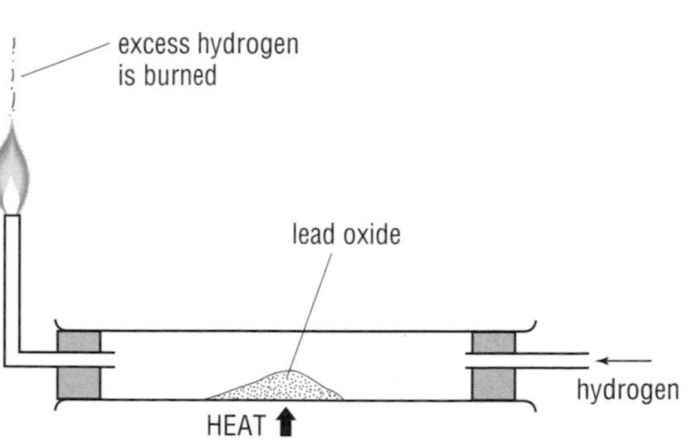

These are the results of the experiment.

Mass of combustion tube	= 51.00 g
Mass of combustion tube and lead oxide	= 54.59 g
Mass of combustion tube and lead	= 54.11 g

Use the results to find the formula of lead oxide. Show all your working.
(A_r(Pb) = 207, A_r(O) = 16.0) *6 marks*

AQA

19 When the Roman army invaded Britain, they found it difficult to light fires to cook their food. Instead they used a self-heating 'meal in a cauldron'. This involved two iron cooking pots called cauldrons being fitted one inside the other with quicklime packed between the two pots. When water was poured into the gap it reacted with the quicklime and the contents became very hot. This heat cooked the food.

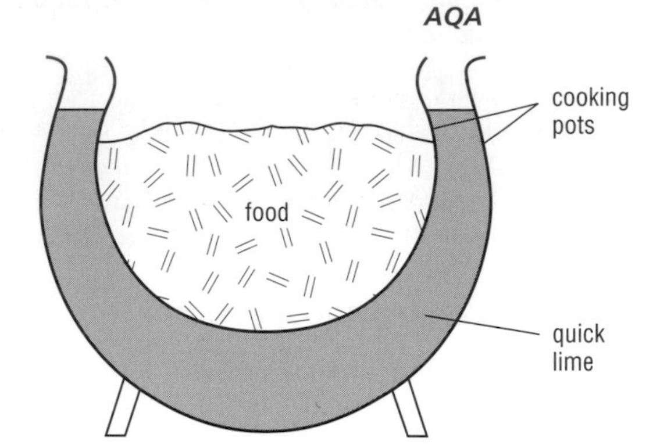

a) What do we call a reaction which gives out heat energy? *1 mark*

b) The chemical reaction involved is:

$$\underset{\text{quicklime}}{CaO} + \underset{\text{water}}{H_2O} \rightarrow \underset{\text{slaked lime}}{Ca(OH)_2}$$

Relative atomic masses: Ca = 40, H = 1, O = 16

(i) What is the mass of water which is needed to react with 56 g of quicklime? Show your working. *2 marks*

(ii) The Romans found they had to add more water than this, even if the water was carefully spread through the quicklime. Suggest an explanation. *1 mark*

(iii) Explain one safety precaution which should be taken when a person carries quicklime. *2 marks*

c) The Romans cooked bread in special ovens. These ovens were made of brick instead of iron. Burning wood was put inside the oven until the bricks were hot enough. The wood was then removed and the bread was placed inside the hot oven to cook for half an hour. Explain one reason why brick ovens cooked the bread better than iron ovens. *2 marks*

OCR

20 a) Acid rain pollutes many lakes. As a result of this pollution, the lakes are a dilute solution of sulphuric acid, H_2SO_4. Lime, calcium hydroxide, $Ca(OH)_2$, can be added to react with the acid.

(i) Give the balanced chemical equation for the reaction between sulphuric acid and calcium hydroxide. *2 marks*

(ii) What name is given to this type of chemical reaction? *1 mark*

b) It was calculated that in a polluted lake there were 4 900 000 kg of sulphuric acid. Relative atomic masses: H = 1, O = 16, S = 32, Ca = 40.

(i) Calculate the relative molecular mass of sulphuric acid, H_2SO_4. *1 mark*

(ii) Calculate the mass of calcium hydroxide, $Ca(OH)_2$, that would be needed to react with all the sulphuric acid in the lake. *4 marks*

(iii) Apart from cost, suggest *two* possible problems of adding this mass of calcium hydroxide to the lake. *2 marks*

AQA

CHAPTER

15

Metals and the reactivity series

SUMMARY

1 The **reactivity series** puts metals **in order of their reactivity** with other substances such as oxygen, water and acid.

Potassium	K	Very reactive metals	YOU MUST KNOW THE ORDER OF METALS IN THE REACTIVITY SERIES
Sodium	Na		
Calcium	Ca		
Magnesium	Mg		
Aluminium	Al	Fairly reactive metals	
Zinc	Zn		
Iron	Fe		
Lead	Pb	Less reactive metals	
Copper	Cu		
Silver	Ag	Unreactive metals	
Gold	Au		

2 **Reaction of metals with air** (oxygen)
Most metals react with air (oxygen)

$$\text{Metal} + \text{oxygen} \rightarrow \text{Metal oxide}$$
$$2M + O_2 \rightarrow 2MO$$

Potassium, Sodium, Calcium, Magnesium	Tarnish in air to form a layer of oxide. Burn vigorously with a bright flame on heating.
Aluminium, Zinc, Iron	Tarnish slowly in air and burn steadily on heating.
Lead, Copper	Only form a layer of oxide on heating.
Silver, Gold	No reaction even on heating.

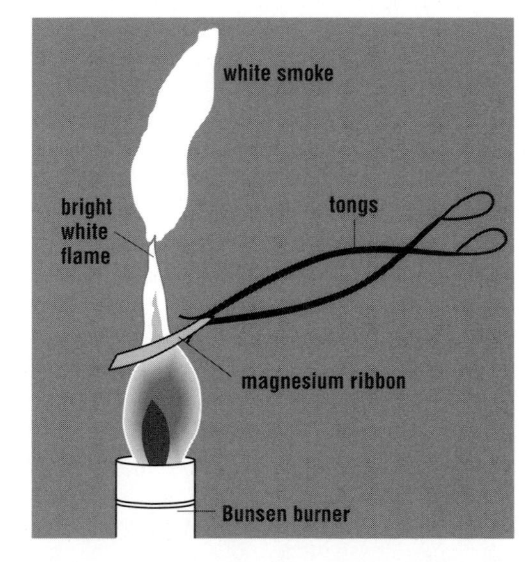

3 Reaction of metals with water and steam

Only the very reactive metals react with cold water although fairly reactive metals will react with steam.

$$\text{Metal} + \text{water (steam)} \rightarrow \text{Metal oxide} + \text{hydrogen}$$
$$M + H_2O \rightarrow MO + H_2$$

The oxides of very reactive metals react with more water to form hydroxides.

$$MO + H_2O \rightarrow M(OH)_2$$

Potassium, Sodium, Calcium, Magnesium	React with cold water with decreasing vigour.
Aluminium, Zinc, Iron	React with steam but not water.
Lead, Copper, Silver, Gold	No reaction even with steam.

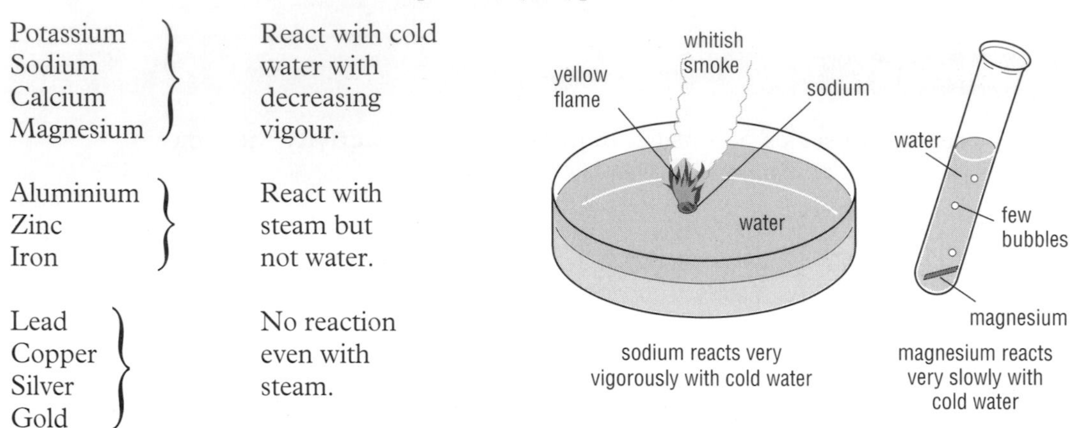

sodium reacts very vigorously with cold water

magnesium reacts very slowly with cold water

4 Reaction of metals with dilute acids

All metals above copper react with dilute acids producing hydrogen.

$$\text{Metal} + \text{acid} \rightarrow \text{salt} + \text{hydrogen}$$
$$M + H_2SO_4 \rightarrow MSO_4 + H_2$$
$$M + 2HCl \rightarrow MCl_2 + H_2$$

Potassium, Sodium, Calcium, Magnesium	violent reactions
Aluminium	– vigorous after a while
Zinc	– steady
Iron	– slow
Lead	– very slow
Copper, Silver, Gold	no reaction

5 Displacement reactions

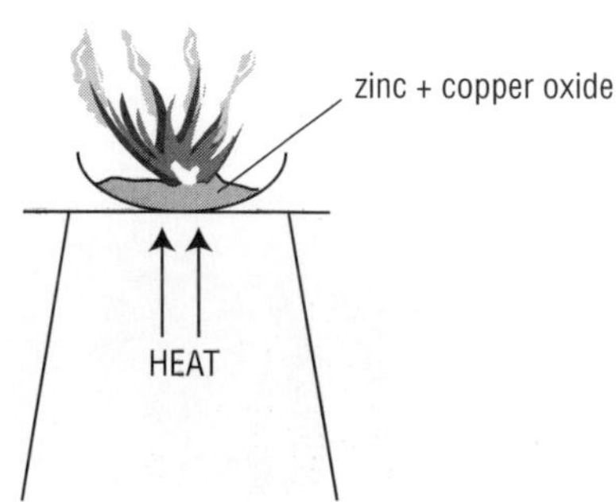

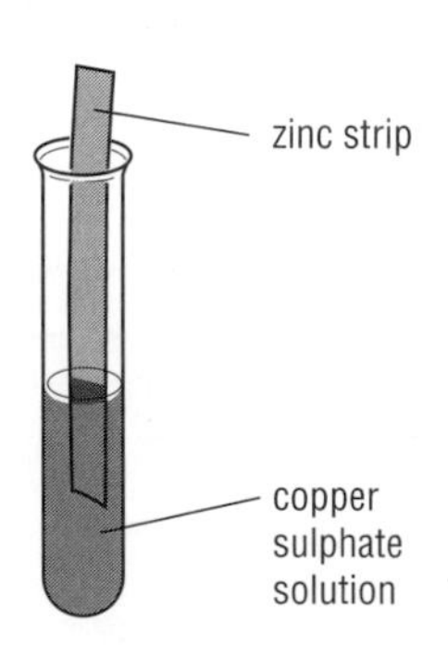

A more reactive metal will displace a less reactive metal from its compounds.

e.g. zinc + copper oxide → zinc oxide + copper
$Zn(s) + CuO(s) \rightarrow ZnO(s) + Cu(s)$

zinc + copper sulphate solution → zinc sulphate solution + copper
$Zn(s) + CuSO_4(aq) \rightarrow ZnSO_4(aq) + Cu(s)$

6 Extracting metals from their ores

Metals	Method
Potassium, Sodium, Calcium, Magnesium, Aluminium	Compounds of metals (e.g. oxides) cannot be reduced to the metal using carbon or carbon monoxide. **Electrolysis of molten compounds** must be used: metal ions are converted to metal atoms at the cathode by electrons in the electric current. e.g. *Cathode*(⁻) $Al^{3+} + 3e^- \longrightarrow Al$ (ions, electrons, atoms)
Zinc, Iron, Lead	**Compounds of metals** (e.g. oxides) **can be reduced using carbon (coke) or carbon monoxide** e.g. $ZnO + C \rightarrow Zn + CO$; $Fe_2O_3 + 3CO \rightarrow 2Fe + 3CO_2$
Copper	**Heat the compound of the metal in a limited supply of air** $CuS + O_2 \rightarrow Cu + SO_2$
Silver, Gold	**Metals occur uncombined in the ground**

7 Advantages and disadvantages of mining for metal ores

Most metals are too reactive to exist on their own in the Earth. They are usually combined with non-metals and found in rocks and minerals as impure substances called **ores**.

Advantages from mining	Disadvantages from mining
Metals are produced and used for thousands of useful articles	Damages the environment
Mining and manufacture create jobs	Destroys wildlife habitats
Sale of products increases the wealth of a community	Subsidence can affect land and buildings

STUDY QUESTIONS

Objective questions

Questions 1 to 5

The figure shows a simplified diagram of a blast furnace.

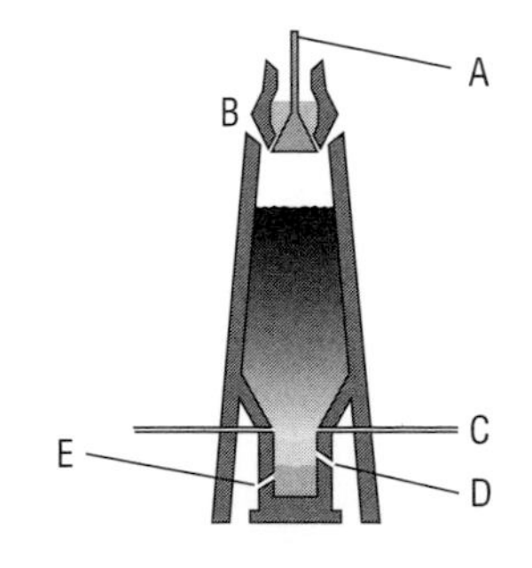

Choose from A, B, C, D and E the point where

1 iron ore is added.

2 hot air is added.

3 molten iron is removed.

4 slag is removed.

5 waste gases come out.

Questions 6 to 9

A lead B tin C tungsten
D uranium E zinc

Choose from A to E the metal which is used

6 as nuclear fuel.

7 as the filament in electric light bulbs.

8 for the electrodes in car batteries.

9 to galvanise iron.

Questions 10 to 16

10 Which *one* of the following substances will not react with magnesium on heating?
A potassium oxide B copper oxide
C iron(III) oxide D water

11 A metal does not react with cold water but reacts slowly with dilute nitric acid. The metal could be
A gold B lead
C silver D sodium

12 A cold water tank is unaffected by cold water, but starts to dissolve when there is some hydrogen chloride in the water. The tank could be made of
A calcium B copper
C magnesium D zinc

13 Red hot carbon will reduce nickel oxide and copper oxide, but it will not reduce strontium oxide. Nickel will reduce copper oxide. Which *one* of the following shows the correct order of increasing reactivity for these elements?
A copper, nickel, carbon, strontium
B nickel, copper, carbon, strontium
C carbon, copper, nickel, strontium
D strontium, carbon, copper, nickel

14 Which *one* of the following equations represents the reaction at the cathode when electroplating nickel with silver?
A $Ag \rightarrow Ag^+ + e^-$
B $Ag + e^- \rightarrow Ag^+$
C $Ag^+ + e^- \rightarrow Ag$
D $Ag^+ \rightarrow Ag + e^-$

15 Aluminium is manufactured by electrolysis of molten aluminium oxide. The aluminium oxide is dissolved in molten cryolite (Na_3AlF_6). The reason for using molten cryolite is to
A allow electrolysis at a lower temperature.
B protect the anode from oxygen.
C help the mixture conduct electricity.
D produce more aluminium.

16 Which *one* of the following metals is manufactured by electrolysis of its molten compounds?
A copper B lead
C lithium D tin

Short questions

17 a) When a piece of sodium is freshly cut, its appearance is typical of a metal.
(i) Explain this statement. *1 mark*
(ii) Describe what you would see if you watched the cut surface of the sodium for a while. Explain what happens. *3 marks*

b) Re-write and balance the following symbol equation for the reaction of sodium with oxygen.

$4Na(s) + \_\_\_\_\_\_(g) \rightarrow \_\_\_Na_2O(s)$

1 mark

18 a) Give *two* physical properties of sodium that are typical of metals. *2 marks*

b) Give *two* physical properties of sodium that are not typical of metals. *2 marks*

19 Cassiterite is an ore of the metal tin from which tin(IV) oxide, SnO_2, can be obtained.

a) What is an ore? *2 marks*
b) Some metals are obtained by removing oxygen from the metal oxide.
(i) What name do we give to this chemical reaction? *1 mark*
(ii) Write a word equation for the reduction of tin(IV) oxide with carbon. *2 marks*
c) Name one metal which must be extracted from its metal ore by electrolysis rather than by using carbon. *1 mark*

20 The table gives some information about the reactions of four metals.

Metal	Reaction with cold water	Reaction with steam	Reaction with dilute sulphuric acid
magnesium	very slow reaction	vigorous reaction	vigorous reaction
copper	no reaction	no reaction	no reaction
sodium	rapid reaction	very vigorous reaction	violent reaction
zinc	no reaction	slow reaction	steady reaction

a) Write the four metals in order of decreasing reactivity. *2 marks*
b) (i) Write a word equation for the reaction of magnesium with steam. (Assume that one product is magnesium oxide.) *2 marks*
(ii) Write a balanced symbol equation for this reaction. *2 marks*
c) (i) Write a word equation for the reaction of zinc with dilute sulphuric acid. *2 marks*
(ii) Write a balanced symbol equation for this reaction. *2 marks*

Further examination questions

21 The table shows information about eight metals.

Metal	Melting point/ °C	Boiling point/ °C	Specific heat capacity/ J/kg °C	Density/ g/cm³	Electrical conductivity	Reaction with water
A	659	2447	900	2.7	0.41	none
B	1083	2582	390	8.9	0.64	none
C	1539	2887	470	7.9	0.11	slight
D	328	1751	130	11.3	0.05	none
E	98	890	1222	0.97	0.20	very reactive
F	183	2500	130	7.3	0.66	none
G	1063	2707	129	19.3	0.49	none
H	3377	5527	135	19.3	0.20	none

Use this information to choose a suitable metal for each of the following uses. Explain which information was important for making your choice.

a) As overhead cables. *2 marks*
b) As a coolant in a nuclear reactor. *2 marks*
c) As electrical solder. *2 marks*
d) As the filament in a light bulb. *2 marks*

EDEXCEL

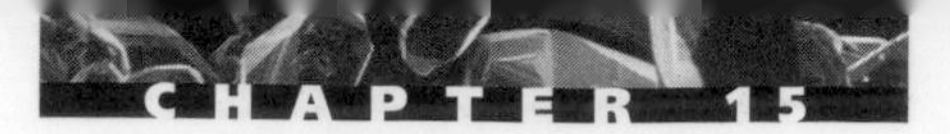

22 a) Describe the observations that you would expect to make when a piece of sodium is placed in water containing some universal indicator solution. *3 marks*

b) Re-write and balance the following symbol equation for the above reaction, by putting the correct numbers in the boxes provided. *2 marks*

$$\square Na(s) + \square H_2O(l) \rightarrow 2NaOH(aq) + H_2(g)$$

c) What does (aq) mean in the above equation? *1 mark*

d) Name the *two* products of the reaction. *2 marks*

e) If you were adding sodium to water, state *two* safety precautions you would take. *2 marks*

f) State and explain the usual method of storing sodium safely in the laboratory. *2 marks*

WJEC

23 Malachite is a green copper ore that is often used to make semi-precious jewellery. It contains copper carbonate.

a) What is an ore? *2 marks*

b) When crushed malachite is heated it produces black copper oxide and carbon dioxide.

Choose the best word or words to describe the reaction which produces copper oxide from copper carbonate.

combustion compound formation decomposition exothermic reaction neutralisation oxidation

1 mark

c) There are different methods which can be used to make copper from copper oxide. One method involves the electrolysis of blue copper sulphate solution. Copper oxide is added to dilute sulphuric acid. They react and form copper sulphate solution. The solid impurities in the ore do not dissolve. The apparatus in the diagram is then used to carry out electrolysis.

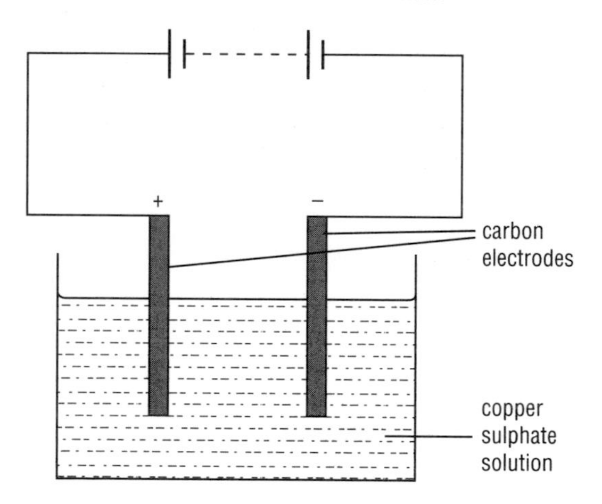

(i) How could the solid impurities be removed from the copper sulphate solution? *1 mark*

(ii) What happens to the colour of the copper sulphate solution when electrolysis takes place? *1 mark*

(iii) Copy and complete the ionic equations for the reactions taking place at the electrodes.

$Cu^{2+} +$ _____ $\rightarrow$ _____

_____ OH^- $\rightarrow$ _____ $H_2O + O_2 +$ _____ *4 marks*

OCR

24 The diagram shows some parts of a boat.

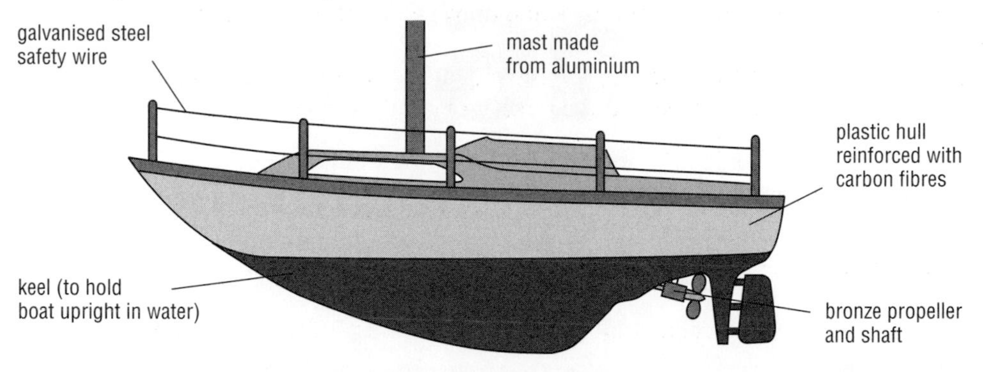

a) The table shows some properties of some metals.

Metal	Density in g/cm³	Resistance to corrosion	Cost
aluminium	2.70	fair	cheap
copper	8.94	very good	cheap
gold	19.32	excellent	very expensive
lead	11.33	excellent	cheap
tin	7.31	very good	expensive

(i) One reason the keel of a boat is there is to stop the boat turning upside down. Which metal in the table should be used to make the keel? State *two* reasons for your answer. *3 marks*

(ii) The hull is made from plastic in which carbon fibres are embedded. What is the purpose of the carbon fibres? *1 mark*

(iii) Why is aluminium a good metal from which to make the mast? *1 mark*

b) Copper has a low tensile strength but when tin is mixed with it, a much stronger metal called bronze is produced. The graph shows how the percentage of tin in bronze affects its strength.

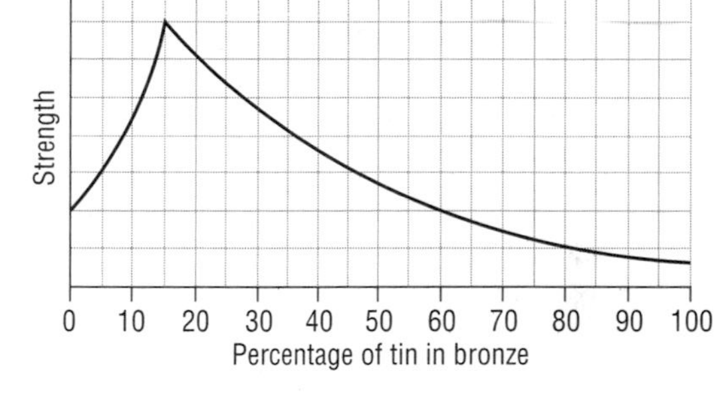

(i) What percentage composition of copper and tin in bronze is suitable for making the propeller and shaft? *1 mark*

(ii) The diagrams show the arrangement of atoms in pure copper and in bronze.

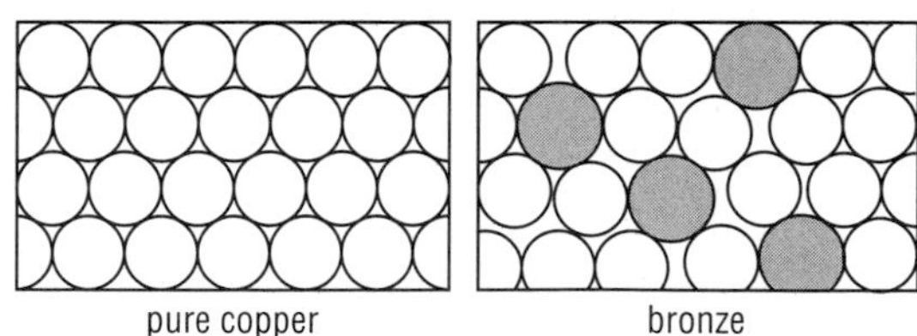

Use the diagrams to explain why:

A copper has a low tensile strength. *1 mark*

B copper has a slightly higher density than bronze. *1 mark*

C bronze is a stronger metal than copper. *1 mark*

c) The steel safety wires on the boat are galvanised to protect them against corrosion. State *one* way in which galvanising protects them against corrosion. *1 mark*

AQA

16

Atomic structure and the periodic table

SUMMARY

1 One of the most useful ways of **classifying elements** is as **metals** and **non-metals**. The most convenient way to decide whether an element is a metal or a non-metal is to check its electrical conductivity.

- Metals are good conductors of electricity.
- Non-metals are poor conductors of electricity.

2 Some elements, like graphite, are difficult to classify as metals or non-metals because they have some properties in common with non-metals and other properties in between those of metals and non-metals. These elements are called **metalloids**.

3 During the 19th century, chemists identified families of similar elements (e.g. chlorine, bromine and iodine). In 1869, Mendeleev found a **pattern** in the **properties of elements** and their **relative atomic masses** which he summarised in a table. Later, chemists realised that the properties of elements were related more closely to their atomic numbers than their relative atomic masses. Mendeleev's table of elements led to the modern periodic table.

The periodic table contains *all* elements.

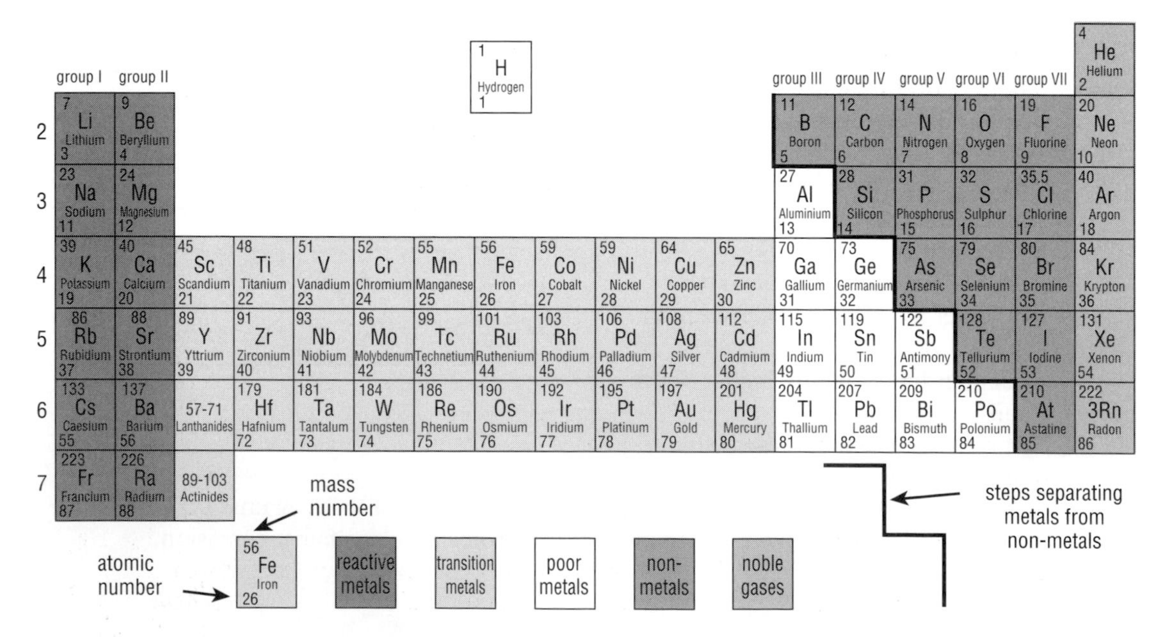

The elements are arranged in order of atomic number. The vertical columns are called **groups**. Each group contains elements with similar properties. The horizontal rows are called **periods**.

4 **Atomic structure**

- All atoms are built up from three tiny particles – **protons**, **neutrons** and **electrons**.
- The proton and the neutron are each assigned a relative mass of one. An electron has a mass of about $\frac{1}{2000}$ of this.
- Protons have a positive charge of +1. Neutrons are uncharged. Electrons have a negative charge of −1.
- Protons and neutrons occupy the **nucleus** at the centre of the atom. Electrons occupy layers or **shells** at different distances from the nucleus.
- Electrons determine the chemical properties of atoms.

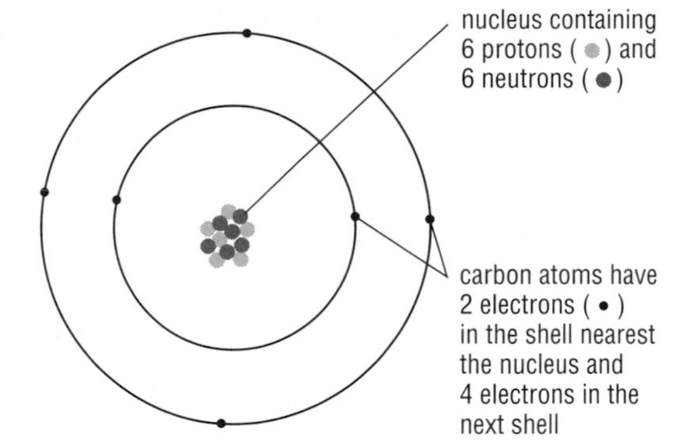
nucleus containing
6 protons (●) and
6 neutrons (●)
carbon atoms have
2 electrons (•)
in the shell nearest
the nucleus and
4 electrons in the
next shell

A carbon atom with six protons, six neutrons and six electrons

5 Protons, neutrons and electrons are the building blocks for all atoms.

e.g. Helium atoms have two protons, two or three neurons and two electrons.
Carbon atoms have six protons, six, seven or eight neutrons and six electrons.
Uranium atoms have 92 protons, 143 or 146 neutrons and 92 electrons.
Uncharged atoms always have the same number of protons as electrons.

6 The **atomic number** of an atom = the number of protons
= the order of the element in the periodic table.

e.g. Fluorine is the ninth element in the periodic table with nine protons.
i.e. its atomic number = 9

The term atomic number is sometimes called **proton number**.

7 The **mass number** of an atom = number of protons + number of neutrons.

Protons and neutrons which occupy the nucleus are sometimes called **nucleons**. So, the term mass number is sometimes called the **nucleon number** of an atom.

8 **Isotopes** are atoms of the same element with the same atomic number but different mass numbers.

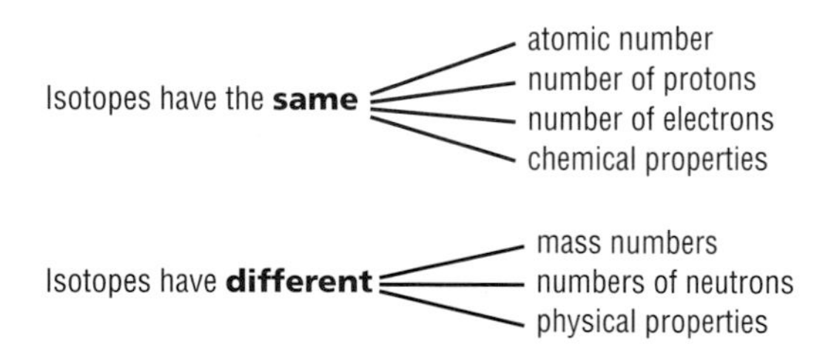
Isotopes have the **same**
atomic number
number of protons
number of electrons
chemical properties
Isotopes have **different**
mass numbers
numbers of neutrons
physical properties

9 **Electron structures**

Electrons occupy layers of shells at different distances from the nuclei of atoms. When these shells are filled, the atoms or ions become stable.

- The first shell nearest the nucleus is full and stable when it contains two electrons.
- The second shell is full and stable when it contains eight electrons.
- The third shell is stable (though not full) when it contains eight electrons.
- The third shell is full and stable when it contains 18 electrons.

Elements in the same group of the periodic table have similar electron structures and this gives them similar properties.
e.g. Lithium atoms have two electrons in the first shell and one in the second. Their electron structure is written as 2,1.
Sodium atoms have an electron structure 2,8,1.
Potassium atoms have an electron structure 2,8,8,1.

10 The alkali metals (Group 1)

- They are reactive metals with very similar properties and reactions.
- They have lower melting points, lower boiling points, lower densities and are softer than other metals.
- They react rapidly with oxygen in the air to form oxides.
- They become more reactive with increase in atomic number.

All the alkali metal atoms have one electron in their outer shell (e.g. Na 2,8,1). When they react, they lose this outermost electron to form stable ions with filled shells and one positive charge (e.g. Na^+ 2,8).

11 The halogens (Group VII)

- They form a group of reactive non-metals.
- They have very similar properties and reactions.
- They form diatomic molecules (e.g. Cl_2, Br_2, I_2).
- They have low melting points and boiling points but vary from F_2 and Cl_2 which are gases at room temperature, through Br_2 which is a liquid to I_2, a solid at room temperature.
- They become less reactive with increase in atomic number.

All the halogen atoms have seven electrons in their outer shell (e.g. Cl 2,8,7). When they react, they try to gain another electron in order to have eight electrons in the outermost shell which is far more stable (e.g. Cl^- 2,8,8).

12 The noble gases (Group O)

- They have very similar properties.
- They form a group of very unreactive non-metals.
- They form monatomic molecules with one atom (e.g. He, Ne, Ar).
- They have very low melting points and boiling points.
- They are all gases at room temperature.

All the noble gas atoms have stable electron structures with eight electrons in their outer shell, apart from helium which has a very stable single shell with two electrons.

13 The transition metals

- They lie between the reactive metals (Groups I and II) and the poor metals (in Groups III and IV) of the periodic table.
- They have similar properties.
- They have high melting points, high boiling points and high densities.
- They are unreactive with cold water.
- They form coloured compounds (e.g. Cu^{2+} compounds are often blue).

- They have catalytic properties as elements and in their compounds (e.g. Fe or Fe_2O_3 is a catalyst for the Haber Process).
- They form more than one ion. Electron structures of the transition elements are more complex than other elements. They can usually form more than one stable ion (e.g. Fe forms both Fe^{2+} and Fe^{3+} ions in its compounds).

STUDY QUESTIONS

Objective questions

Questions 1 to 10

The diagram shows part of the periodic table divided into five sections labelled A, B, C, D and E.

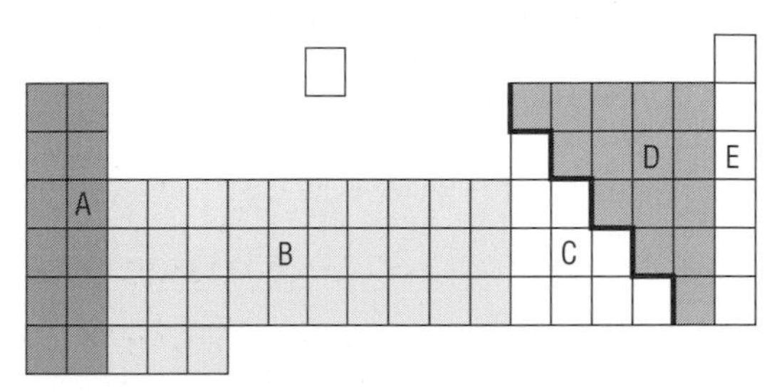

In which of the sections A to E would you find

1 a gas used in advertising signs?

2 an element used to purify water supplies?

3 a metal used in making tanks for hot water?

4 an element that has replaced wood and steel in window frames?

5 the most unreactive element?

6 an element whose oxide is used to colour glass?

7 the hardest naturally-occurring element?

8 the most abundant metal in the atmosphere?

9 an element which floats on water and reacts with it?

10 a gas which is used to kill bacteria in swimming pools?

Questions 11 to 21

In questions 11 to 21, choose the letter A, B, C, or D which is the correct answer.

11 Elements in the same group of the periodic table have
A the same reactivity.
B atoms of the same size.
C chlorides with similar properties.
D the same number of electrons.

12 The atomic number of sulphur is 16. How many electrons are there in the outer shell of a sulphur atom?
A 2
B 4
C 6
D 8

13 Which two atoms in the following table are isotopes of the same element?

Atoms	Mass number	Atomic number
W	19	12
X	19	10
Y	20	10
Z	20	11

A W and X
B X and Y
C Y and Z
D W and Z

14 In the periodic table, rubidium is below potassium in Group I and iodine is below bromine in Group VII. Which of the following pairs of elements will react most vigorously under the same conditions?
A potassium and bromine
B potassium and iodine
C rubidium and bromine
D rubidium and iodine

15 Caesium is in Group I of the periodic table. Which *one* of the following sets of properties is it likely to have?
A soft, low density, low melting point
B soft, low density, high melting point
C hard, high density, low melting point
D hard, high density, high melting point

16 A coloured element, X, reacted with sodium to form a compound with a formula NaX. Which *one* of the following elements could be X?
A argon
B chlorine
C sulphur
D potassium

17 Which *one* of the following elements will not react with oxygen?
A neon
B nickel
C nitrogen
D neptunium

18 Which *one* of the following is the best test for chlorine?
A It turns damp litmus paper blue.
B It turns damp litmus paper red.
C It turns damp litmus paper neutral.
D It turns damp litmus paper white.

19 As the halogens increase in relative atomic mass, there is a decrease in
A melting point.
B atomic number.
C molecular size.
D reactivity with water.

20 Which *one* of the following oxides reacts with water to form a solution with the highest pH?
A magnesium oxide, MgO
B phosphorus(V) oxide, P_2O_5
C potassium oxide, K_2O
D silicon dioxide, SiO_2

21 Moving across the periodic table, the elements in a period change from
A solid to liquid to gas.
B reactive metal to transition metal to non-metal.
C metal to non-metal to metalloid.
D reactive elements to unreactive elements.

Short questions

22 Complete the second row in the following table for the halogens. *3 marks*

Name of halogen	Symbol	State at room temperature	Colour
chlorine	Cl	gas	pale green
bromine	________	________	________
iodine	I	solid	dark grey

23 The first 20 elements in the periodic table are shown here.

							H									He
Li	Be											B	C	N	O	F	Ne
Na	Mg											Al	Si	P	S	Cl	Ar
K	Ca																

a) Group I contains the elements lithium, sodium and potassium. Why do these three elements have similar chemical properties? *2 marks*

b) Look at the part of the periodic table and write down
(i) the first element in Group IV.
(ii) an element which consists of molecules containing a pair of atoms.
(iii) an element which forms an ion of the type X^{2-}. *3 marks*

24 Magnesium oxide is used in some antacid medicines. Magnesium will react with oxygen to form magnesium oxide.

a) Name an element with similar properties to magnesium. *1 mark*

b) Write a word equation for the reaction of magnesium with oxygen. *2 marks*

c) The diagrams are incomplete. Draw completed diagrams to show the electronic structure of a magnesium atom and an oxygen atom. *2 marks*

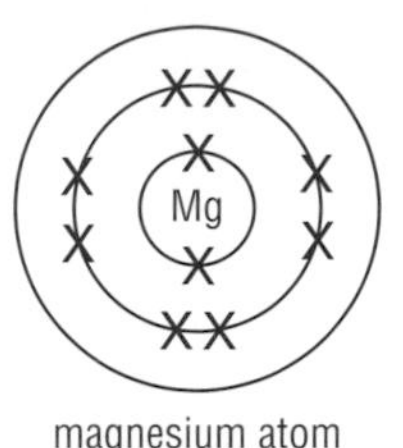

magnesium atom

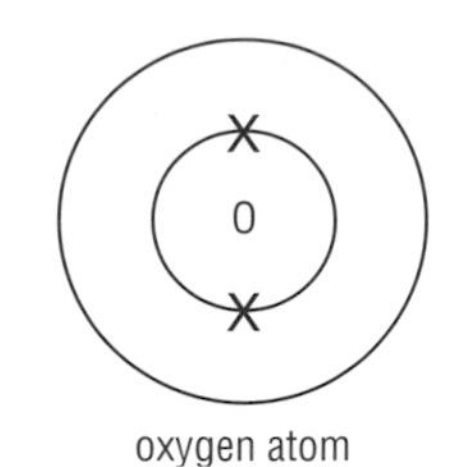

oxygen atom

25 In the nucleus of an aluminium atom there are 13 protons and 14 neutrons.

a) What is the mass number of aluminium?

b) How many electrons are there in an aluminium atom?

c) How many electrons are there in an Al^{3+} ion?

d) How many protons are there in an Al^{3+} ion?

e) Why is an aluminium atom electrically neutral? *5 marks*

26 Neon exists as two different isotopes. The structures of the two neon isotopes are shown here.

$$^{20}_{10}Ne \text{ and } ^{22}_{10}Ne$$

a) How are these isotopes similar in atomic structure?

b) How are these isotopes different in atomic structure?

c) The relative atomic mass of neon is 20.2. What does this tell you about the relative amounts of the two different isotopes in a sample of neon? Explain your answer.

4 marks

27 Suppose you have discovered a new element with the following properties.

A shiny grey in colour
B boiling point 400°C
C sinks in water
D conducts electricity
E burns easily in oxygen

a) Which property suggests that the new element is a metal?

b) The oxide of the new element dissolves in water and universal indicator is added. What colour will the universal indicator produce?

c) When the new element is dipped into copper(II) sulphate solution, a brown substance forms on its surface. What is the brown substance?

d) How does the reactivity of the new element compare with that of copper?

e) What happens to the new element when it is dipped in the copper(II) sulphate solution?

6 marks

Further examination questions

28 Use the periodic table on page 88 and your knowledge of more familiar elements to suggest answers to the following questions. You are not expected to know about the chemistry of rubidium, iodine and selenium.

a) Rubidium, Rb, has the atomic number 37.
(i) In what group of the periodic table is rubidium?

(ii) Name the *two* products you would expect when rubidium reacts with water.
(iii) Suggest *two* observations which you might see during the reaction.

5 marks

b) Iodine, I, has the atomic number 53. It reacts with hydrogen to form a gas, hydrogen iodide.

$$H_2 + I_2 \rightarrow 2HI$$

Hydrogen iodide is very soluble in water.
(i) Name a common laboratory liquid you would expect to behave like hydrogen iodide solution.
(ii) What *two* products would you expect to be formed when magnesium reacts with hydrogen iodide solution?
(iii) Suggest *two* observations which you might see during the reaction.

5 marks

c) Selenium, Se, has the atomic number 34. Selenium trioxide reacts with water to form a new compound.

$$H_2O(l) + SeO_3(s) \rightarrow H_2SeO_4(aq)$$

(i) What effect would the solution have on universal indicator (full-range indicator)?
(ii) Suggest *two* observations you might see when this solution was added to sodium carbonate.
(iii) Name *one* of the compounds formed in the reaction with sodium carbonate.

4 marks

EDEXCEL

29 a) Lithium has an atomic number of 3 and a mass number of 7. This is often represented by the symbol $^{7}_{3}Li$.
(i) State the number of neutrons, protons and electrons which make up a neutral atom of this element.
(ii) Sometimes atoms of the same element occur with a different number of neutrons in the nucleus. What do we call such atoms?
(iii) Using the symbolism at the start of this question, how would you represent an atom of lithium which contained only three neutrons.
(iv) Atoms of lithium form ions by the loss of one electron. Write the formula of a lithium ion.

7 marks

b) (i) Give the electronic structure of a single, uncombined atom of chlorine.
1 mark
(ii) Draw, showing only the outer electrons of each atom, the structure of a chlorine molecule, Cl_2. *2 marks*

c) Show what happens to the electronic structure of an atom of chlorine when it forms a chloride ion and say whether the atom has been oxidised, reduced or neither of these. *2 marks*

EDEXCEL

30 a) Lithium, sodium and potassium belong to Group I of the periodic table. The atomic number of lithium is 3, sodium 11 and potassium 19.
(i) Which of these elements – lithium, sodium or potassium – is the most reactive? *1 mark*
(ii) Explain your choice in terms of electron structure. *2 marks*
b) Sodium hydroxide and sodium carbonate are important compounds. Both sodium hydroxide and sodium carbonate react with dilute hydrochloric acid. Both reactions produce products. Name *two* products that are common to both reactions. *2 marks*
c) (i) What name is given to the type of reaction that occurs between sodium hydroxide and dilute hydrochloric acid? *1 mark*
(ii) Complete the following symbol equation.

NaOH(aq) + HCl(aq) → __________ + __________ *2 marks*

WJEC

31 Use the periodic table on page 88 to help you answer this question. The structure of an atom of an element is shown.

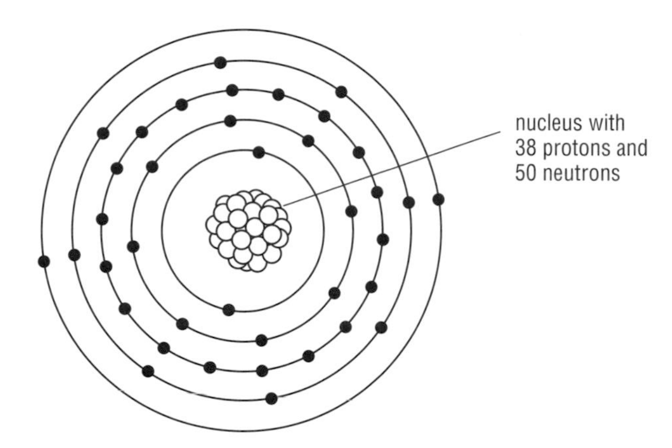

a) (i) To which group of the periodic table does the element belong? *1 mark*
(ii) To which period of the periodic table does the element belong? *1 mark*
(iii) What is the atomic number of this element? *1 mark*
(iv) What is the mass number of this element? *1 mark*
b) (i) What is the name and chemical symbol of this element? *2 marks*
(ii) This element forms ions. What will be the charge on these ions? *1 mark*
(iii) Name one element which will have similar chemical properties to the one named in b)(i). *1 mark*
c) The element, an atom of which is shown in the diagram, is classified as a metal. Give *two* chemical properties of a metal. *2 marks*

AQA

CHAPTER

17

The structure and bonding of materials

SUMMARY

1 Studying structures

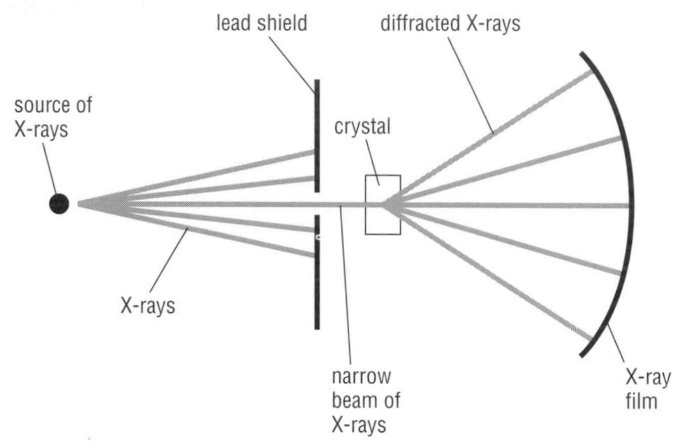

The structures of materials can be studied using X-rays.

- A narrow beam of X-rays is directed at a well-formed crystal of the material. Some of the X-rays are diffracted by the particles in the crystal onto X-ray sensitive film.
- From the diffraction pattern on the X-ray film, it is possible to work out the arrangement of particles in the crystal.
- The regular arrangement of particles in a crystal is called a **lattice**.

2 The structure of substances

- The **structure and bonding** of a substance determine its **properties**. The **properties** of a substance determine its **uses**.

  i.e. **structure and bonding** → **properties** → **uses**
- All substances are made up from only three different types of particle. **atoms** / **ions** / **molecules**
- There are four different kinds of solid structures:
  - **giant metallic**: composed of metal atoms
  - **giant covalent**: composed of non-metal atoms joined together in large molecules
  - **giant ionic**: composed of ions
  - **simple molecular**: composed of non-metal atoms in small molecules.
- The table summarises the particles, bonding and properties of the four different solid structures.

Type of structure	Particles in the structure	Type of substance	Bonding	Properties	Structure
giant metallic	**atoms** close-packed	metals e.g. Na, Fe, Cu and alloys such as steel	atoms are held in a close-packed giant structure by the attraction of positive ions for mobile electrons	• high melting points and boiling points • conduct electricity • high density • hard but malleable	
giant covalent	**very large molecules** containing thousands of atoms (giant molecules)	a few non-metals (e.g. diamond, graphite) and some non-metal compounds (e.g. polythene, PVC, sand)	large numbers of atoms are joined together by strong covalent bonds to give a giant structure as a 3D network or a very long, thin molecule	• high melting points and boiling points • do not conduct electricity • hard but brittle (3D networks) or flexible (long, thin structures)	3D network long, thin structures
giant ionic	**ions**	metal/non-metal compounds e.g. Na^+Cl^-, $Ca^{2+}O^{2-}$, $Mg^{2+}(Cl^-)_2$	positive and negative ions are held together by the attraction between their opposite charges	• high melting points and boiling points • conduct electricity when molten and in aqueous solution • hard but brittle • often soluble in water	
simple molecular	**small molecules** containing a few atoms	most non-metals and non-metal compounds e.g. O_2, S_8, H_2O, CO_2, sugar	atoms are held together in small molecules by strong covalent bonds The bonds between separate molecules are weak	• low melting points and boiling points • do not conduct electricity • soft when solid	

3 Chemical bonding

- When elements react, they try to gain, lose or share electrons in order to get a more stable electron structure, very often like that of a noble gas.
- **Ionic bonding** involves **transfer of electrons** from metal atoms to non-metal atoms forming positive and negative ions. The attractions between the oppositely-charged ions produce strong ionic bonds.

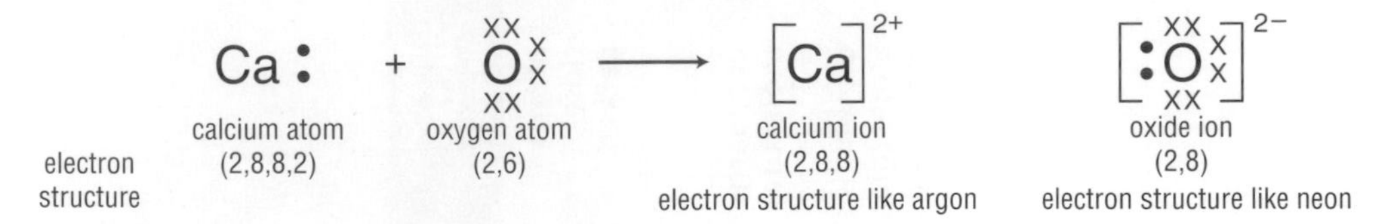

- **Covalent bonding** involves the **sharing of electrons** by two atoms. Each atom contributes one electron to the bond.

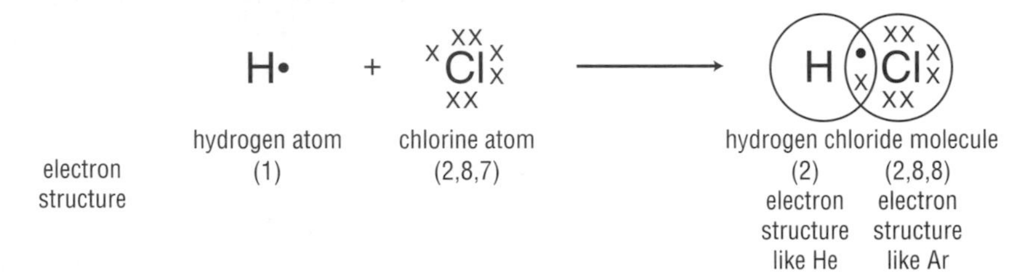

- The energy change in a chemical reaction can be calculated from strengths (bond energies) of the bonds involved. Bond breaking is an endothermic process. Bond making is an exothermic process.

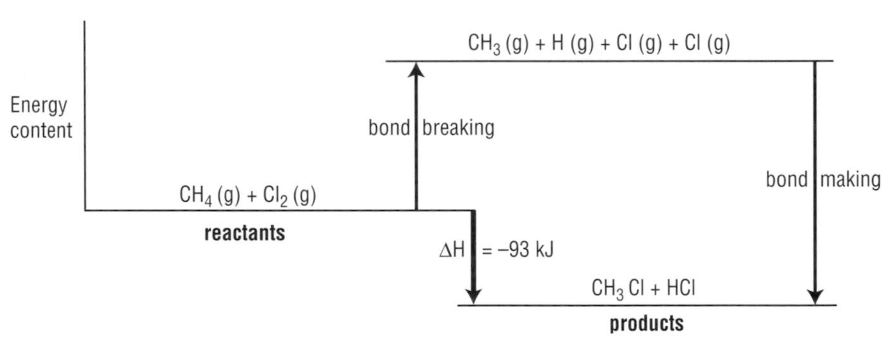

Bonds broken

1 mole of C — H bonds	= +435 kJ
1 mole of Cl — Cl bonds	= +242 kJ
	= +677 kJ

Bonds made

1 mole of C — Cl bonds	= −339 kJ
1 mole of H — Cl bonds	= −431 kJ
	= −770 kJ

$\therefore$ Energy change of reaction, $\Delta H = (+677 - 770)$ kJ $= -93$ kJ

STUDY QUESTIONS

Objective questions

Questions 1 to 4

Questions 1 to 4 concern the following substances:

A butane (C_4H_{10}) used in camping GAZ
B potassium nitrate (KNO_3) used as a fertiliser
C polythene (polyethene) used for clingfilm
D brass, an alloy of zinc and copper.

Choose from A to D the substance which

1 contains ions.

2 contains uncombined atoms.

3 has a giant covalent structure.

4 contains molecules with a small number of atoms.

Questions 5 to 11

Questions 5 to 11 concern the following structures:

A giant ionic
B giant metallic
C giant covalent
D simple molecular.

Choose from A to D the structure

5 which has freely moving electrons.

6 which contains molecules with a low relative molecular mass.

7 which has anions and cations.

8 which has a low boiling point.

9 for petrol.

10 for gold.

11 for limestone (calcium carbonate).

Questions 12 to 15

In questions 12 to 15 choose from A, B, C or D which is the correct answer.

12 The diagram shows a simplified structure for asbestos.

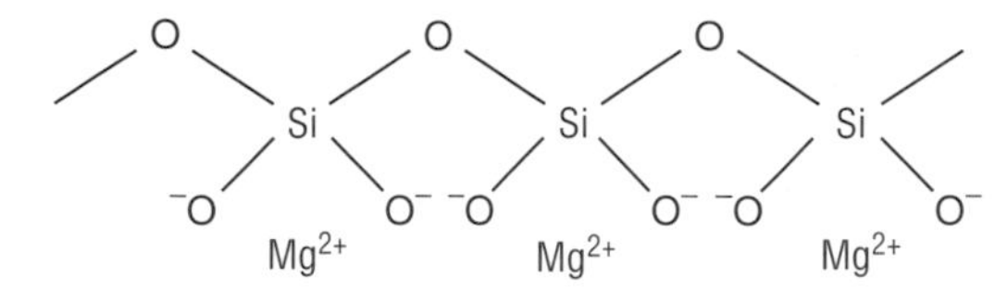

The bonding in asbestos is
A purely covalent.
B purely ionic.
C both covalent and ionic.
D neither covalent nor ionic.

13 Solid copper sulphate will
A conduct electricity.
B contain mobile electrons.
C contain ions.
D have a low boiling point.

14 Which *one* of the following processes is exothermic?
A photosynthesis
B ice melting
C petrol evaporating
D mortar setting.

15 When rubidium in Group I reacts with sulphur in Group VI
A rubidium atoms lose one electron and sulphur atoms gain one electron.
B rubidium atoms lose one electron and sulphur atoms gain two electrons.
C rubidium atoms lose two electrons and sulphur atoms gain one electron.
D rubidium atoms lose two electrons and sulphur atoms gain two electrons.

Short questions

16 The compound magnesium oxide, MgO, is made up of positive and negative ions.

a) Magnesium is in Group II. How are positive magnesium ions formed from magnesium atoms? *1 mark*
b) Oxygen is in Group VI. How are negative oxide ions formed from oxygen atoms? *1 mark*
c) How are the ions held together in magnesium oxide? *1 mark*
d) What is the type of bonding in magnesium oxide? *1 mark*

17 Hydrogen reacts with oxygen to form water. The reaction can be represented by these symbol and structural equations.

$$2H_2 + O_2 \longrightarrow 2H_2O$$
$$2[H-H] + [O=O] \longrightarrow 2[H-O-H]$$

a) What type of bonding is present in (i) H_2 (ii) H_2O? *2 marks*
b) Draw an electron dot/cross diagram to show all the electrons and the bonding in an H_2 molecule. *2 marks*

18 Phosphorus and fluorine react to form a simple molecular compound, phosphorus trifluoride.

Write down the words to fill the spaces labelled a), b) and c) in the following sentences.

Phosphorus trifluoride is made up of phosphorus and fluorine . . . (a) . . . . These are joined together by sharing pairs of . . . (b) . . . to form . . . (c) . . . bonds.
3 marks

19 Calcium oxide (lime) is an ionic compound with a high melting point, whereas paraffin wax (candle wax) is a molecular compound which melts easily. Why is there a difference? *4 marks*

20 a) Iron is a typical metal. Describe the structure and bonding in a metal like iron. *4 marks*

b) Steel, which is mainly iron, is used to make the bendable yet strong body panels for cars. How does the structure and bonding of iron allow the body panels to be both strong and bendable? *4 marks*

Diagrams may help your answers to both parts of this question.

21 The diagrams represent atoms of potassium and bromine. Only the outer shell electrons are shown.

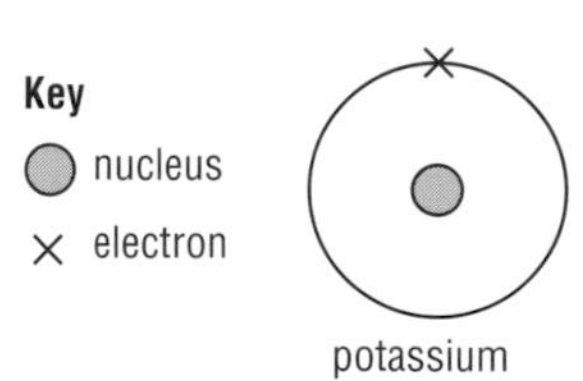

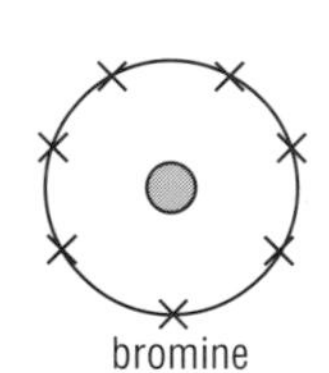

a) Potassium bromide is an ionic compound. Draw a diagram to show the arrangement of electrons in potassium bromide. *2 marks*

b) Atoms in a bromine molecule, Br_2, are covalently bonded. Draw a diagram to show the arrangement of electrons in a bromine molecule.
2 marks

22 Three bond energies are listed in the table.

Bond	Bond energy in kJ/mole
I — I	151
Cl — Cl	242
I — Cl	202

Iodine will react with chlorine to form iodine monochloride. The structural equation for this reaction is:

$$\mathrm{I—I + Cl—Cl \rightarrow 2I—Cl}$$

a) What is the total energy required to break one mole of I — I and 1 mole of Cl — Cl bond? *1 mark*

b) What is the energy given out when two moles of I — Cl bonds are formed? *1 mark*

c) What is the overall energy change for the reaction represented in the equation above? *1 mark*

d) What can you deduce about the reaction from your answer to part c)?
1 mark

Further examination questions

23 The diagram shows the chemical bond changes that occur as hydrogen and oxygen react.

H—H
      + O=O ⟶ H–O–H
H—H   H–O–H

The table gives the relative amounts of energy needed to break specific bonds.

Bond	Amount of energy needed to break the bond/kJ
H — H	436
O = O	498
H — O	464

Note: The amount of energy needed to make a bond is equal and opposite to that needed to break a bond.

a) Calculate the total energy needed to break the bonds in hydrogen and oxygen in the above reaction. *2 marks*

b) Calculate the total energy released when the bonds in water are formed in the above reaction. *2 marks*

c) Find the difference between answers b) and a) and hence state whether the reaction is exothermic or endothermic. *2 marks*

WJEC

24 Calor gas is a convenient energy source for boating, camping and caravanning. It is available as butane Calor gas or propane Calor gas. This label has been taken from a cylinder containing butane Calor gas, C_4H_{10}.

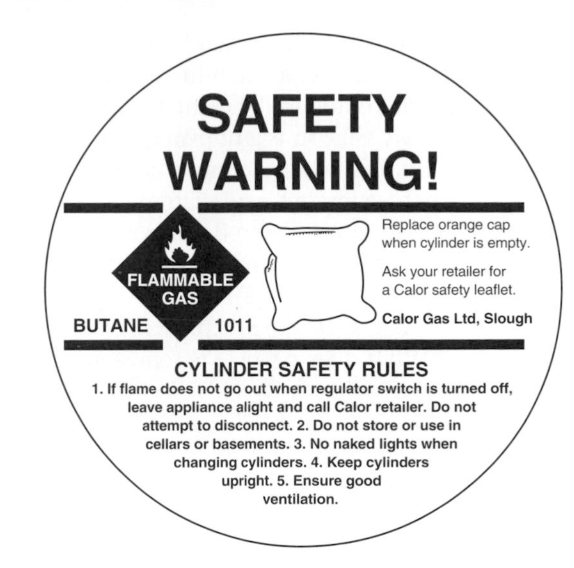

a) Suggest why these cylinders should not be (i) stored or used in cellars or basements, *1 mark*
(ii) changed near naked lights. *1 mark*

b) The pressure inside a cylinder full of gas increases on a hot day. Explain, in terms of the particles present, why this happens. *3 marks*

c) The diagram below shows the structure of a butane molecule.

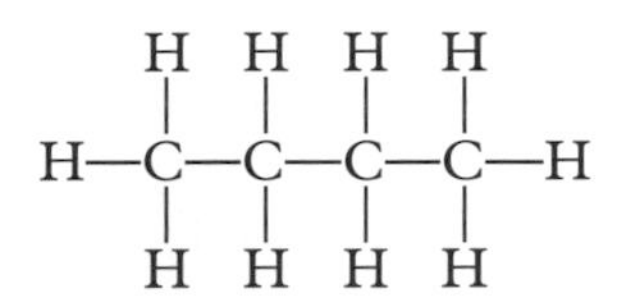

The carbon and hydrogen atoms in a butane molecule are joined by covalent bonds. What is a covalent bond? *2 marks*

d) From your knowledge of hydrocarbons, suggest why propane Calor gas, C_3H_8, is preferred as a fuel in winter rather than butane Calor gas even though it is more expensive and gives out less heat per gram than butane Calor gas. *2 marks*

e) The equation represents the complete combustion of the formula mass in grams of butane in air. The structural formulae of the chemicals involved are shown.

$$(C_4H_{10}) + 13\,(O{=}O) \longrightarrow 8\,(O{=}C{=}O) + 10\,(H{-}O{-}H)$$

Use the information in the table to help you to answer this question.

Bond	Bond energy in kJ/mole
C—H	413
C—C	346
O=O	497
C=O	803
H—O	463

(i) Calculate the energy needed to break all the bonds in the reactants.
3 marks
(ii) Calculate the energy released as new bonds are formed in the products.
2 marks
(iii) Calculate the energy change. *1 mark*

AQA

25 a) Paving stones can be laid on a layer of sand and cement. When water is added to the mixture, a reaction takes place between silicon dioxide in the sand and calcium oxide in the cement. From the reaction, calcium silicate, $CaSiO_3$, is formed. Silicon is a non-metal and calcium is a metal.

(i) Explain why calcium is classified as a metal. Use the idea of electronic structure in your explanation. *2 marks*

(ii) Why does silicon dioxide react with calcium oxide? *2 marks*

b) Part of the structure of silicon dioxide is shown in the diagram.

(i) What does line A represent? *1 mark*

(ii) What does particle B represent? *2 marks*

(iii) Silicon dioxide has a high melting point. Explain why in terms of its molecular structure. *1 mark*

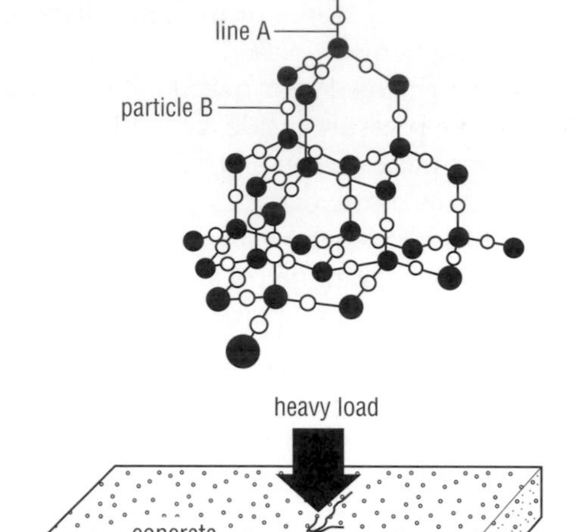

c) Concrete is a composite material made from sand, cement, gravel and water. Concrete is very strong when compressed, but not when under tension.

(i) Explain why concrete is a composite material. *2 marks*

(ii) How can concrete be prevented from cracking under tension? *1 mark*

AQA

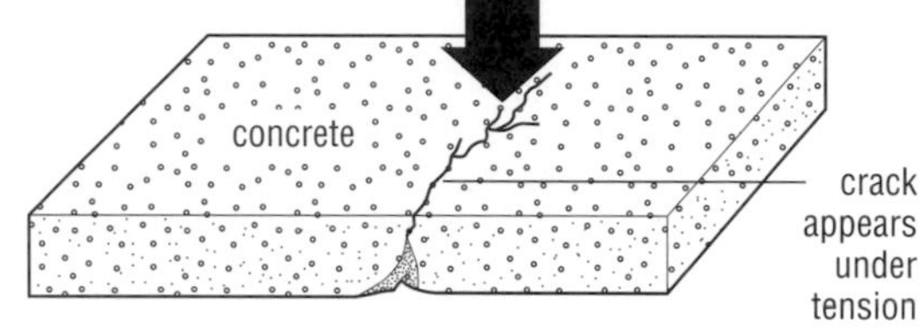

26 a) The compound sodium chloride is obtained from rock salt. Rock salt is a mixture of sodium chloride and sand. Sodium chloride is made from the elements sodium and chlorine.

(i) Why is sodium called an element? *1 mark*

(ii) Why is sodium chloride called a compound? *1 mark*

(iii) Explain why rock salt is called a mixture. *2 marks*

b) The diagrams represent the electronic structures of an atom of sodium and an atom of chlorine.

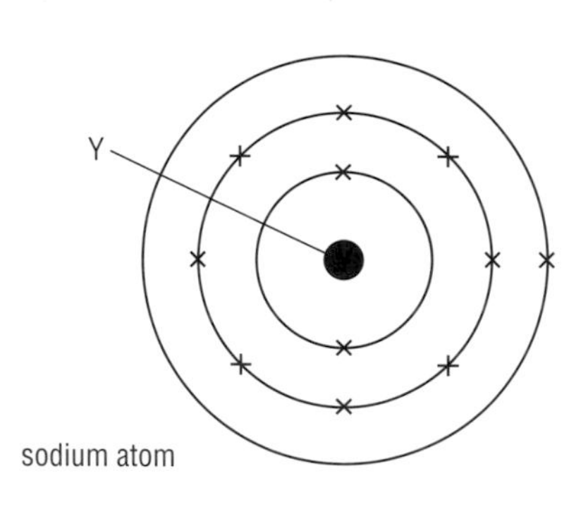

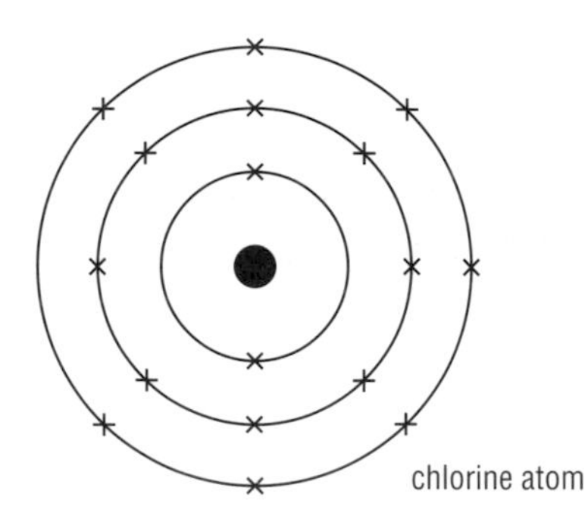

(i) Name the part of the atom labelled Y on the diagram of the sodium atom. *1 mark*

(ii) Give the chemical formula for the compound sodium chloride. *1 mark*

(iii) Explain how a sodium atom and a chlorine atom bond to form the compound sodium chloride. Use the idea of electronic structure in your answer. *5 marks*

AQA

CHAPTER

18

Reaction rates

SUMMARY

1 Different reaction rates

Chemical reactions can occur at different speeds or rates.

- Weathering of rocks such as marble and limestone may take decades, possibly centuries.
- Rusting and cooking take place at a steady rate.
- Burning and explosions are incredibly fast reactions.

2 What affects reaction rates?

Look at these examples then decide which of the factors below is responsible.

Concentration   Temperature   Surface area   Light   Catalysts

- Twigs burn faster than logs.
- It takes one minute to fry an egg, but five minutes to boil an egg.
- Bathroom cleaner works faster when it's not mixed with water.
- Plants will not grow in the dark.
- Biological detergent will clean badly stained clothes.

3 Measuring reaction rates

During a reaction, reactants are being used up and products are forming. We can use these changes to measure the reaction rate by calculating how much reactant is used up or how much product forms in a given time.

$$\text{i.e. Reaction rate} = \frac{\text{change in amount of reactant or product}}{\text{time taken}}$$

Here are the results for the experiment shown over the page.

Time/min	Mass of O_2 given off /g
2.0	1.5
4.0	2.5
6.0	3.1
8.0	3.4
10.0	3.6

Calculation of average reaction rate:

i) in the first 2 minutes $= \dfrac{\text{change in mass}}{\text{time taken}} = \dfrac{1.5}{2} = 0.75$ g/min.

ii) from 8 to 10 minutes $= \dfrac{\text{change in mass}}{\text{time taken}} = \dfrac{0.2}{2} = 0.10$ g/min.

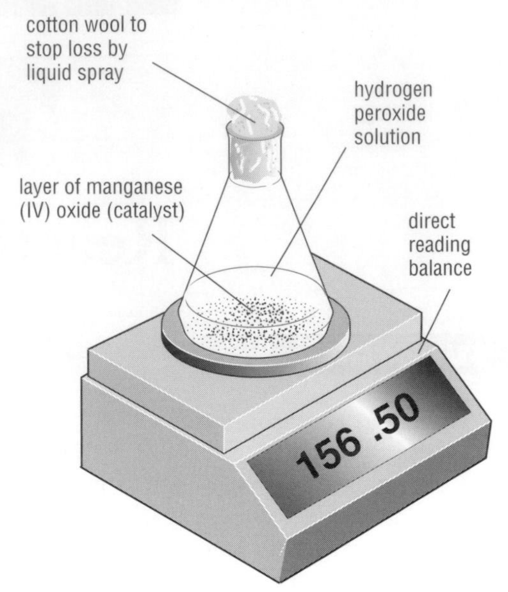

Studying the rate of reaction when manganese (IV) oxide catalyses the decomposition of hydrogen peroxide

$$2H_2O_2(aq) \rightarrow 2H_2O(l) + O_2(g)$$

4 Explaining the effect of different factors on reaction rates

The kinetic theory can be used to explain how different factors affect the rates of reactions. **Chemical reactions** occur when **particles of the reacting substances collide** with each other. This is called the **collision theory**.

The diagram uses the collision theory to explain how different factors affect the reaction rate.

SURFACE AREA	CONCENTRATION	TEMPERATURE
In general, reactions go faster when there is more surface area to react.	In general, reactions go faster when the concentration of reactants is greater.	In general, reactions go faster at higher temperatures.
Surface area of whole potato is low – few collisions with potato	Concentration of bathroom cleaner is low – few collisions with grease blob	At lower temperature, water particles move, occasionally hitting the potato
hot water particle; water at 100°C	blob of grease	water at 100°C
Surface area of four quarters of potato is greater – more surface area exposed for collisions	**Concentration of bathroom cleaner is higher – more collisions with grease blob**	**At higher temperature, water particles move faster, colliding with the potato more often and with more energy**
water at 100°C	blob of grease	water at 120°C under pressure; trivet

5 Catalysts and enzymes

- **Catalysts** are substances which **change the rate of a reaction without being used up**. They can therefore take part in the reaction over and over again.
- Different reactions need different catalysts and catalysts play an important part in the chemical industry.
- Petrol, ammonia, margarine and sulphuric acid are all produced by processes involving catalysts. The catalysts for many important industrial processes are transition metals or their compounds.
- Catalysts speed up reactions by allowing substances to react more readily. They do this by allowing bonds in the reactants to break more readily and new bonds to form more easily.
- The catalysts for biological processes are called **enzymes**. Almost every chemical reaction in living things has its own specific enzyme.
- More and more industrial processes are being developed which use enzymes. These processes include baking, brewing, yogurt- and cheese-making and the manufacture of fruit juices, vitamins and pharmaceuticals.

STUDY QUESTIONS

Objective questions

Questions 1 to 6

In questions 1 to 6 choose from A, B, C or D which is the correct answer.

1 Which *one* of the following will start reacting the fastest?

A 1 g of limestone chips in 100 cm^3 of acid at 25°C
B 1 g of limestone chips in 100 cm^3 of acid at 35°C
C 1 g of limestone powder in 100 cm^3 of acid at 25°C
D 1 g of limestone powder in 100 cm^3 of acid at 35°C

2 When powdered chalk is added to hydrochloric acid, it reacts and disappears in seconds. When a marble chip is added to the same acid, it takes several minutes to react and disappear. This is because

A acids react fast with all powders.
B chalk is more soluble in the acid.
C chalk is purer than marble.
D the chalk has a larger surface area.

3 A piece of magnesium ribbon was placed in excess sulphuric acid. The table shows the volume of gas produced at minute intervals.

Time/minutes	0	1	2	3	4	5	6	7	8	9	10
Volume/cm^3	0	8	13	17	20	23	25	27	29	30	30

How long did it take for half the metal to react?

A 2.5 minutes B 4 minutes
C 4.5 minutes D 5 minutes

4 The main reason for using catalysts in industry is that

A they increase the yield of products.
B they increase the rate at which products form.
C they reduce the temperature for a reaction.
D they remove products from the reaction mixture.

5 Catalysts
- A are chemically changed at the end of a reaction.
- B are changed in mass at the end of a reaction.
- C change the products in a reaction.
- D react with the reactants in a reaction.

6 A catalyst can change
- A the speed of a reaction.
- B the heat of a reaction.
- C the amount of product in a reaction.
- D the direction of a reaction.

Questions 7 to 9

Questions 7 to 9 concern the following graphs in which a variable, X, is plotted against time.

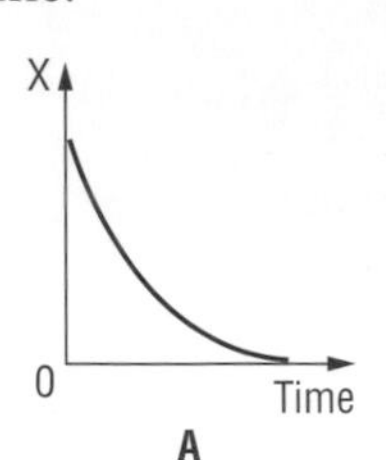

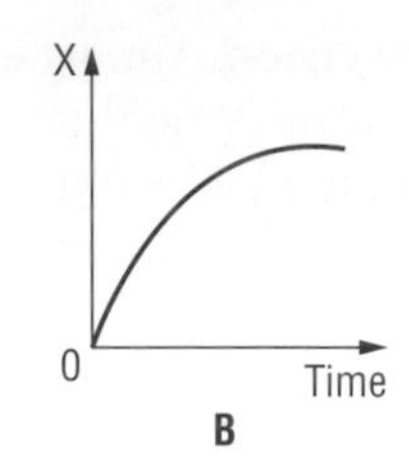

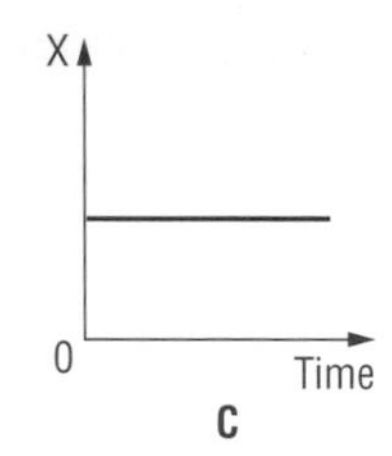

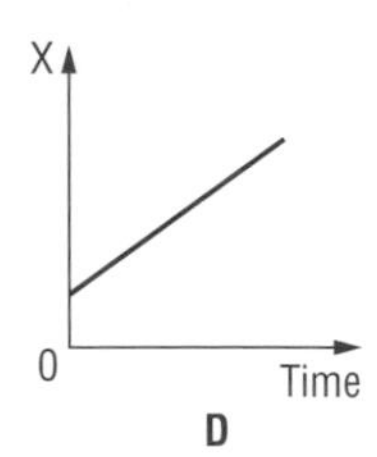

Choose from A to D, the graph which shows the change in

7 the volume of hydrogen produced (X), when excess acid is added to magnesium.

8 the mass of magnesium oxide (X), when magnesium oxide is heated in an open crucible.

9 the reaction rate (X), when excess acid is added to marble chips (calcium carbonate).

Questions 10 to 13

The curves in the graph show the volume of hydrogen produced during different experiments to investigate the reaction between zinc and hydrochloric acid. Curve X is obtained when 1 g of zinc pieces react with 100 cm^3 (excess) hydrochloric acid at 30°C.

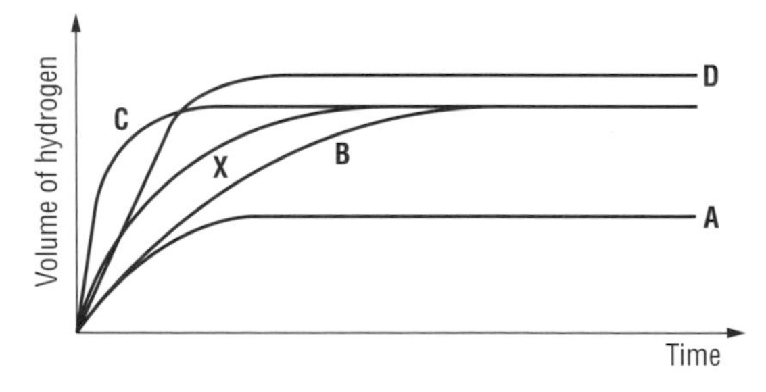

Which of the curves A, B, C or D would you expect to obtain when

10 1 g of zinc pieces reacts with 100 cm^3 of the same acid at 50°C?

11 1 g of zinc pieces reacts with 100 cm^3 of the same acid at 15°C?

12 0.5 g of zinc pieces reacts with 100 cm^3 of the same acid at 30°C?

13 1.2 g of zinc pieces reacts with 100 cm^3 of the same acid at 30°C?

14 Manganese(IV) oxide catalyses the decomposition of hydrogen peroxide to form water and oxygen. As a catalyst, the manganese(IV) oxide
- A causes the decomposition to begin.
- B increases the mass of oxygen which is formed.
- C increases the rate at which oxygen is produced.
- D increases the temperature of the manganese(IV) oxide.

Short questions

15 Hydrogen peroxide solution decomposes to produce water and oxygen. Either gold powder or copper powder can be used as a catalyst for this reaction. The apparatus shown can be used to investigate the reaction and find out whether gold or copper is the better catalyst.

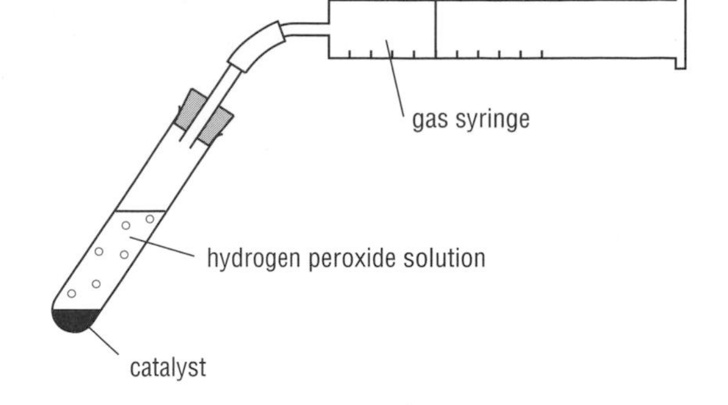

a) Write a word equation for the reaction involved. *1 mark*
b) State *three* things you would do to make the investigation fair. *3 marks*
c) Powdered glass is also a catalyst for the decomposition of hydrogen peroxide solution. Why does hydrogen peroxide solution in a glass bottle with rough inner walls keep badly compared with that in a glass bottle which has smooth inner walls? *2 marks*

16 A student reacted 100 cm^3 of dilute hydrochloric acid with excess marble chips. The mass of carbon dioxide produced changed with time and was noted every 20 seconds. The results obtained are shown in the table.

Time/s	0	20	40	60	80	100	120	140	160
Mass of carbon dioxide/g	0	0.36	0.58	0.72	0.85	0.92	0.94	0.94	0.94

a) Plot a graph of these results showing the mass of carbon dioxide on the vertical axis and time along the horizontal axis. *5 marks*
b) From your graph, deduce the mass of gas produced after 30 seconds. *1 mark*

17 The sketch graph below shows how the volume of hydrogen changes when a small piece of magnesium reacts with dilute hydrochloric acid.

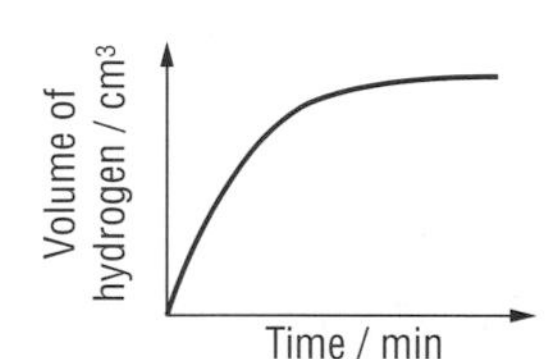

a) How does the rate of the reaction change with time? *1 mark*
b) What happens to the acid concentration during the reaction? *1 mark*
c) Why does the change in the magnesium affect the reaction rate? *2 marks*
d) Explain in terms of particle collisions why the rate of this reaction changes with time. *2 marks*

18 The sketch graph below shows how the reaction rate of one particular enzyme-catalysed reaction varies with temperature.

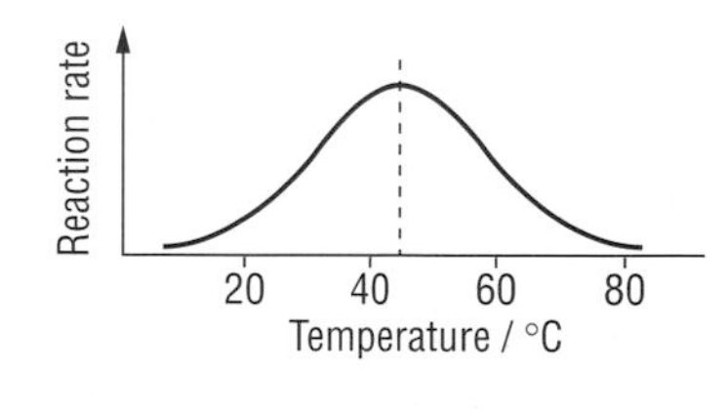

a) What are enzymes? *1 mark*
b) At what temperature is the enzyme used in the experiments shown in the sketch graph most effective? *1 mark*
c) Why does the curve in the sketch graph rise at first, reach a peak and then fall? *5 marks*
d) Why is it better to use water at 50°C with biological washing powders and not boiling water? *2 marks*

Further examination questions

19 A student placed some marble in some dilute hydrochloric acid and a gas was produced.

$$CaCO_3(s) + 2HCl(aq) \rightarrow CaCl_2(aq) + H_2O(l) + CO_2(g)$$

a) Give the name of the gas produced. *1 mark*

b) The student repeated the experiment with a small cube of marble (excess) and 25 cm^3 of the acid and measured the volume of gas evolved at 10 second intervals. This table of results was obtained.

Time/s	0	10	20	30	40	50	60
Volume/cm^3	0	20	30	36	39	40	40

(i) Draw a graph of these results. Plot the time on the horizontal (x) axis. Use the graph, where necessary, to answer the questions which follow. *4 marks*
(ii) After how many seconds did the reaction stop? *1 mark*
(iii) From what simple observation would the student know that the reaction had stopped? *1 mark*
(iv) What would the student see which shows that all of the acid, rather than all of the marble, had been used by the end of the reaction? *1 mark*
(v) What volume of gas had been collected at the end of the experiment? *1 mark*

c) The experiment was repeated with three pieces of marble, each of the same size as before, and 25 cm^3 of the same acid.
(i) At the start of the experiment, would the gas be produced more quickly, less quickly or at the same rate as in the first experiment? *1 mark*
(ii) Give a reason for your answer to (i). *1 mark*
(iii) Sketch, on the grid, using a broken line (---) the sort of curve which might have been produced in this experiment. *2 marks*

EDEXCEL

20 The table gives the results from the reaction of excess marble chips (calcium carbonate) with (A) 50 cm^3 of dilute hydrochloric acid, (B) 25 cm^3 of the dilute hydrochloric acid + 25 cm^3 of water.

Time/minutes	Total mass of carbon dioxide produced/g	
	(A) 50 cm^3 dilute HCl	(B) 25 cm^3 dilute HCl + 25 cm^3 water
0	0.0	0.0
1	0.9	0.4
2	1.4	0.7
3	1.8	0.9
4	2.1	1.0
5	2.2	1.1
6	2.2	1.1

a) Plot the two curves on the same graph. Label the first curve A and the second curve B. *4 marks*

b) Sketch, on the same grid, the curve you might have expected for the reaction A if it had been carried out at a higher temperature. Label this curve C. *2 marks*

c) Sketch, on the same grid, the curve you might have expected for reaction B if the marble had been ground to a powder. Label this curve D. *2 marks*

d) Complete the word equation:

hydrochloric acid + calcium carbonate → carbon dioxide + ________ + ________ *1 mark*

e) Carbon dioxide can also be obtained by heating limestone (calcium carbonate). [$A_r(Ca) = 40, A_r(C) = 12, A_r(O) = 16$]

$$CaCO_3 \rightarrow CaO + CO_2$$

Calculate:
(i) the relative formula mass (M_r) of limestone (calcium carbonate); *1 mark*
(ii) the relative formula mass (M_r) of lime (calcium oxide); *1 mark*
(iii) the maximum mass of lime which could be obtained from 10 g of limestone (show your working). *2 marks*
(iv) the maximum volume of carbon dioxide, at 1 atmosphere pressure and 25°C, which could be obtained from 10 g of limestone. (Assume that 1 mole of gas occupies 24 dm^3 at 1 atmosphere pressure and 25°C. Show your working.) *3 marks*

WJEC

21 A compound called zinc iodide is formed when zinc is added to a solution of iodine in ethanol. A lump of zinc is weighed. It is then hung in 25 cm^3 of a solution of iodine. At intervals the zinc is taken out of the solution, thoroughly washed with ethanol, dried and reweighed. It is then put back into the solution. The results are shown in the table.

Number of days	0	0.5	1	2	3	4	5	6
Mass of lump of zinc/g	2.33	2.18	2.11	2.05	2.02	2.01	2.00	2.00
Loss of mass/g		0.15	0.22	0.28	0.31	0.32	0.33	0.33

a) (i) Plot these points on a graph, showing the loss of mass of zinc vertically and the time in days horizontally. *4 marks*
(ii) Finish the graph by drawing the line of best fit. *1 mark*

b) The experiment is repeated with a more concentrated iodine solution. The graph obtained is steeper. A greater mass of zinc is used up.
The diagrams represent some of the particles present in the two solutions.

less concentrated iodine solution

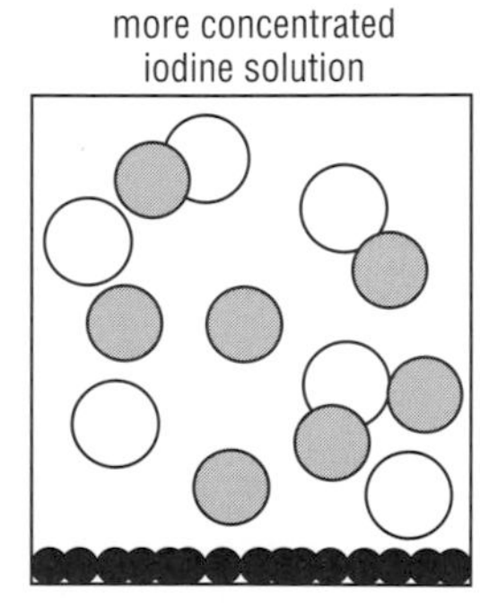

more concentrated iodine solution

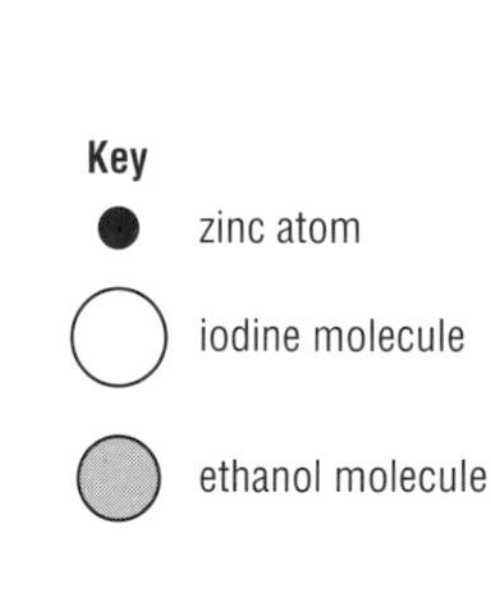

Key
zinc atom
iodine molecule
ethanol molecule

(i) Explain why the graph in this experiment is steeper than the original graph. *2 marks*
(ii) Explain why a greater mass of zinc is used up. *2 marks*

OCR

22 a) Some students investigated the rate of decomposition of 50 cm^3 of hydrogen peroxide solution, using 1.0 g of powdered manganese(IV) oxide catalyst. During the reaction, oxygen and water were produced.

$$2H_2O_2(aq) \rightarrow 2H_2O(l) + O_2(g)$$

The oxygen was collected and measured at regular time intervals. The results are shown on the graph.

(i) Mark with a cross (x) a part of the curve at which the rate of reaction is fastest. *1 mark*

(ii) The experiment was repeated using 1.0 g of powdered manganese(IV) oxide and 50 cm^3 of less concentrated hydrogen peroxide solution. Draw a sketch graph showing the curve you might expect to get using this less concentrated solution with a copy of the graph above on the same axes. *2 marks*

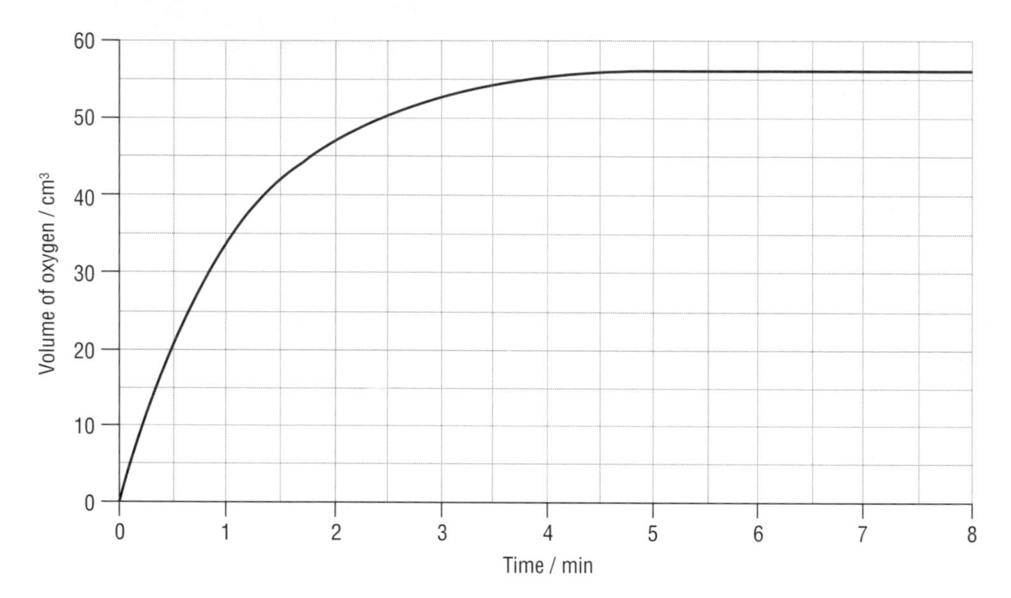

b) Explain, in terms of the particles taking part in a reaction, how heating can increase the rate of reaction. *3 marks*

c) Catalysts are used in many industrial processes including the catalytic cracking and polymerisation of oil fractions. Explain what is meant by:

(i) catalytic cracking, *2 marks*

(ii) polymerisation. *2 marks*

AQA

CHAPTER

19

Earth science

SUMMARY

1 **The Earth – our source of raw materials:**

- minerals, rocks and fossil fuels from the **Earth's crust**
- water, salt and other minerals from the **seas**
- important gases from the **air**
- timber, food and fibres from **living things**.

Rocks are made up from **minerals** which are **compounds** containing different **elements**. Chalk, limestone and marble are different rocks all of which contain calcium carbonate, $CaCO_3$, made up of the elements calcium, carbon and oxygen.

2 **Layers of the Earth**
The Earth has four clear layers. The temperature and density of these layers increases from the atmosphere to the core.

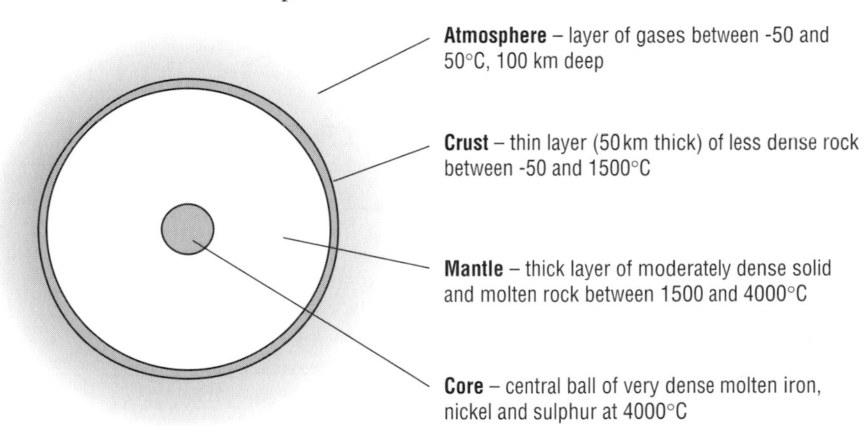

Atmosphere – layer of gases between -50 and 50°C, 100 km deep
Crust – thin layer (50 km thick) of less dense rock between -50 and 1500°C
Mantle – thick layer of moderately dense solid and molten rock between 1500 and 4000°C
Core – central ball of very dense molten iron, nickel and sulphur at 4000°C

3 **Weathering** is the breaking up of rocks. There are two main kinds:

- **Physical weathering**
  - by expansion and contraction of the rock as the temperature changes
  - by the freezing of water in cracks.
- **Chemical weathering**
  - by the action of rainwater (or more rapidly acid rain) on carbonates (e.g. chalk, marble and limestone)
  - by the action of water itself on feldspar in granite.

Very often, the weathered material from rocks is carried away by rivers, ocean currents and winds. This is called **transport**.

Erosion = Weathering + Transport

4 Rock types

There are three different rock types:

		Igneous	Sedimentary	Metamorphic
FORMED FROM	→	Cooling of **molten magma** in mantle through volcanoes or underground	Layers of **sediment** in lakes or seas over millions of years	**Heat and pressure** on existing rocks over long periods
STRUCTURE	→	Various minerals in interlocking crystals	Grains cemented by salt crystals	Grains, layers or small crystals
EXAMPLES	→	Granite, basalt	Sandstone, limestone, chalk	Marble, slate

5 The **rock cycle** takes millions of years.

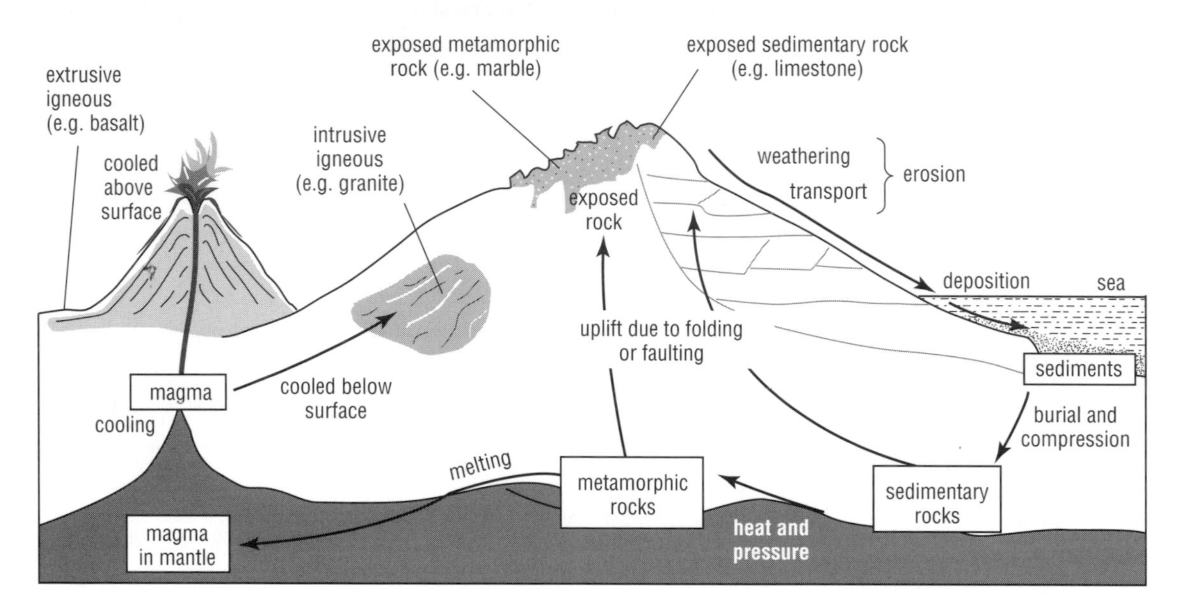

6 Plate tectonics

The Earth's crust is cracked and broken into vast sections called **plates** which float on the dense liquid mantle. The continents and oceans sit on top of the plates which move very, very slowly due to **convection currents** in the liquid mantle. The study of the movement and interaction of the giant plates is called **plate tectonics**.

- *When plates slide past each other*, tear faults and earthquakes can occur.
- *When plates move apart*, normal faults, rift valleys and volcanoes occur.
- *When plates move towards each other* and collide, reverse faults and folds can occur.

STUDY QUESTIONS

Objective questions

Questions 1 to 5

The diagram on the next page shows a simplified rock cycle. Some of the labels, circled and numbered 1 to 5, have been removed.

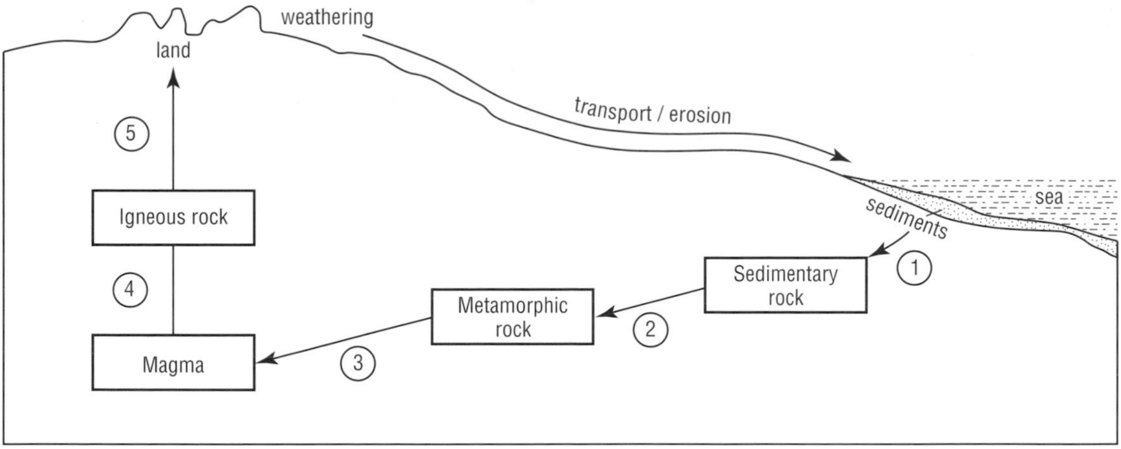

weathering
land
transport / erosion
sea
sediments
5
Igneous rock
4
Magma
3
Metamorphic rock
2
Sedimentary rock
1

Which of the following labels, A to E, correspond to each of the labels 1 to 5?

A burial and compacting
B cooling and solidifying
C exposure by erosion
D heat and pressure
E melting.

Questions 6 to 14

In questions 6 to 14, choose from A, B, C, D or E which is the correct answer.

6 Volcanoes are most likely to occur when
 A one tectonic plate moves on top of another.
 B tectonic plates move alongside each other.
 C two tectonic plates move apart.
 D two tectonic plates collide.
 E two tectonic plates move in the same direction.

7 The conditions below the Earth's crust have been studied by measuring
 A global warming.
 B magnetic fields.
 C ocean temperatures.
 D tidal changes.
 E wind speeds.

8 Rocks exposed to the weather are slowly broken down. When water enters the cracks in rocks, it causes them to splinter when the water
 A cools.
 B freezes.
 C heats up.
 D is acidic.
 E evaporates.

9 Fossils are found in
 A igneous and metamorphic rocks.
 B igneous and sedimentary rocks.
 C metamorphic and sedimentary rocks.
 D metamorphic rocks only.
 E sedimentary rocks only.

10 When limestone (calcium carbonate, $CaCO_3$) is heated, it decomposes losing carbon dioxide and forming lime. The formula of lime is
 A CaC.
 B CaCO.
 C $CaCO_2$.
 D CaO.
 E CaO_2.

11 When limestone is weathered chemically, it reacts to form calcium hydrogencarbonate ($Ca(HCO_3)_2$) solution. This decomposes to form stalagmites and stalactites in caves. The formula of the cave deposits is
 A CaO.
 B $Ca(OH)_2$.
 C CaC.
 D $CaCO_3$.
 E $CaHCO_3$.

12 Slate is formed by the action of heat and high pressure on
A chalk.
B gravel.
C mud.
D sand.
E shale.

13 Slate is
A an igneous rock.
B a lava.
C a metamorphic rock.
D a sediment.
E a sedimentary rock.

14 The Earth's crust is divided into plates. When two plates collide, they form
A extrusions.
B folds.
C hot spots.
D rift valleys.
E volcanoes.

Short questions

15 The diagrams show three stages in the formation of an island.

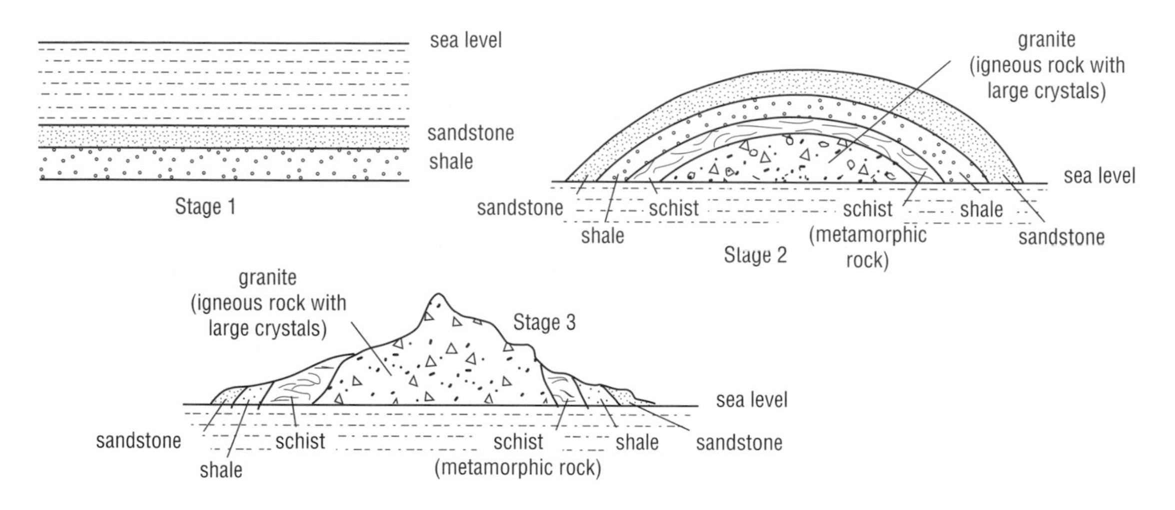

Explain how the shape of the island has been produced by referring to the formation of the rocks present and any changes which have happened since their formation. *5 marks*

EDEXCEL

16 Basalt and granite are both igneous rocks which are formed when molten magma cools. Basalt is formed when magma cools relatively quickly, but granite is formed when the magma cools more slowly.

a) State *two* ways in which granite and basalt will have similar appearances. *2 marks*
b) In what ways will the appearance of basalt and granite differ? *1 mark*
c) Explain how the different rates of cooling cause the different appearances of basalt and granite. *2 marks*

17 a) What is the difference between 'weathering' and 'erosion'? *2 marks*
b) Explain how temperature changes alone can lead to the physical weathering of rocks. *2 marks*
c) How and why is limestone weathered chemically by rain water? *4 marks*

Further examination questions

18 a) Outline the formation of
(i) igneous rock, (ii) metamorphic rock. *4 marks*
b) Explain how igneous rocks can become sedimentary rocks. Give an indication of the time scale involved. *4 marks*
c) Explain how sedimentary rock can be recycled as igneous rock. *4 marks*

EDEXCEL

19 The diagrams A, B and C show microscopic slides of the rocks granite, sandstone and slate, although not necessarily in that order.

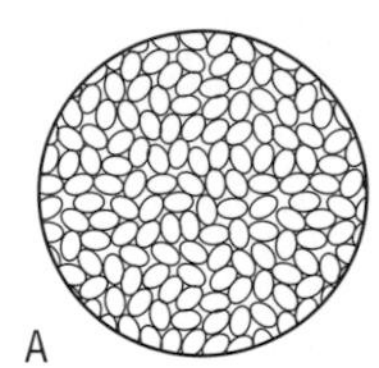
A

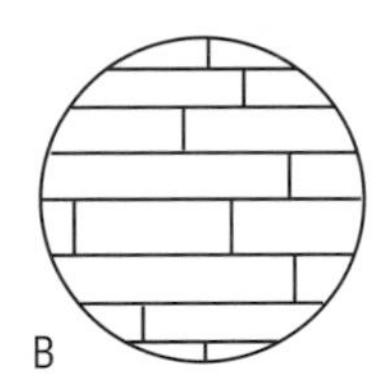
B

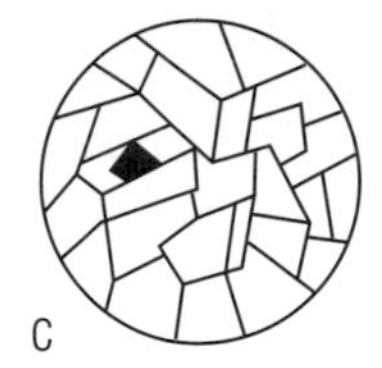
C

a) Copy out and complete the table by
(i) identifying each slide, A, B and C, as granite, sandstone or slate *3 marks*
(ii) naming the type of rock shown in each slide. *3 marks*

	Granite	Sandstone	Slate
Slide A, B or C			
Rock type			

b) Explain the formation of each rock, using evidence from the slides. *6 marks*
c) How could a geologist find the age of the rock sample C? *1 mark*

WJEC

20 a) The diagram shows part of the rock cycle. Identify the material present at each of the positions 1 to 5 in the diagram. *5 marks*

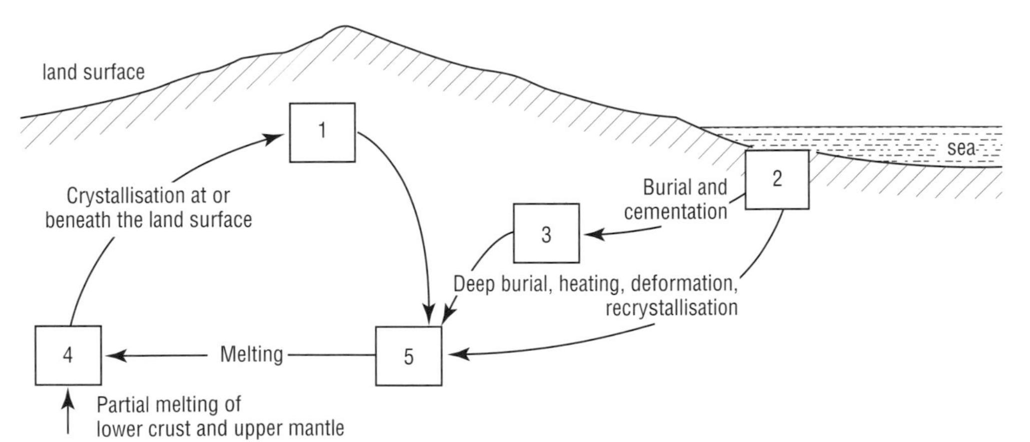

b) Both igneous and metamorphic rocks can form during mountain building. These rock types have a similar appearance.
(i) In what way is their appearance similar? *1 mark*
(ii) Describe how you could distinguish between these two rock types by their appearance. *2 marks*
c) Explain how the age of an igneous rock can be estimated. *3 marks*

AQA

CHAPTER

20

Chemicals from crude oil

SUMMARY

1 Crude oil and its products

- Crude oil is a **fossil fuel**, like coal and natural gas.
- Crude oil contains hundreds of different carbon compounds. These carbon compounds are often called **organic compounds**.
- Most of the compounds in crude oil are **hydrocarbons** – compounds containing only hydrogen and carbon.
- Hydrocarbons are typical simple molecular compounds. They have low melting points and boiling points. They do not conduct electricity and are insoluble in water.

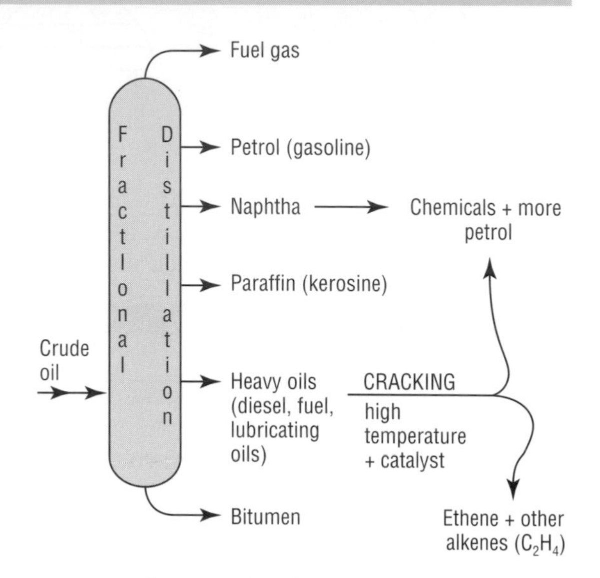

Processes from crude oil

2 Alkanes

- The simplest hydrocarbons are **alkanes**.
- The alkanes are a **homologous series** of compounds with similar properties in which the formulas differ by CH_2 e.g. methane, CH_4; ethane, C_2H_6; propane, C_3H_8 etc.
- Alkanes have strong C—C and C—H bonds so they have few reactions. They do, however, combine with reactive non-metals.

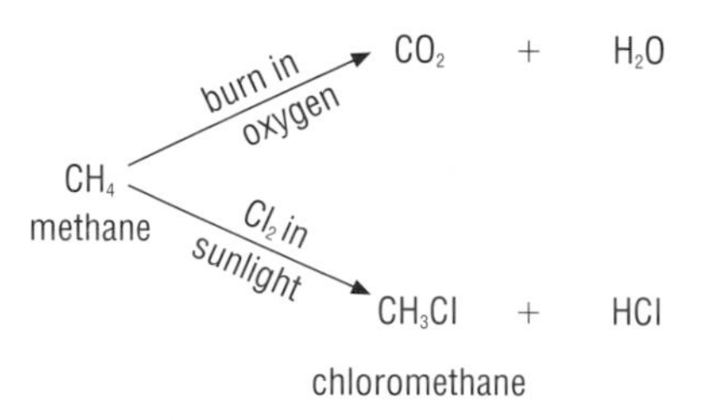

3 Alkenes

- **Cracking** is used to produce petrol from the heavier fractions from crude oil. During cracking, alkane molecules are split into smaller, more combustible molecules (alkanes and alkenes).

$$\text{e.g.}\quad \underset{\text{decane}}{C_{10}H_{22}} \rightarrow \underset{\text{octane}}{C_8H_{18}} + \underset{\text{ethene}}{C_2H_4}$$

- Compounds, like ethene, which contain a carbon–carbon double bond are called **alkenes**.

```
    H  H
    |  |            H       H
 H—C—C—H             \     /
    |  |               C=C
    H  H             /     \
                    H       H
   ethane              ethene
```

- Alkenes are much more reactive than alkanes because of their reactive double bonds.
- Alkenes readily undergo **addition reactions** as shown below.

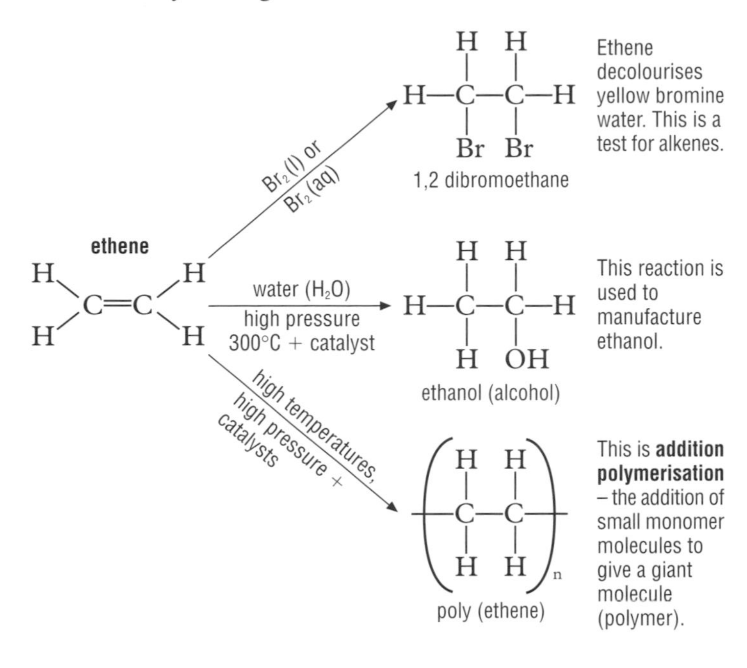
ethene
Br_2(l) or Br_2(aq)
1,2 dibromoethane
Ethene decolourises yellow bromine water. This is a test for alkenes.
water (H_2O)
high pressure
300°C + catalyst
ethanol (alcohol)
This reaction is used to manufacture ethanol.
high temperatures, high pressure + catalysts
poly (ethene)
This is **addition polymerisation** – the addition of small monomer molecules to give a giant molecule (polymer).

STUDY QUESTIONS

Objective questions

Questions 1 to 5

This is a simplified diagram of the distillation column at an oil refinery.

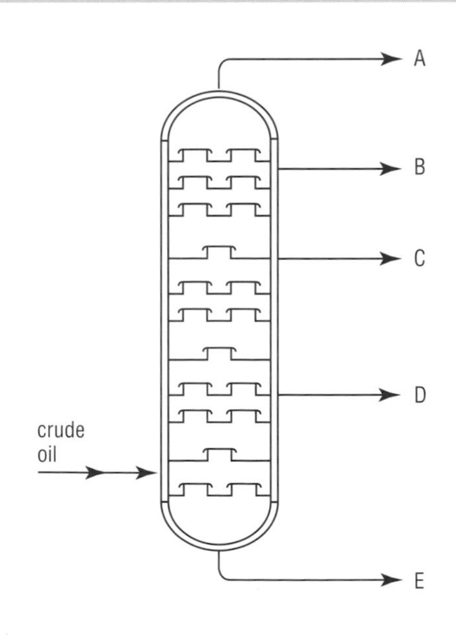
A
B
C
D
crude oil
E

Choose from A, B, C, D and E the fraction which

1 is used as petrol.

2 is used for lubricating oils and waxes.

3 is used to produce liquefied petroleum gas (LPG).

4 is used as kerosine (paraffin) for central heating and jet engines.

5 provides tar (bitumen) for roads and roofing.

Questions 6 to 8

A butane B ethane C methane
D propane E octane

Choose from A to E the substance which is

6 the main constituent of natural gas.

7 used in GAZ for camping and caravanning.

8 a major constituent of petrol.

Questions 9 to 15

In questions 9 to 15 choose from A, B, C or D which is the correct answer.

9 How many different structural formulae are there with the molecular formula C_3H_7Cl?
A 1 B 2 C 3 D 4

10 Crude oil can be separated by fractional distillation because each fraction has a different
A boiling point B colour C density
D flammability

11 When bromine reacts with ethane, the main products are
- A bromoethane and hydrogen.
- B bromoethane and hydrogen bromide.
- C dibromoethane and hydrogen.
- D dibromoethane and hydrogen bromide.

12 Alkanes and alkenes both react with
A chlorine B conc. sulphuric acid
C sodium D sodium hydroxide solution

13 Which one of the following substances will decolourise bromine water?
A petrol B poly(ethene) C sugar
D vegetable oil

14 Plastics are a pollution problem because they
- A are manufactured from crude oil.
- B are not attacked by air, water or bacteria.
- C burn easily producing carbon dioxide.
- D contain large polymer molecules.

15 Polythene (poly(ethene)) and ethene have
- A the same molecular formula.
- B unsaturated molecules.
- C the same ratio of C to H atoms.
- D similar properties.

Short questions

16 Petrol for cars contains up to 5% benzene, a hydrocarbon with the formula, C_6H_6.

a) What is a hydrocarbon? *1 mark*

b) Name the products formed when benzene undergoes complete combustion. *2 marks*

c) Some modern cars contain catalytic converters in their exhaust systems. These catalyse the combustion of any unburnt benzene.

(i) State *two* factors (other than the use of a catalyst) which will affect the rate of combustion of this unburnt benzene. *2 marks*

(ii) Why is the concentration of unburnt benzene in the exhaust gases highest just after the car has been started up? *1 mark*

17 The diagram shows two processes that take place in a petrochemical plant.

Kerosine (paraffin) fraction —*Reactor 1*→ Ethene —*Reactor 2*→ Poly(ethene)

a) Name the type of reaction which takes place in
(i) Reactor 1, (ii) Reactor 2. *2 marks*

b) Ethene, C_2H_4, is said to be an unsaturated hydrocarbon. What does unsaturated mean when used to describe hydrocarbons? *1 mark*

c) Butene, C_4H_8, is also an unsaturated hydrocarbon. Draw *two* possible structural formulae for butene.
2 marks

18 Liquid paraffin can be 'cracked' on a small scale by passing its vapour over strongly heated broken porous pot.

a) Describe how you would test for the ethene produced. *2 marks*
b) What would the result of the test be? *1 mark*
c) Why would ethene be produced at a faster rate if the porous pot was broken up into smaller pieces? *1 mark*
d) The formula for ethene is C_2H_4. Draw a diagram to show how the atoms are arranged in a molecule of ethene. *1 mark*
e) What chemical is produced when ethene is polymerised? *1 mark*

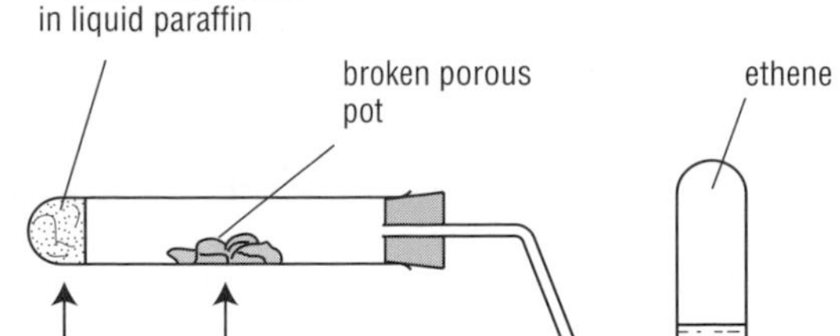

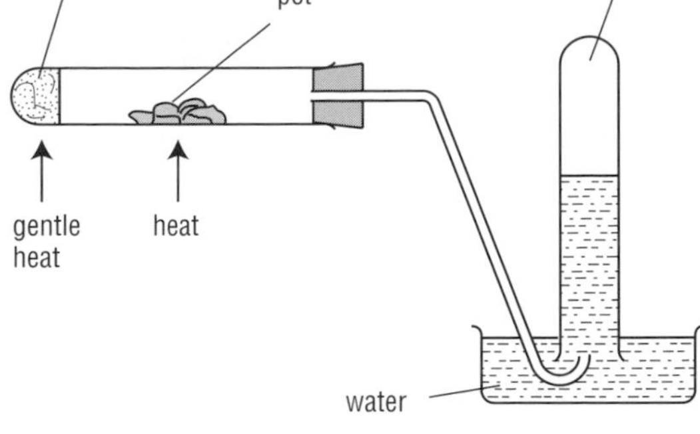

19 The main component of natural gas is methane, CH_4.

a) Name the products formed when methane burns completely. *2 marks*
b) Write a balanced chemical equation for the complete combustion of methane. *2 marks*
c) Use the equation in b) to calculate how many grams of oxygen are needed to react with 10 g of methane. (C = 12, O = 16, H = 1) *3 marks*

20 The most commonly used fossil fuels are coal, oil and natural gas.

a) Why are coal, oil and natural gas called fossil fuels? *1 mark*
b) What materials have led to the formation of oil and natural gas? *2 marks*
c) What conditions have been necessary for the formation of oil and natural gas? *3 marks*

Further examination questions

21 At an oil refinery, fractional distillation is used to separate crude oil into fractions.

a) The separation happens in a fractionating tower. The vapourised crude oil enters at the bottom of the tower. Describe what happens in the fractionating tower to produce the separate fractions as shown in the diagram. *3 marks*
b) (i) Large amounts of the heavier fractions are 'cracked'. What is meant by cracking? *1 mark*
(ii) Cracking can only happen at high temperatures. Explain why high temperatures are needed. *2 marks*
(iii) The conditions needed for cracking make the process expensive. Give *two* reasons why heavy fractions are cracked, even though the process is expensive. *2 marks*

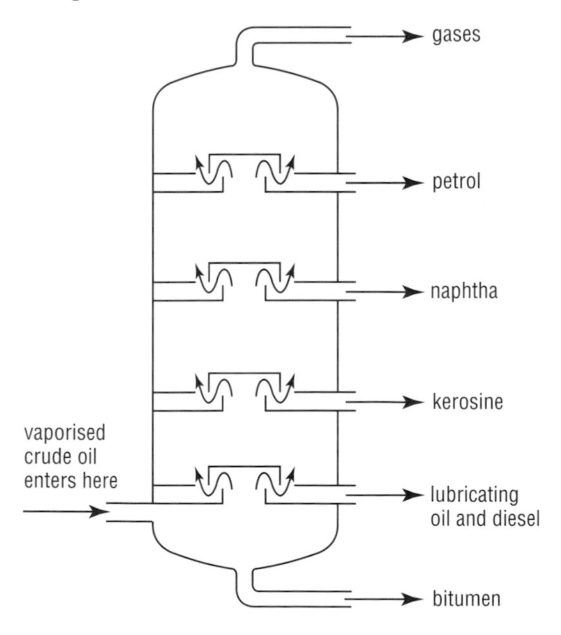

c) Ethene is one chemical obtained by cracking. It can be used to make poly(ethene) – polythene.

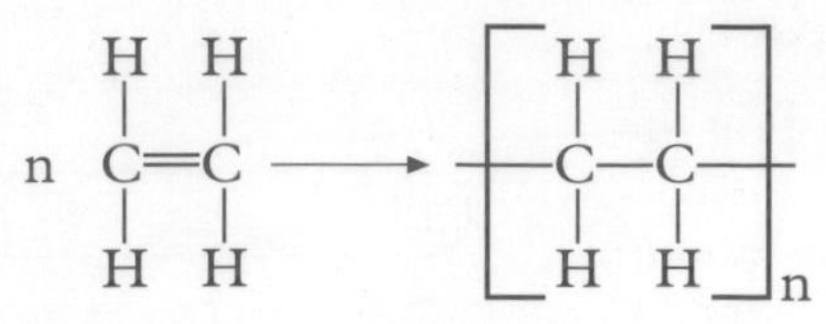

(i) Name the type of reaction taking place. *1 mark*

(ii) Describe how the ethene molecules join together to form poly(ethene). *2 marks*

EDEXCEL

22 Crude oil can be separated into fractions by a process called fractional distillation. Here is some information on these fractions.

Name of fraction	Boiling range	Number of carbon atoms per molecule
gas fraction	below 25°C	1–4
petrol fraction	25–100°C	5–10
paraffin fraction	100–250°C	11–15
diesel fraction	250–350°C	16–20
lubricant fraction	350–500°C	21–35
bitumen fraction	above 500°C	above 35

a) Why is a range of boiling points given for each fraction? *2 marks*

b) In which fraction would you find the hydrocarbon with the formula $C_{17}H_{36}$? *1 mark*

c) The paraffin fraction has a higher boiling range than the petrol fraction. Write down *two* further ways in which it is different. *2 marks*

d) The hydrocarbon $C_{16}H_{34}$ was heated strongly in the absence of air. This is one of the reactions that took place:

$$C_{16}H_{34} \rightarrow C_6H_{14} + C_6H_{12} + 2C_2H_4$$

This type of reaction is carried out because there is a greater demand for the products than for the original hydrocarbon. Suggest *two* reasons for this. *2 marks*

e) A molecule of the compound methane, CH_4, can be shown like this:

```
    H
    |
H — C — H
    |
    H
```

Draw a molecule of the compound ethene, C_2H_4. *2 marks*

f) Small molecules of substances called monomers can be joined together in polymerisation e.g. ethene → poly(ethene).

(i) Complete the equation to show the formation of the polymer from the monomer propene. *1 mark*

```
   ⎧ H   H  ⎫
   ⎪ |   |  ⎪
n  ⎨ C = C  ⎬ ——→
   ⎪ |   |  ⎪
   ⎩ H  CH3 ⎭
```

(ii) Suggest the name of the polymer formed. *1 mark*

AQA

23 Chloroethene is a liquid which is important in industry. It is produced by the reaction of one molecule of hydrogen chloride with one molecule of ethyne in the presence of a catalyst.

$$C_2H_2 + HCl \rightarrow C_2H_3Cl$$

a) When a solution of bromine is added to chloroethene, the red-brown colour of the bromine disappears. What does this tell you about the structure of chloroethene? *1 mark*

b) The reaction of hydrogen chloride with ethyne can be shown as:

$$H-C\equiv C-H \; + \; H-Cl \longrightarrow \begin{matrix} H & & H \\ & C=C & \\ H & & Cl \end{matrix}$$

Write down the molecular formula of the compound formed when two molecules of hydrogen chloride react with one molecule of ethyne. *1 mark*

c) Much of the chloroethene produced is used to make a polymer called PVC.
   (i) Draw a part of a PVC chain clearly showing bonds between atoms. *2 marks*
   (ii) This polymer is used to produce guttering and drainpipes. Previously, they were often made of cast iron. Suggest *one* advantage, apart from cost, that PVC has when making guttering and drainpipes. *1 mark*

d) (i) Household waste contains PVC, also known as poly(chloroethene). This can cause problems when waste is burnt in an incinerator. Explain how possible combustion products could have effects on the local environment. *3 marks*
   (ii) Some local councils prefer to put waste into landfill sites where the action of microbes can rot down some of the organic parts of the waste, such as carbohydrates, proteins and fats. Discuss the advantages and disadvantages of depositing PVC in landfill sites. *3 marks*

OCR

24 The table shows the fractions in a sample of crude oil.

Name of fraction	Percentage
gases	3
petrol	15
paraffin	17
diesel oil	15
lubricating oil and bitumen	50

a) Name the process used to separate crude oil into these fractions. *1 mark*

b) The lubricating oil and bitumen fraction is not as useful as the other fractions.
   (i) Name the process used to convert this fraction into more useful fractions. *1 mark*
   (ii) One of the products of this process is ethene, C_2H_4. Give a chemical equation, using structural formulae, to show how ethene can form a polymer. *2 marks*

c) The gases fraction contains methane CH_4. Methane burns in air to form carbon dioxide and water.

$$CH_4 + 2O_2 \rightarrow CO_2 + 2H_2O$$

Bond	Approximate bond energies/ kilojoules per mole
C—H	+415
O═O	+495
O—H	+465
C═O	+740

   (i) Calculate the total energy transferred from the surroundings to break the bonds of the reactants, methane and oxygen. *2 marks*
   (ii) Calculate the total energy transferred to the surroundings from the formation of the bonds of the products, carbon dioxide and water. *2 marks*
   (iii) Use your answers from (i) and (ii) to explain what type of chemical reaction occurs. *2 marks*

AQA

21

Ammonia and fertilisers

SUMMARY

1 Reversible reactions and equilibria

Reversible reactions can go in both directions by changing the conditions or adding or removing reagents.

e.g.	$CuSO_4.5H_2O$	*heat* $\rightleftharpoons$	$CuSO_4$	$+\ 5H_2O$
	blue hydrated copper(II) sulphate	*add water*	white anhydrous copper(II) sulphate	

In some reversible reactions, the reactants are completely changed to the products. In other reversible reactions, the reactants are not completely changed to the products and the substances form an equilibrium in which the concentrations of the reactants and products stay constant. When equilibrium is reached, both the forward and the backward reactions are still taking place, but at the same rate. So, there is no change in the overall amount of any substance. This is described as a **dynamic equilibrium** with equilibrium arrows in the equation.

$$N_2(g) + 3H_2(g) \rightleftharpoons 2NH_3(g)$$

2 Manufacturing ammonia

Raw materials

				Reactants		**Product**
Crude oil	$\xrightarrow[\textit{distillation}]{\textit{fractional}}$	Naphtha fraction	$\xrightarrow{\textit{cracking}}$	Hydrogen	$\xrightarrow[\textit{Process}]{\textit{Haber}}$	**Ammonia**
Air	$\xrightarrow{\textit{liquefy}}$	Liquid air	$\xrightarrow[\textit{distillation}]{\textit{fractional}}$	Nitrogen		

Notice:

- The raw materials are chosen so that ammonia is produced as economically as possible.
- The conditions for the Haber Process ensure a *fast reaction rate*
  - a pressure of 200 atm
  - a temperature of 400°C
  - a catalyst of iron.

- The *yield* of ammonia increases when
  - the pressure is increased
  - the temperature is reduced.
- Ammonia is important in industry and agriculture. It is used to make fertilisers, nitric acid and nylon.

3 Ammonia – a base and an alkali

- *Ammonia is a gas* at room temperature – colourless, pungent and toxic.
- *Ammonia is very soluble in water*, reacting with it to form ammonia solution, containing ammonium hydroxide.

$$NH_3(g) + H_2O(l) \rightleftharpoons \underset{\text{ammonium hydroxide (ammonia solution)}}{NH_4^+(aq) \quad + OH^-(aq)}$$

- *Ammonia is a soluble base* (alkali) so it neutralises acids to form salts. This is how fertilisers are made.

$$\underset{\text{base (alkali)}}{NH_4OH} + \underset{\text{acid}}{HNO_3} \rightarrow \underset{\text{salt (ammonium nitrate)}}{NH_4NO_3} + \underset{\text{water}}{H_2O}$$

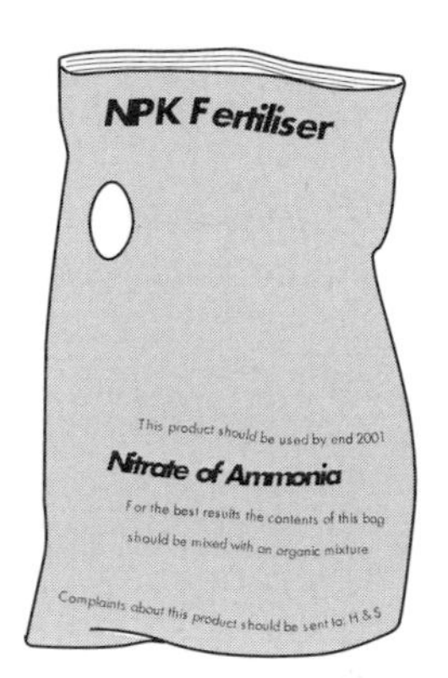

NPK Fertiliser
This product should be used by end 2001
Nitrate of Ammonia
For the best results the contents of this bag should be mixed with an organic mixture
Complaints about this product should be sent to: H & S

- Test for ammonia – *ammonia turns damp red litmus paper blue.*

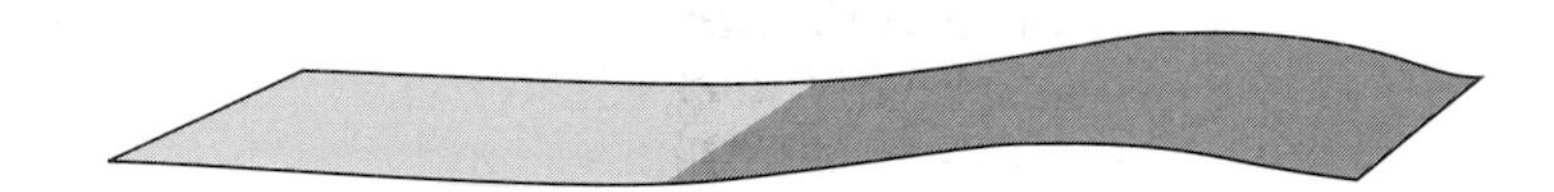

4 Fertilisers

Carbon dioxide and water provide the carbon, oxygen and hydrogen which plants need for growth. After these three elements, the most important elements for plant growth are nitrogen, phosphorus and potassium.

In heavily cultivated areas, nitrogen, phosphorus and potassium are rapidly depleted from the soil so they must be replaced by using fertilisers. The chemicals used in fertilisers containing these three essential elements (nutrients), their role in plant growth and the effect of shortages are summarised in the table.

Nutrient	Chemicals providing the nutrient	Role in plant growth	Effect of shortage
nitrogen	ammonium salts, nitrates	essential for: • protein synthesis • chlorophyll synthesis	• stunted plants • yellow leaves
phosphorus	phosphates	essential for: • nucleic acid synthesis	• plants' growth slowed • small fruit and seeds
potassium	potassium salts	important for: • protein synthesis • carbohydrate synthesis	• yellow leaves • curled up leaves

STUDY QUESTIONS

Objective questions

Questions 1 to 5

A absorption B catalysis C fixation D neutralisation E oxidation

Choose from A to E the name of the process in which

1 nitrates in the soil pass into plants through their roots.

2 ammonia reacts with nitric acid.

3 ammonia is manufactured from nitrogen and hydrogen.

4 sulphur burns to form sulphur dioxide.

5 atmospheric nitrogen is turned into nitrogen compounds in plants.

Questions 6 to 9

The flow diagram shows important stages in the manufacture of sulphuric acid.

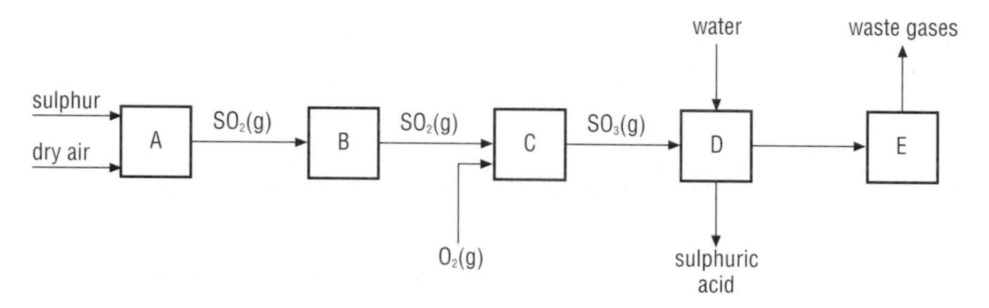

Which of the boxes A to E would be labelled

6 absorption tower?

7 chimney?

8 contact process converter?

9 furnace?

Questions 10 to 14

In questions 10 to 14, choose from A, B, C or D which is the correct answer.

10 Which one of the following reactions will not produce ammonia?
A heating ammonium chloride
B heating ammonium chloride with sodium hydroxide solution
C heating ammonium chloride with hydrochloric acid
D passing a mixture of nitrogen and hydrogen over hot iron

11 The main reason for using catalysts in industry is that
A they increase the rate of reaction.
B they increase the amount of products.
C they reduce the temperature of the reaction.
D they remove the products of the reaction.

12 Some minced beef was heated with calcium oxide. A gas was produced which turned red litmus blue. This suggests that minced beef contains

A a base.
B an alkali.
C ammonia.
D nitrogen.

13 In an NPK fertiliser, the P is present as

A phosphate.
B phosphorus.
C potash.
D potassium.

14 The nitrate concentration in a river was 10 times larger at a point 10 miles further downstream than at the first testing point. Which *one* of the following areas has the river probably flowed through?

A arable farmland
B deserted moorland
C thick forest
D undrained fenland

Short questions

15 a) List *three* important properties for a good fertiliser in addition to its improvement of plant growth. *3 marks*

b) Which *one* of the following fertilisers will provide the greater mass of nitrogen for each kilogram of the fertiliser added to the soil? Show your working. (N = 14, H = 1, O = 16, C = 12)

Fertiliser	Formula	Relative formula mass
ammonium nitrate	NH_4NO_3	80
urea	$(NH_2)_2CO$	60

3 marks

16 Write balanced chemical equations for the following reactions.

a) Ammonia reacts with nitric acid to form ammonium nitrate. *2 marks*
b) Ammonium chloride decomposes into ammonia and hydrogen chloride on heating. *2 marks*
c) Ammonium chloride reacts with sodium hydroxide on warming to produce ammonia, water and sodium chloride. *2 marks*

17 When a mixture of nitrogen and hydrogen are passed repeatedly over a heated iron wool catalyst, they come to an equilibrium with ammonia.

a) Write an equation for the equilibrium. *3 marks*
b) What is meant by the term equilibrium? *1 mark*
c) The equilibrium is described as dynamic. What does this mean? *2 marks*

18 Two of the most important elements for plant growth are nitrogen and phosphorus.

a) State *one* important role for nitrogen in plant growth. *1 mark*
b) State *one* effect on plants of a shortage of nitrogen. *1 mark*
c) State *one* important role for phosphorus in plant growth. *1 mark*
d) State *one* effect on plants of a shortage of phosphorus. *1 mark*
e) Suggest *two* problems caused by the over use of fertilisers. *2 marks*

Further examination questions

19 Ammonia is manufactured by reacting nitrogen with hydrogen.

$$N_2 + 3H_2 \rightleftharpoons 2NH_3$$

The graph shows the percentage yield of ammonia under various conditions of temperature and pressure.

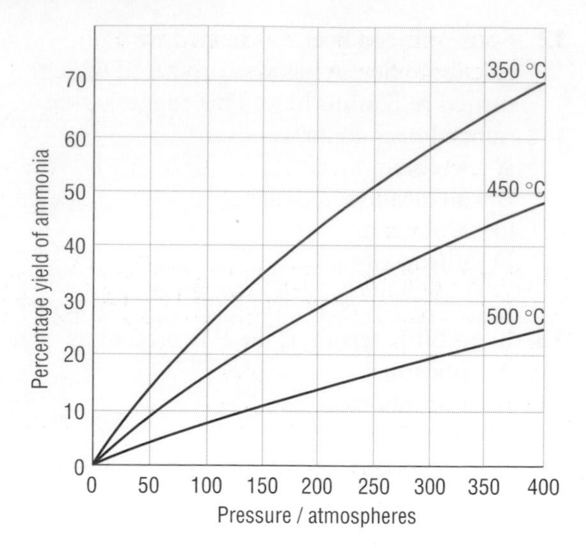

a) Why is the yield always less than 100%? *1 mark*

b) The graph shows that the maximum yield of ammonia would be obtained by using a pressure of 400 atmospheres and a temperature of 350°C. However, a chemical plant uses a pressure of 200 atmospheres and a temperature of 500°C. Explain why:

   (i) a pressure of 200 atmospheres is used, rather than 400 atmospheres *3 marks*

   (ii) a temperature of 500°C is used, rather than 350°C. *3 marks*

c) Suggest why the yield of ammonia is greater at lower temperatures. *1 mark*

d) The table shows the gases in the resulting mixture.

Gas	Boiling point/°C
nitrogen	−196
hydrogen	−253
ammonia	−33

The mixture leaving the plant contains 15% ammonia. Suggest what can be done with this mixture to improve the overall efficiency of the plant. *3 marks*

EDEXCEL

20 Ammonia is manufactured by the Haber Process in which nitrogen and hydrogen are combined over an iron catalyst at, for example, 250 atmospheres pressure and 500°C.

$$N_2(g) + 3H_2(g) \rightleftharpoons 2NH_3(g)$$

a) What mass of ammonia could be manufactured from 1 tonne of hydrogen (assuming 100% yield was obtained). (Relative atomic masses: H = 1, N = 14) *4 marks*

b) Explain why a high pressure is used in this process. *3 marks*

c) Ammonia is used to manufacture fertilisers.

   (i) Why do most fertilisers need to be soluble in water? *1 mark*

   (ii) How can excessive use of such fertilisers cause environmental problems? *2 marks*

EDEXCEL

21 Ammonia is manufactured by the Haber Process, where nitrogen and hydrogen react together as follows:

$$N_2 + 3H_2 \rightleftharpoons 2NH_3$$

The reaction is reversible. A balance is eventually reached when ammonia is being formed at the same rate at which it is decomposing. This point is called 'equilibrium'.

	Percentage of ammonia at equilibrium		
Pressure/atm	**100°C**	**300°C**	**500°C**
25	91.7	27.4	2.9
100	96.7	52.5	10.6
400	99.4	79.7	31.9

a) (i) What is meant by a 'reversible reaction'? *1 mark*
(ii) Which substances are present in the mixture at equilibrium? *1 mark*

b) (i) Under what conditions shown in the table is the maximum yield of ammonia obtained? *2 marks*
(ii) The Haber Process is usually carried out at a higher temperature than that which would produce the maximum yield. Suggest why. *2 marks*

c) Ammonia can be converted into nitric acid in three stages:

Stage 1: Ammonia reacts with oxygen from the air to form nitrogen monoxide and water.

$$4NH_3 + 5O_2 \rightarrow 4NO + 6H_2O$$

Stage 2: On cooling, nitrogen monoxide reacts with oxygen from the air to form nitrogen dioxide.

Stage 3: Nitrogen dioxide reacts with water to form nitric acid and nitrogen monoxide.

(i) Suggest the conditions under which the reaction in Stage 1 takes place. *2 marks*
(ii) Balance the equation for the reaction at Stage 2.

$NO + O_2 \rightarrow \quad NO_2$ *1 mark*

(iii) Balance the equation for the reaction at Stage 3.

$NO_2 + H_2O \rightarrow \quad HNO_3 + NO$ *2 marks*

d) The chemical plant for manufacturing ammonia is often on the same site as plants manufacturing nitric acid and fertilisers.
(i) What advantages will this have for the manufacturing company? *2 marks*
(ii) Briefly describe *two* important ways in which it is possible to reduce the environmental impact of such plants on the surrounding area. *2 marks*

AQA

22

Forces and motion

SUMMARY

1 The extension of a spring, a rubber band and other elastic materials is proportional to the stretching force. This is known as **Hooke's Law**.

2 Provided the **elastic limit** is not exceeded, elastic materials return to their original shape and size when the stretching force is removed. If the elastic limit is exceeded, materials remain permanently deformed.

3 **Pressure** is the force per unit area;

$$\text{pressure} = \frac{\text{force (N)}}{\text{area (m}^2\text{)}}$$

Units of pressure are N/m^2 or pascals (Pa). 1 Nm^2 = 1 Pa

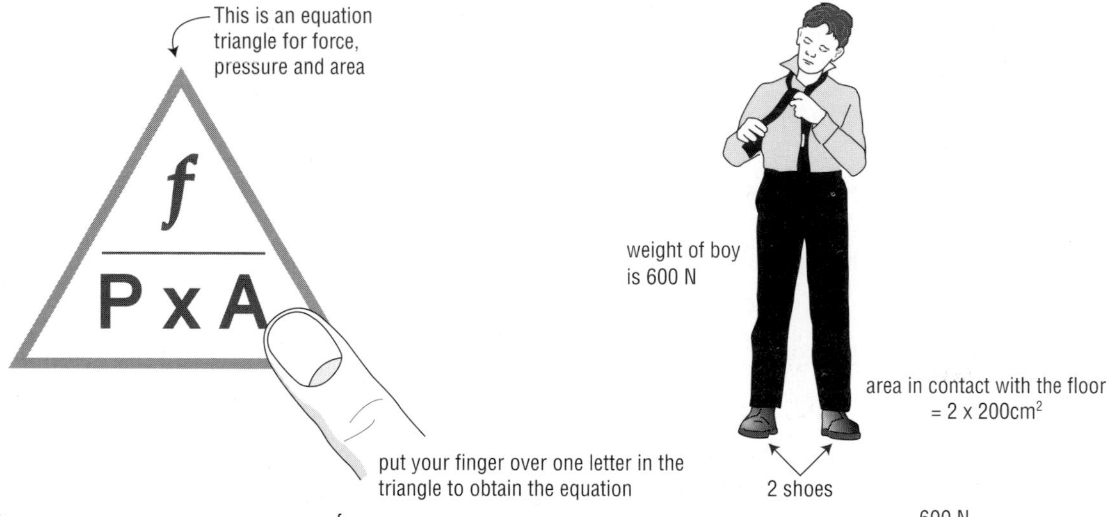

This is an equation triangle for force, pressure and area
f
P x A
put your finger over one letter in the triangle to obtain the equation
weight of boy is 600 N
area in contact with the floor = 2 x 200cm²
2 shoes

e.g. Put your finger over A, which leaves $\frac{f}{p}$

i.e. area = $\frac{\text{force}}{\text{pressure}}$

pressure of boy on floor = $\frac{600\ \text{N}}{400\ \text{cm}^2}$

= $1.5\ \text{N/cm}^2$

4 The pressure from a gas is caused by the bombardment from tiny, rapidly-moving molecules.

5 For a fixed mass of gas at constant temperature, the volume is inversely proportional to the pressure.

$$\text{i.e. pressure} \propto \frac{1}{\text{volume}} \text{ or pressure} \times \text{volume} = \text{constant}$$

This is **Boyle's Law**.

6 **Work** = force × distance
(J) (N) (m)
Work is measured in **joules** (J). 1 J = 1 Nm.

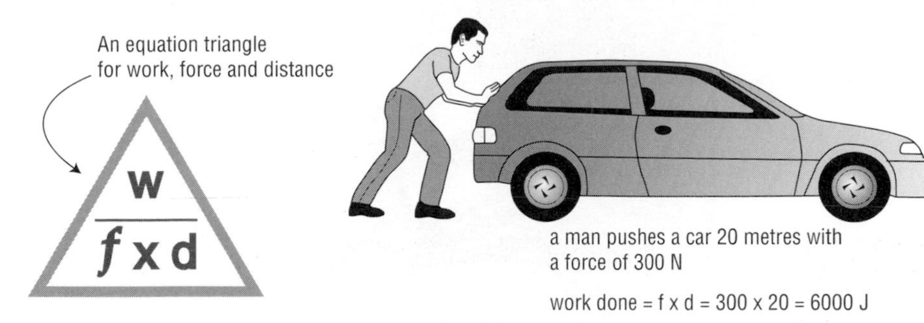

An equation triangle
for work, force and distance
w
f x d
a man pushes a car 20 metres with
a force of 300 N
work done = f x d = 300 x 20 = 6000 J

7 **Power** is the rate of working.

$$\text{Power (W)} = \frac{\text{work done (J)}}{\text{time taken (s)}}$$

Power is usually measured in **watts**. 1 W = 1 J/s

8 **Distance** is the total path taken. **Displacement** is the distance moved in a particular direction.

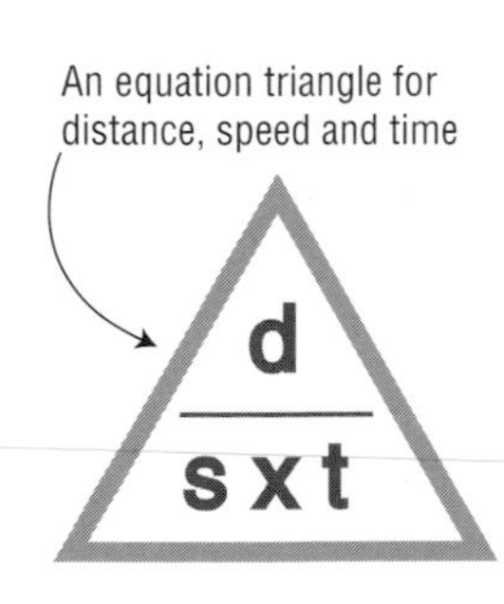

a bus travels 0.6 km between stops in 1 minute

$$\text{speed in m/s} = \frac{\text{d (m)}}{\text{t (s)}} = \frac{600 \text{ m}}{60 \text{ s}} = 10 \text{ m/s}$$

9 $$\textbf{Speed} \text{ (m/s)} = \frac{\text{distance (m)}}{\text{time (s)}}$$

$$\textbf{Velocity} \text{ (m/s in a given direction)} = \frac{\text{displacement (m in a given direction)}}{\text{time (s)}}$$

10 Speed = the gradient of a distance–time graph.

11 **Acceleration** $(m/s^2) = \dfrac{\text{change in speed (m/s)}}{\text{time (s)}}$

Acceleration = the gradient of a speed–time graph.

An equation triangle for change in speed (Δs), acceleration and time

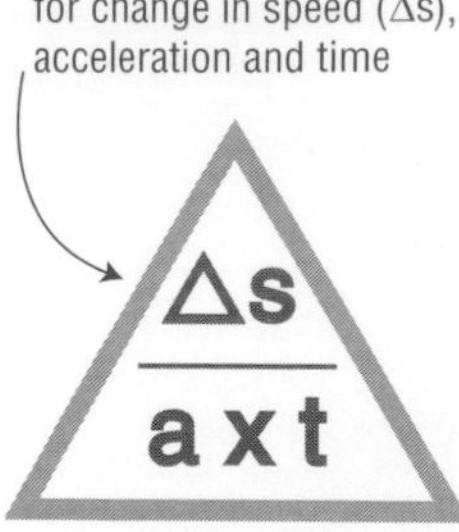

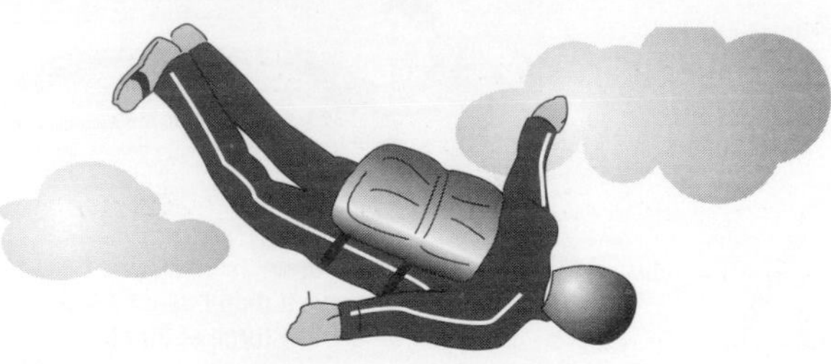

a freefall skydiver accelerates from 10 m/s to 70 m/s in 6 seconds as he falls.

his acceleration $= \dfrac{\Delta s}{t} = \dfrac{60 \text{ m/s}}{6\text{s}} = 10 \text{ m/s}^2$

12 **Newton's first law of motion** says:
If an object is stationary it will remain so. Or, if it is moving, it will continue to move at the same speed and in the same direction unless there are unbalanced forces acting on it.

13 **Newton's second law of motion** says:
The acceleration of an object is directly proportional to the force acting on it and inversely proportional to its mass.

$$\text{i.e. acceleration } (m/s^2) = \frac{\text{force (N)}}{\text{mass (kg)}} \text{ or force} = \text{mass} \times \text{acceleration}$$

An equation triangle for force, mass and acceleration

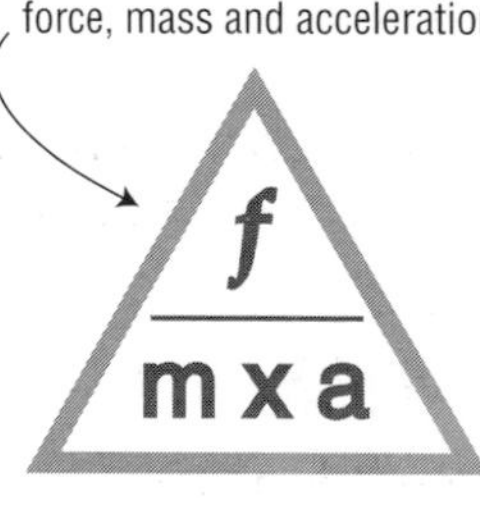

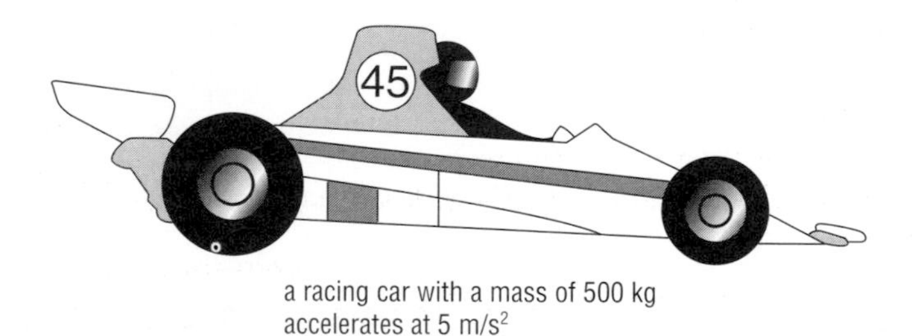

a racing car with a mass of 500 kg accelerates at 5 m/s^2

the force causing this acceleration
f = m x a = 500 x 5 = 2500 N

14 In the absence of friction or air resistance, all objects fall to the Earth with the same acceleration due to gravity of 10 m/s^2.

15 **Weight** is the force of gravity on an object.

$$\therefore \text{ weight (N)} = \text{mass (kg)} \times \text{acceleration due to gravity } (m/s^2)$$
$$w = m \times g$$

16 Energy enables humans and machines to do work.
Potential energy (gravitational potential energy),

$$\text{P.E.} = m \times g \times h$$

Kinetic energy (movement energy),

$$\text{K.E.} = \tfrac{1}{2} mv^2$$

STUDY QUESTIONS

Objective questions

Questions 1 to 4

In questions 1 to 4, consider the following five units:

A N B N/m^2 C N/m D Nm E Nm/s

Which of the above units could be correctly used in measuring

1 the pressure of the air?

2 the power of a lawnmower?

3 the work done in lifting a bale of hay?

4 the weight of a cake?

Questions 5 to 7

The figure shows five graphs of distance against time labelled A to E.

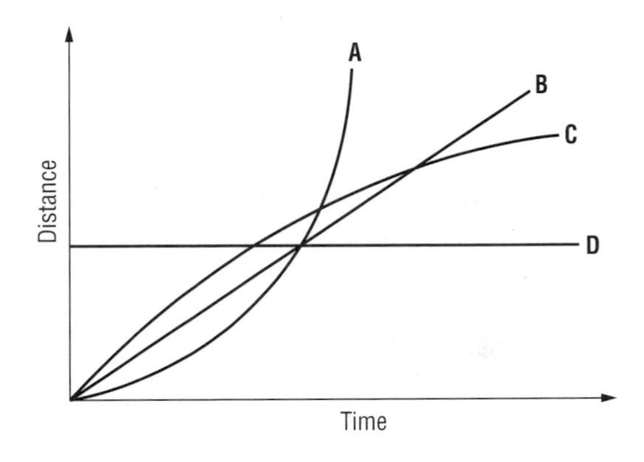

Which graph represents an object which is

5 stationary?

6 accelerating?

7 moving at constant speed?

In questions 8 to 15, choose from A, B, C or D which is the correct answer.

Questions 8 to 11

A 10 kg mass is moving with a speed of 5 metres per second. The mass is brought to rest in 4 seconds.

8 What is the average speed of the 10 kg mass during the 4 seconds it is brought to rest?
A $\frac{4}{5}$ m/s B $\frac{5}{4}$ m/s C $\frac{5}{2}$ m/s D 5 m/s

9 How far does the 10 kg mass travel during the 4 seconds it is brought to rest?
A 4 m B 5 m C 10 m D 20 m

10 What is the deceleration of the mass during the 4 seconds it is brought to rest?
A 20 m/s^2 B $\frac{5}{4}$ m/s^2 C $\frac{4}{5}$ m/s^2
D $\frac{1}{20}$ m/s^2

11 What is the force acting on the 10 kg mass to bring it to rest?
A 200 N B 100 N C 50 N
D 12.5 N

Questions 12 and 13

A stone with a mass of 2 kg is dropped from a cliff onto the beach 20 m below ($g = 10\ m/s^2$)

12 The weight of the stone is
A 20 N B $\frac{10}{2}$N C 2 N D $\frac{2}{10}$ N

13 The change in potential energy of the stone is
A 800 J B 400 J C 200 J D 20 J

Questions 14 and 15

A girl of mass 64 kg balances evenly on two stilts. Each stilt has an area of 8 cm^3 in contact with the pavement ($g = 10\ m/s^2$).

14 What is the force exerted by the girl on the pavement?
A 64 N B 64 kg C 640 N D 640 kg

15 What is the pressure exerted by the girl on the pavement?
A 640 N/cm^2 B 160 N/cm^2
C 80 N/cm^2 D 40 N/cm^2

Short questions

16 In the 1996 Atlanta Olympic Games, Michael Johnson won the 200 metres in a new world record time of 19.32 seconds.

a) Calculate Michael Johnson's speed over the full 200 metres. *2 marks*
b) After 4 seconds from the start, Johnson was sprinting at a speed of 11 m/s. What was his average acceleration between 0 and 4 seconds? *3 marks*
c) Johnson's mass was 86 kg. Calculate his kinetic energy when his velocity was 11 m/s. *3 marks*

17 In the shot put at the 1996 Atlanta Olympics, one competitor threw a 7.0 kg shot by applying a force of 112 N. What was the acceleration of the shot? *3 marks*

18 These results show the extension of a spring with different loads.

Extension/cm	0	4	10	16	22	30
Load/N	0	4	10	16	20	24

a) Use the results to plot a graph of extension (vertical) against load (horizontal). *3 marks*
b) Use your graph to estimate the load required for an extension of 8 cm. *1 mark*
c) The length of the spring is 10 cm with no load on it. What is the length of the spring when the load on it is 6 N? *2 marks*
d) What is the load when the spring reaches its elastic limit? *1 mark*

19 The diagram shows the forces on an aircraft during flight.

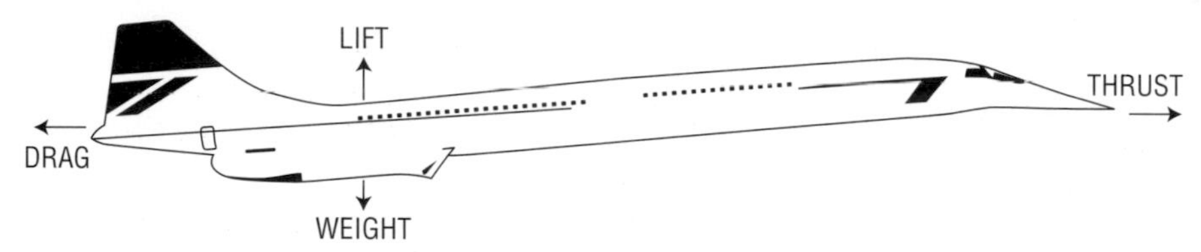

a) Which forces affect:
(i) the speed of the aircraft? *1 mark*
(ii) the height of the aircraft above the ground? *1 mark*
b) What are the relative sizes of these forces for the aircraft to:
(i) cruise at constant speed and at a steady height? *2 marks*
(ii) accelerate and ascend as it travels along a runway at take-off? *2 marks*

20 Part of a motorist's journey is shown on the speed–time graph below.

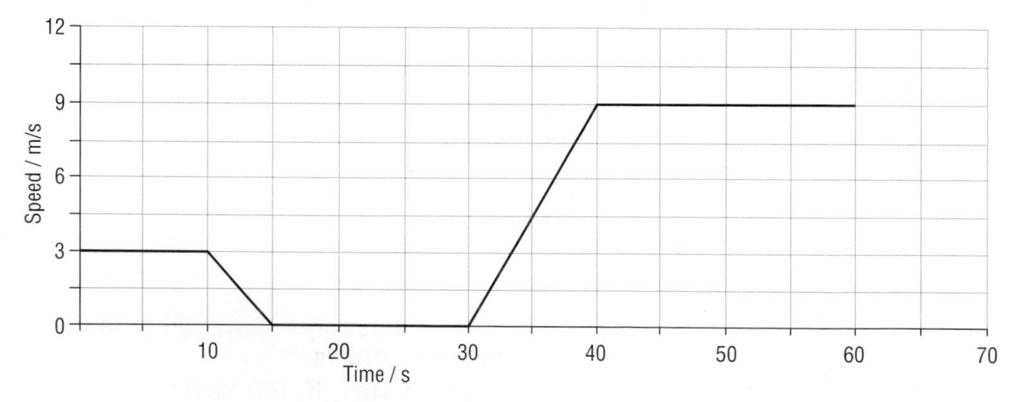

Describe the motion of the vehicle fully. *7 marks*

21 A man's car will not start, so two friends push him to help him to get going. By pushing as hard as possible for 10 seconds, the car reaches a speed of 3 metres per second.

a) What acceleration do they give the car? *1 mark*

b) Another motorist has the same problem with his car. The two friends push the car along the same road with the same force as before. They take 15 seconds to get the second car moving at 3 metres per second. What can you conclude about the mass of the second car? *3 marks*

22 a) A car has a mass of 750 kg. Once the engine has started, it accelerates from 3 m/s to 8 m/s in 10 seconds. What is the driving force of the engine? *4 marks*

b) On a flat stretch of motorway, a lorry driver changes into top gear. He then makes the lorry go as fast as possible. The graph shows what happens to the speed of the lorry. Why does the speed of the lorry increase at first but then reach a constant speed? *4 marks*

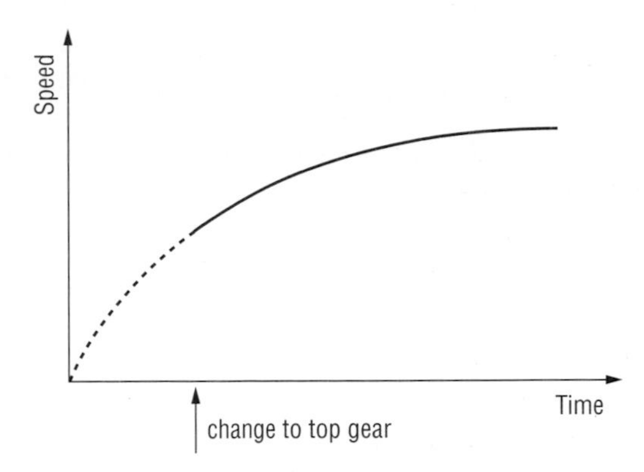

23 The work done in stopping a car of mass 800 kg is 90 000 J. The braking force is 6000 N.

a) What is the braking distance? *3 marks*

b) The kinetic energy of the car before braking is 90 000 J and it is travelling at 15 m/s. What is the kinetic energy when it travels at 30 m/s? *3 marks*

c) What is the braking distance of a car from a speed of 30 m/s? (Assume that the braking force is 6000 N as in part a)). *2 marks*

d) The Highway Code recommends that on fast roads, in good conditions, drivers should allow a 2 second gap between their car and the one in front. Is this good advice? (Look at your answer to part c)). *3 marks*

Further examination questions

24 The diagram shows part of the new Blackpool roller coaster.

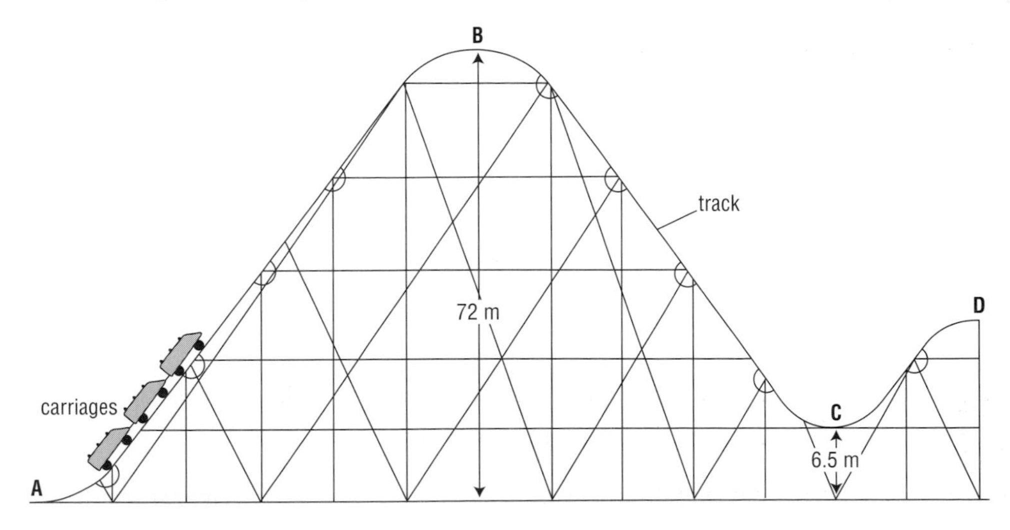

a) The carriages are pulled up to point B by an electric motor. Once a carriage is at point B, it is released and free-wheels down the track towards point C.
   (i) The total mass of carriage and passengers is 3100 kg. How much gravitational potential energy will be gained in moving from point A to point B? (Take g = 10 N/kg.) *3 marks*
   (ii) The power rating of the electric motor is a constant 50 kW. Calculate the time it would take for the carriage and passengers to move from point A to point B. *3 marks*
   (iii) In practise, the time taken to reach point B will be longer than you have calculated. Explain why. *2 marks*

b) (i) On release from point B, the carriage moves down towards point C. Describe the main energy change taking place as it does so. *1 mark*
   (ii) By using energy considerations, calculate the maximum possible speed of the carriage as it passes through point C. *4 marks*
   (iii) Give *two* reasons why the height of the next peak at D has to be less than that at B. *2 marks*

EDEXCEL

25 Peter cycles from home to school. The graph represents his journey.

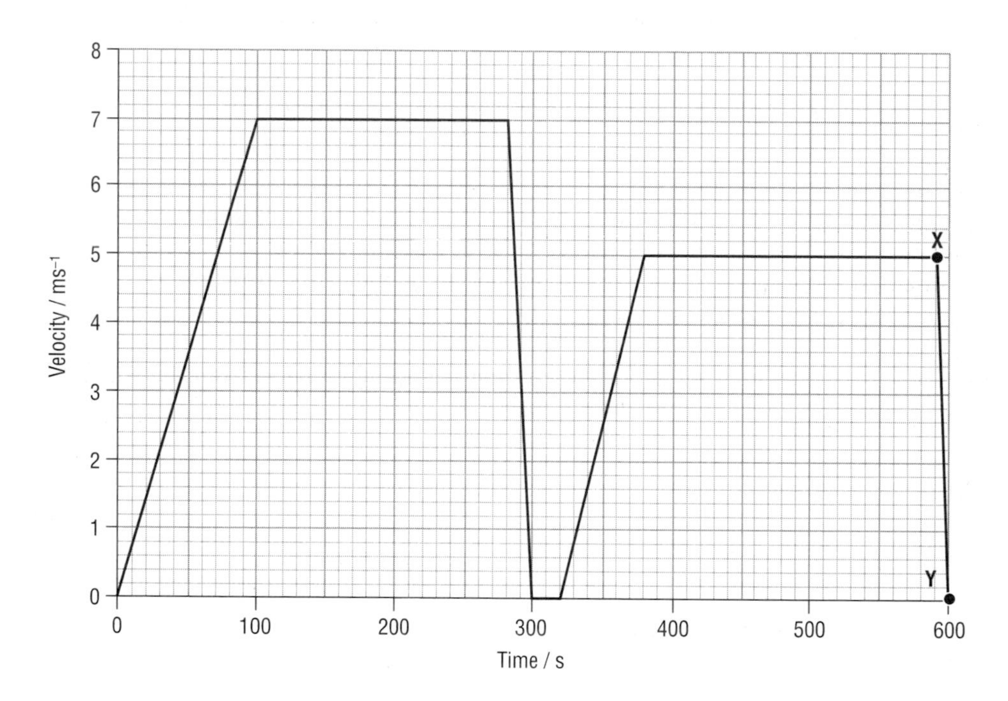

a) (i) What is Peter's velocity after 50 seconds? *1 mark*
   (ii) After how many seconds does Peter stop at some traffic lights? *1 mark*
   (iii) Calculate Peter's deceleration, in m/s^2, between points X and Y. *2 marks*
   (iv) Peter and his bicycle have a combined mass of 60 kg. Calculate the resultant force exerted on Peter and his bicycle as he decelerates between points X and Y. *2 marks*

b) (i) After his journey, Peter noticed that his bicycle tyres and the air they contained were warm. Explain why. *1 mark*
   (ii) Explain what effect, if any, this has on the force exerted on the walls of the tyres. *2 marks*

EDEXCEL

26 A type of toy catapult consists of a movable plunger which has a spring attached. The handle was pulled down to fully compress the spring and on release, the metal ball of mass 0.1 kg (weight 1 N) was projected 0.75 m vertically.

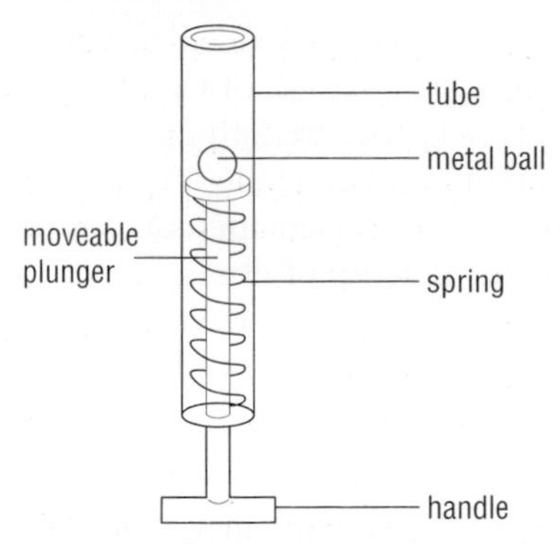

a) (i) What type of energy is stored in a compressed spring? *1 mark*
   (ii) What happens to this stored energy when the handle of the plunger is released? *2 marks*

b) Calculate the maximum potential energy acquired by the metal ball from the catapult. Write down the formula that you use and show your working. Take the acceleration due to gravity to be 10 m/s^2. *3 marks*

c) Explain why the maximum potential energy gained by the metal ball is less than the original stored energy of the spring. *3 marks*

WJEC

27 A ball bearing was projected horizontally from rest at a point P above the ground. The diagram shows five positions of the ball P, Q, R, S and T at consecutive 0.2 second time intervals. The distances marked are the actual distances moved.

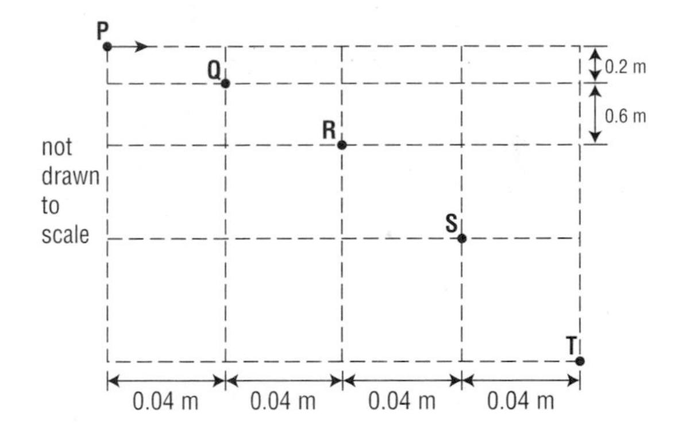

a) Explain what the data given show about the ball bearing's
   (i) horizontal speed *1 mark*
   (ii) vertical speed. *1 mark*

b) (i) Calculate the horizontal speed of the ball bearing. *2 marks*
   (ii) Calculate the average vertical speed of the ball bearing between P and Q. *1 mark*
   (iii) State the ball bearing's vertical speed at P. *1 mark*
   (iv) What is the ball bearing's vertical speed at Q? *1 mark*

WJEC

28 The diagram shows how you can compare the strength of different materials when they are being stretched. Here are the results of some tests:

Sample	Force needed to break sample/kN
aluminium	7.0
concrete	0.4
wood (across the grain)	0.3
wood (along the grain)	10.0

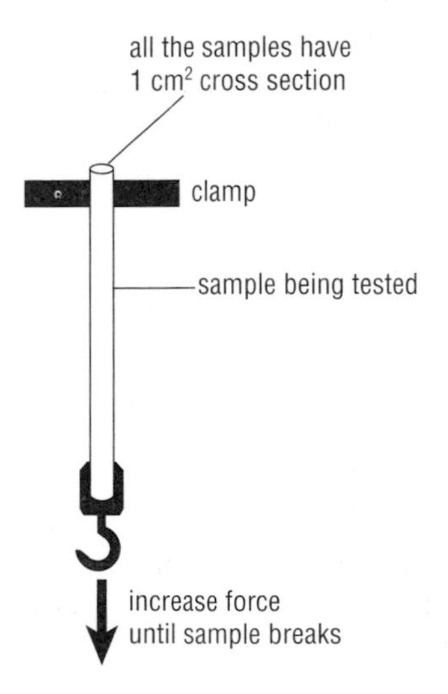

a) Describe, in as much detail as you can, how the strength of wood compares to the strength of aluminium and concrete. (You do not need to do any calculations.) *4 marks*

b) Wood is made up of fibres. Suggest why wood is stronger along the grain than it is across the grain. *2 marks*

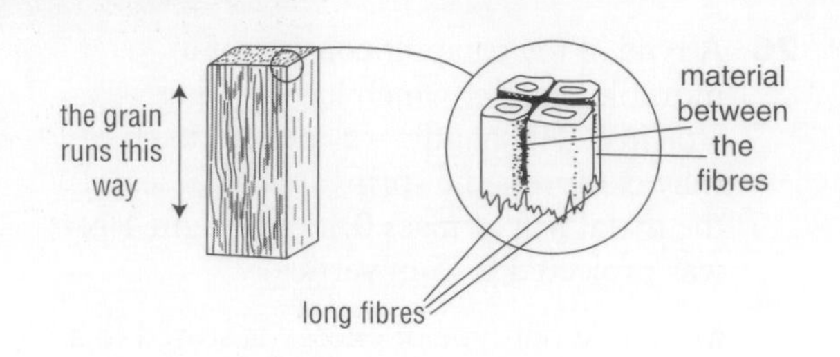

AQA

29 Some cars are fitted with airbags to protect the occupants in case of a collision. The airbag is filled from a canister of sodium azide which decomposes very quickly to produce sodium and large amounts of nitrogen gas which then inflates the bag.

a) Explain why it is so important that the bag must inflate very quickly. *2 marks*

b) The gas is produced at high pressure. Then the bag expands and the pressure of the gas inside the bag changes.
   (i) What happens to the pressure of the gas inside the bag as it expands? *1 mark*
   (ii) What equation links the pressure and volume of a gas under different conditions? *1 mark*
   (iii) When the gas pressure is 6 atmospheres, the bag occupies 22 litres and all the gas has been produced. Calculate the pressure once the bag occupies 66 litres. Assume there is no change in temperature. *2 marks*
   iv) 22 litres of nitrogen at 6 atmospheres are needed to fill the bag properly. Experiments show that 44 g of sodium azide can produce 4 litres of nitrogen at this pressure.
      1. Use this information to calculate the minimum mass of sodium azide needed to fill the air bag. *2 marks*
      2. Suggest *one* reason why twice this amount of sodium azide is normally used. *1 mark*

OCR

CHAPTER

23

Energy transfers

SUMMARY

1 Types of energy

There are eight different types of energy:
electrical energy, **heat** or thermal energy, **light**, **sound**, **kinetic** energy, **potential** energy, **elastic** energy and **chemical** energy.
Energy is continually being changed from one form to another.

- When floodlights illuminate a pitch,
  ELECTRICAL ENERGY → light energy and heat energy.
- When playing a guitar,
  CHEMICAL ENERGY → elastic energy → sound.
- When ski-ing downhill,
  POTENTIAL ENERGY → kinetic energy.

2 Heat energy can be transferred from one place to another by three different processes.

- **Conduction** – when hotter, vibrating particles pass on their vibrational (kinetic) energy to cooler neighbouring particles with the material(s) moving.

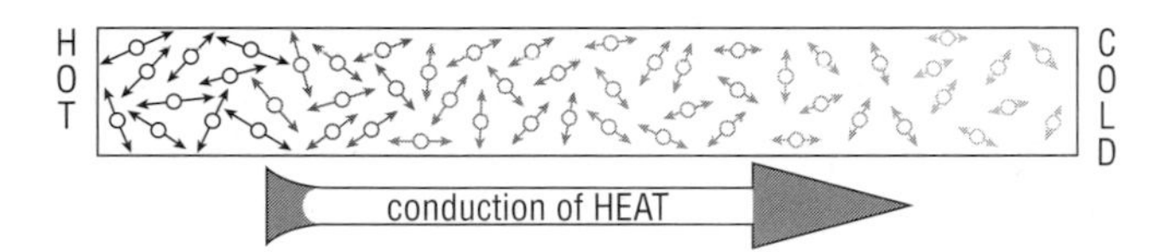

- **Convection** – when heated liquid or gas becomes less dense than cooler surrounding material. This causes the hot, less dense material to rise and the cooler, more dense material to sink in convection currents.

convection currents of air carry hang gliders higher

- **Radiation** – when invisible, infra-red electromagnetic waves travel outwards from hot objects. Radiation involves waves and not particles so it can take place through a vacuum.

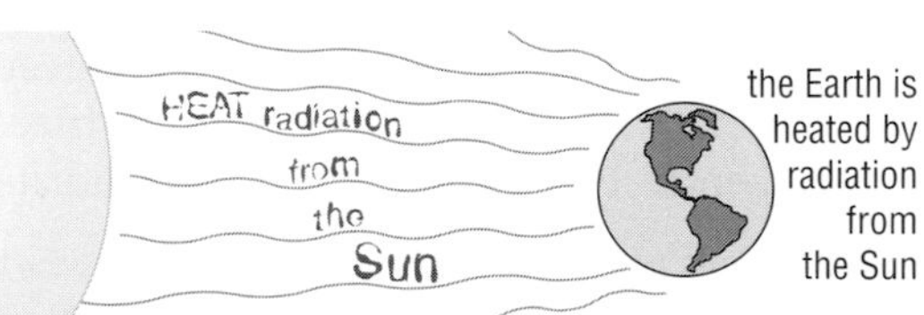

the Earth is heated by radiation from the Sun

3 Emitting (giving out) and absorbing (taking in) radiation
Dark, dull surfaces are good emitters and good absorbers of radiation.
Light, shiny surfaces are poor emitters and poor absorbers of radiation.

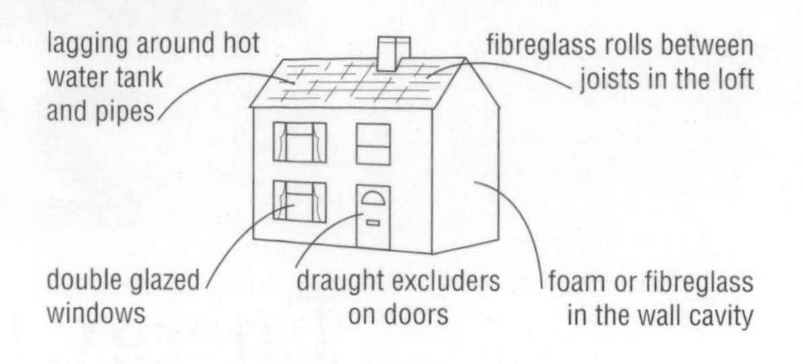

4 Reducing energy loss from our homes (see diagram right)

5 Conservation of energy
Energy can be changed from one form to another, but it cannot be created or destroyed.
Machines (e.g. motor cars, kettles) are often used to convert energy from one form into another. Unfortunately, these and all other energy transfers are never perfect. Some energy is always lost during the transfer, nearly always as heat.

e.g. petrol engine in a car

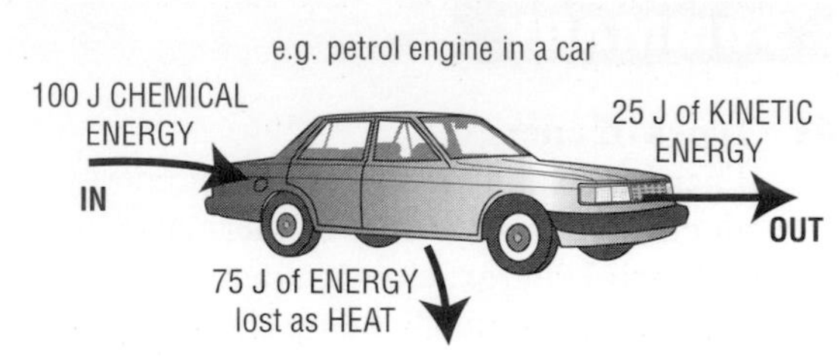

∴ Total energy INPUT = Useful energy OUTPUT + WASTED energy

$$\therefore \text{Efficiency} = \frac{\text{useful energy (work) output}}{\text{total energy input}}$$

STUDY QUESTIONS

Objective questions

Questions 1 to 10
In questions 1 to 10 choose from A, B, C or D which is the correct answer.

1 Electricity can be generated from different sources of energy. Which *one* of the following energy sources can be used to generate electricity without being converted into heat energy first?
A coal B natural gas C water waves D oil

2 If you walk barefooted on a woollen carpet and a ceramic tiled floor, the carpet feels warmer even when the carpet and the tiles are at the same temperature. This is because
A ceramic tiles are better insulators than wool.
B wool is a better insulator than ceramic tiles.
C ceramic tiles are colder than wool.
D ceramic tiles conduct cold into your feet.

3 Chefs use thick, padded gloves to remove dishes from a hot oven to protect their hands from heat transferred by
A conduction B convection C radiation D absorption

4 Steelworkers wear dark glasses to protect themselves from heat transferred from hot materials in high temperature furnaces by
A conduction B convection C radiation D absorption

5 Energy can be transferred from a hot radiator into a room. Which two properties of the radiator's painted surface will allow it to give out energy fastest?
A dark and matt B dark and shiny C light and matt D light and shiny

6 Radiators in a room transfer their energy mainly by
A convection with some conduction.
B convection with some radiation.
C radiation with some conduction.
D radiation with some convection.

7 Which of the following would provide the best insulation?
A 2 sheets of 10 cm thick chipboard glued together
B a single sheet of 20 cm thick chipboard
C 2 sheets of 10 cm thick chipboard separated by an air gap of 10 cm
D 2 sheets of 10 cm thick chipboard separated by a sheet of 10 cm thick copper

8 Heat can travel through a vacuum by
A conduction and convection.
B convection and radiation.
C convection only.
D radiation only.

9 The double glazing of windows improves heat insulation because
A air trapped between sheets of glass is a poor conductor.
B the glass sheets prevent convection currents.
C radiation cannot pass through the gap between the glass sheets.
D radiation cannot pass through two sheets of glass.

10 When heat is transferred through a metal by conduction, the energy is transferred mainly by
A electrons B neutrons C protons D waves

Short questions

11 a) Name *four* renewable energy sources that are used to generate electricity in Britain. *4 marks*
b) The main advantage of using renewable sources to generate electricity is that there are no fuel costs. Give *two* major disadvantages of using renewable energy. *2 marks*

12 The diagram shows what happens to each 100 joules of chemical energy from the coal burnt in a power station when the electricity generated is used for lighting.

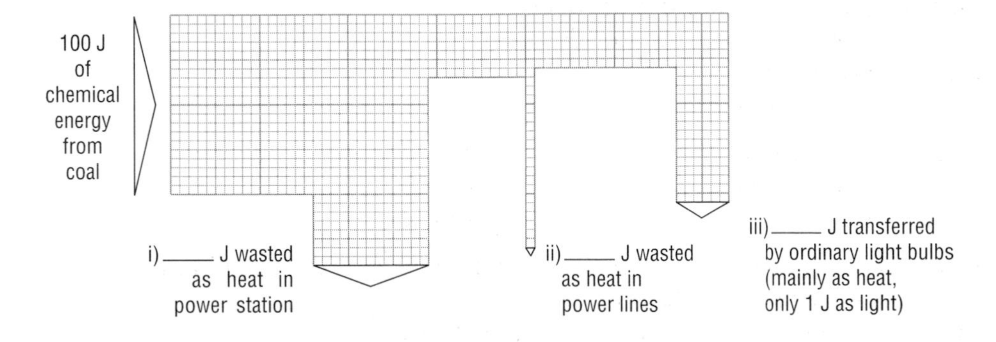

a) What figures replace (i), (ii) and (iii) in the diagram? *3 marks*
b) What is the efficiency of ordinary light bulbs in transferring electrical energy into light? *2 marks*

13 This is the inside of a refrigerator.

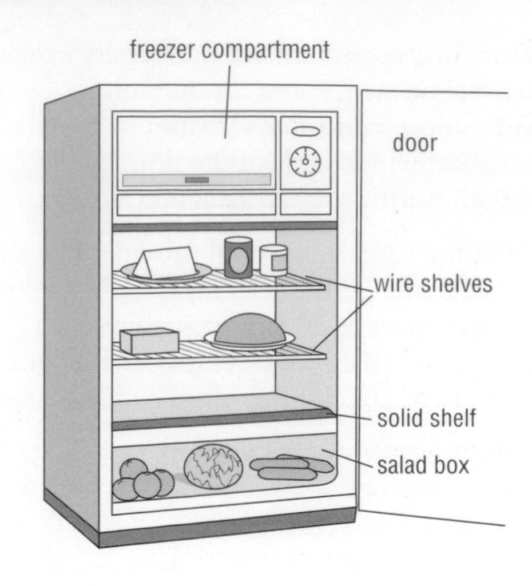

a) How do convection currents keep the contents of the refrigerator cool? *4 marks*

b) The shelves in the middle of the refrigerator are made of plastic-coated wire. The shelf over the salad box is a sheet of glass.

(i) Why are wire shelves used in the middle of the refrigerator? *1 mark*

(ii) When food at room temperature is placed in the salad box, heat is conducted through the glass shelf as the food cools. Explain how heat is conducted through the glass shelf using the idea of particles. *4 marks*

14 The table compares the energy flow through a petrol car with that through an electric car.

	Petrol car	Electric car
Energy from each kilogram petrol or battery/kJ	45 000	700
Energy wasted as heat in exhaust and cooling system/kJ	30 000	0
Energy wasted in drive mechanism/kJ	4500	210
Energy available to drive car/kJ	10 500	490

a) Calculate the efficiency of the petrol car. *2 marks*
b) Calculate the efficiency of the electric car. *1 mark*
c) Which car is the more efficient? *1 mark*

Further examination questions

15 a) The table contains information about developed countries, such as Britain, and developing countries, such as India. It shows the proportion of energy obtained from different energy sources.

Energy source	Developed countries (%)	Developing countries (%)
biomass	3	35
coal	25	28
hydroelectric	6	6
natural gas	23	7
nuclear	5	1
oil	37	23

(i) What percentage of energy is obtained from renewable sources in:
A developed countries; B developing countries? *2 marks*

(ii) Suggest *two* reasons why the proportion of energy obtained from renewable sources is significantly different in developed and developing countries. *2 marks*

b) The diagram shows a solar panel used to heat water.

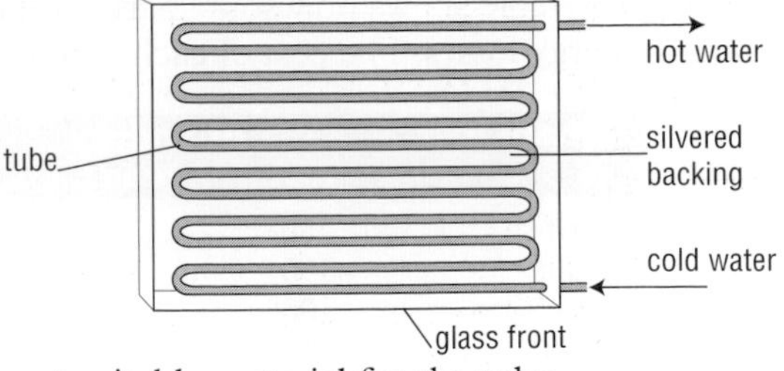

(i) Name the process by which energy is transferred:

A from the Sun to the tube in the panel;

B from the tube to the water.

2 marks

(ii) Which of the following would be the most suitable material for the tube containing the water?

black plastic   copper painted black   frosted glass

polished stainless steel *1 mark*

(iii) Give *two* reasons for your choice in part (ii). *2 marks*

(iv) Discuss *one* advantage and *one* disadvantage of heating water with a solar panel rather than a gas boiler. *2 marks*

EDEXCEL

16 A 3 kW immersion heater is used to heat a tank of water. It is switched on for 1800 seconds. The tank contains 20 kg of water and the water temperature rises by 50°C in the time that the heater is switched on.

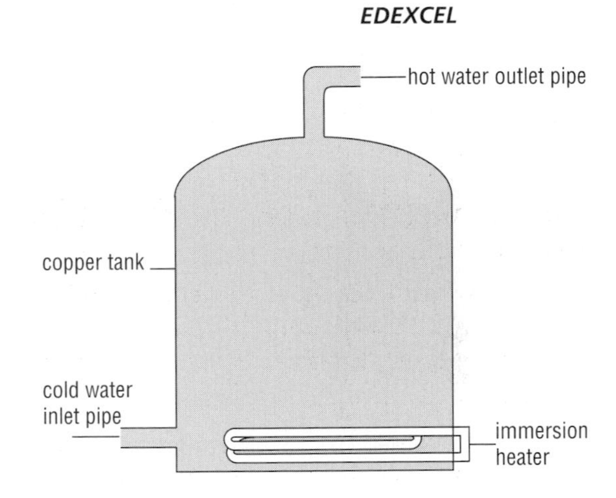

a) Calculate the energy in joules transferred by the immersion heater. Write down the formula that you use and show your working. *3 marks*

b) Calculate the heat energy gained by the water, given that the specific heat capacity of water is 4200 J/kg°C. Write down the formula that you use and show your working. *3 marks*

c) (i) Calculate the efficiency of this system in heating the water given that

$$\text{efficiency} = \frac{\text{useful energy transferred}}{\text{total energy supplied}} \times 100\%$$

(ii) Give *one* reason, apart from the energy loss to the surroundings, why this energy transfer is not 100% efficient. *1 mark*

d) The hot water tank loses heat energy by radiation, convection and conduction.

(i) Explain how the tank loses heat energy by convection. *3 marks*

(ii) Explain how the energy lost by radiation can be reduced. *1 mark*

WJEC

17 The diagram shows a house which has not been insulated. The cost of the energy lost from different parts of the house during one year is shown.

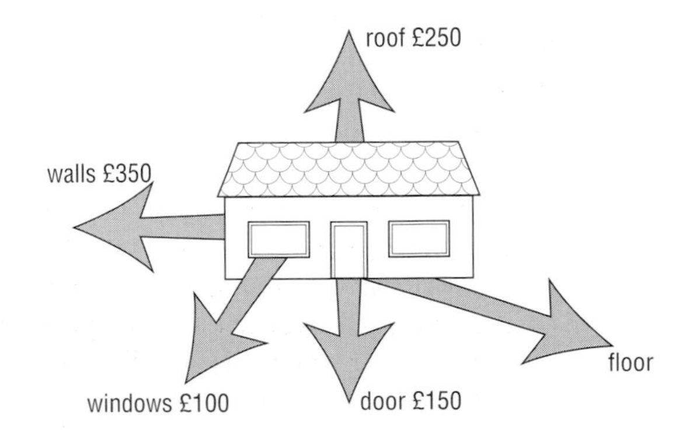

a) The total cost of the energy lost during one year is £1000.

(i) What is the cost of the energy lost through the floor? *1 mark*

(ii) Suggest one way of reducing this loss. *2 marks*

b) The table shows how some parts of the house may be insulated to reduce energy losses. The cost of each method of insulation is also given.

Where lost	Cost of energy lost per year/£	Method of insulation	Cost of insulation/£
roof	250	fibre glass in loft	300
walls	350	foam-filled cavity	800
windows	100	double glazing	4500
doors	150	draught proofing	5

(i) Which method of insulation would you install first? Explain why. *3 marks*
(ii) Which method of insulation would you install last? Explain why. *3 marks*

AQA

18 Hikers often wear jackets made with different layers to help them keep warm.

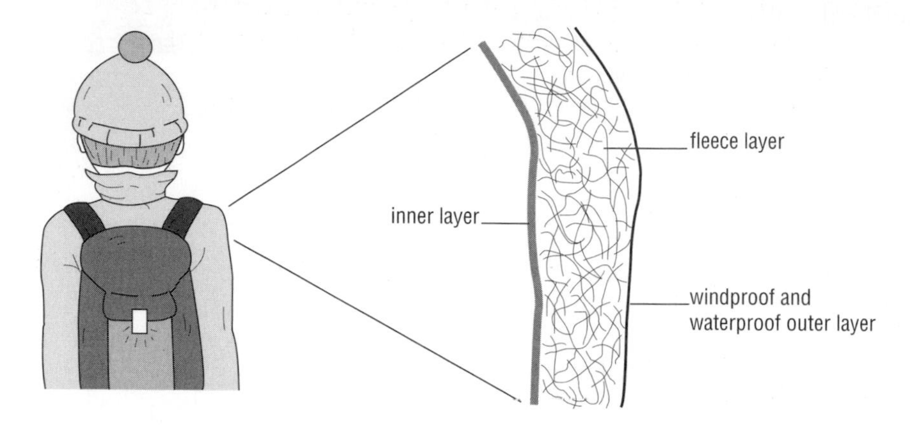

a) (i) Sometimes the jacket has a reflective inside surface. Which process of heat transfer is this designed to reduce? *1 mark*
(ii) Which process of heat transfer is the fleece layer designed to reduce? *1 mark*
(iii) Explain why the fleece layer is good at reducing heat loss from a hiker's body. *2 marks*

b) The drawing shows a stove used by hikers to heat water.

(i) The specific heat capacity of water is 4200 Joules per kilogram degree Celsius (J/kg°C). Calculate the heat energy required to raise the temperature of 0.75 kg of water from 10°C to its boiling point at 100°C. Include in your answer the equation you are going to use. Show clearly how you get to your final answer and give the unit. *4 marks*

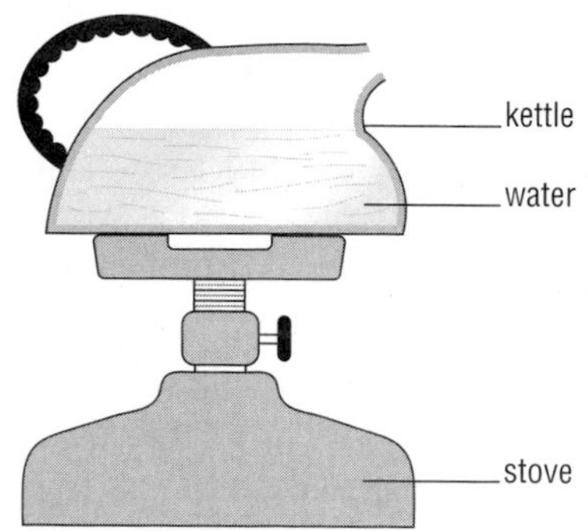

(ii) The hikers use the stove to raise the temperature of 0.75 kg of water from 10°C to 100°C. Give *three* reasons why the stove will need to give out more heat than the quantity calculated in part b) (i). *3 marks*

AQA

CHAPTER

24

Currents and circuits

SUMMARY

1 Circuits and flow of charge

- An **electric current** is a flow of charge around a circuit. The **actual charges** which move are **tiny, negatively-charged electrons** flowing from the negative terminal of the battery to the positive.
- The negative terminal of the cell gives up electrons which flow round the circuit to the positive terminal which is keen to take electrons.
- The **conventional current** is shown as an arrow on the connecting wires in circuit diagrams pointing from positive to negative.
- A **battery** is two or more cells connected together.
- The different pieces of equipment in a circuit (e.g. cells, bulbs, ammeters) are called **components**. Each component has a standard symbol for use in circuit diagrams.

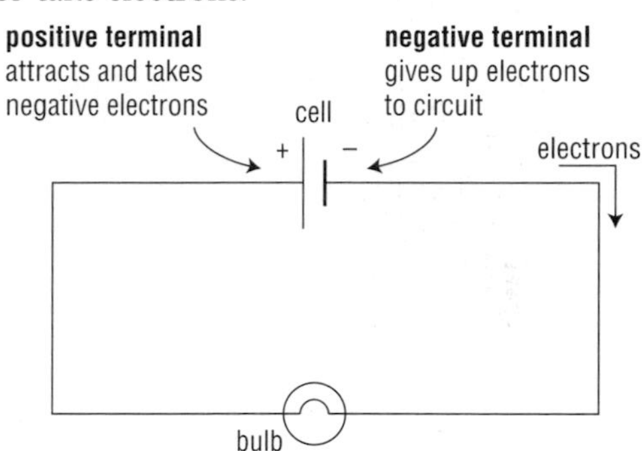

2 Series and parallel circuits

- In a series circuit, the current has no choice of route.
- The current flowing through a component is measured in **amperes** (A) by an **ammeter** in series with the component.
- Current is not used up as it passes round a circuit.
- In a series circuit, the current is the same at all points.
- In a parallel circuit, the current has a choice of routes. More current takes the easier path at a junction. The total current going into a junction equals the total current coming out of the junction.

Make sure that you understand the information in the diagram.
(A_1 shows current at that point, A_2 shows current at that point, etc. All the bulbs are identical.)

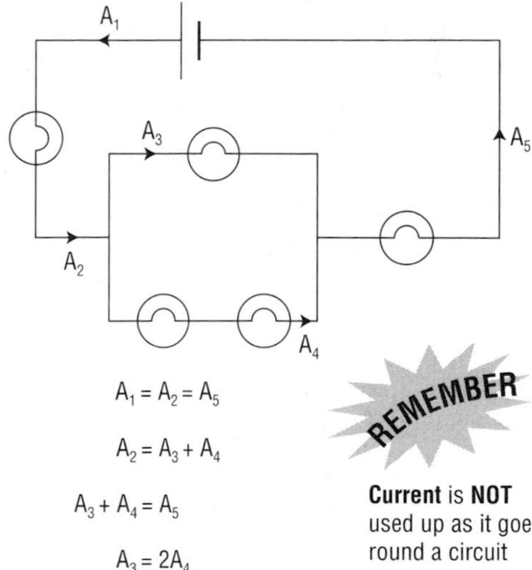

$A_1 = A_2 = A_5$

$A_2 = A_3 + A_4$

$A_3 + A_4 = A_5$

$A_3 = 2A_4$

REMEMBER

Current is **NOT** used up as it goes round a circuit

3 Cells and bulbs

- Cells in a circuit provide an electrical driving force on the electrons. They act like a pump pushing (forcing) the electrons round the circuit.
- If there are more cells in the circuit, there is a bigger electrical driving force and the current increases.
- This electrical driving force in a circuit is called the **voltage** or **potential difference** (p.d.) which is measured in **volts** (V).
- Bulbs and other components in a circuit hinder the flow of electrons and cause **resistance** measured in ohms (Ω).
- As the number of bulbs and other components in a circuit increases, the resistance increases and the current gets smaller.

Make sure you understand the information in the diagram. (All the cells are identical and all the bulbs are identical.)

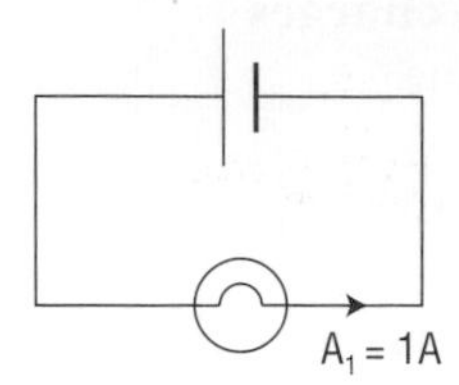

$A_1 = 1A$

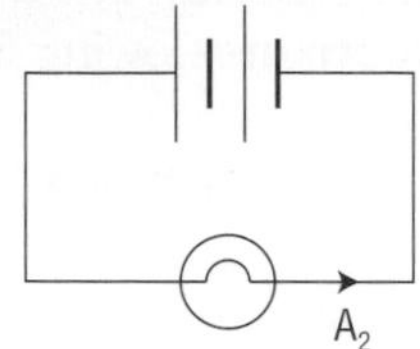

A_2

2 cells ∴ twice the electrical driving force
$\therefore A_2 = 2A$

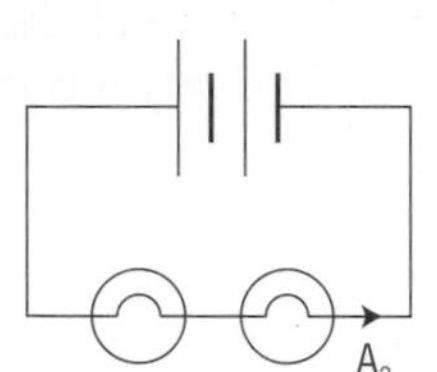

A_3

2 cells and 2 bulbs
Twice the electrical driving force, but twice the resistance
$\therefore A_3 = 1A$

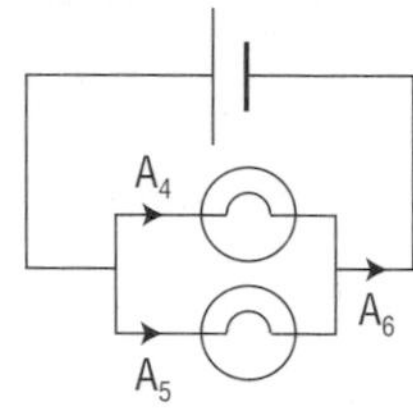

$A_4 + A_5 = A_6$

Each bulb has an electrical driving force from 1 cell
$\therefore A_4 = 1A$
$A_5 = 1A$
and $A_6 = 2A$

4 **Ohm's Law** says:

- The voltage across a metal resistor is proportional to the current through it, provided its temperature is constant.
  i.e. $V \propto I$

$$\Rightarrow \frac{V}{I} = \text{voltage per unit current} = \text{Resistance, R}$$

$$\text{So } R = \frac{V}{I}, \text{ Resistance (ohms)} = \frac{\text{Voltage (volts)}}{\text{Current (amps)}}$$

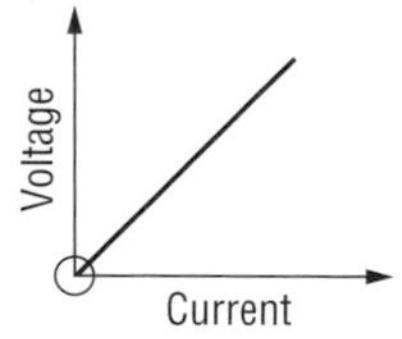

- Filament lamps and thermistors do not obey Ohm's Law. These are known as **non-ohmic conductors**.
- In a filament lamp, the temperature rises sharply as the current increases. This causes the resistance to increase and so the gradient of the voltage–current graph rises (Figure a).

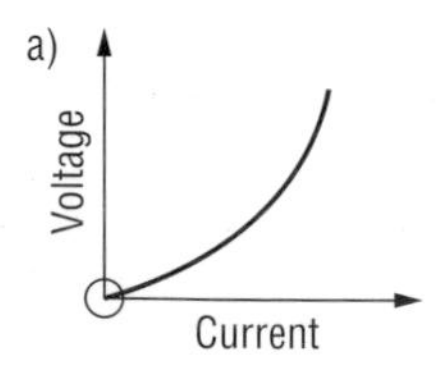

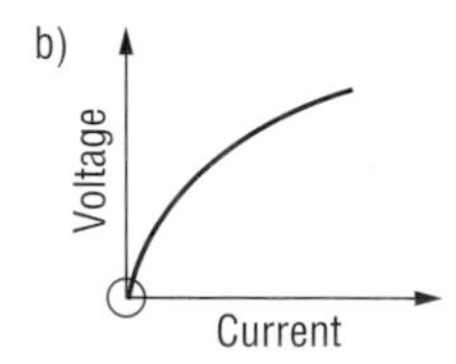

- Thermistors and light-dependent resistors (LDRs) are made from semiconductors in which the resistance falls as the temperature increases. This means that the gradient of the voltage–current graph falls as the current increases and the temperature rises (Figure b).

STUDY QUESTIONS

Objective questions

Questions 1 to 4

In questions 1 to 4, choose from A, B, C, D or E which is the correct answer.

Look carefully at the circuit diagrams in which all the cells are identical and all the bulbs are identical.

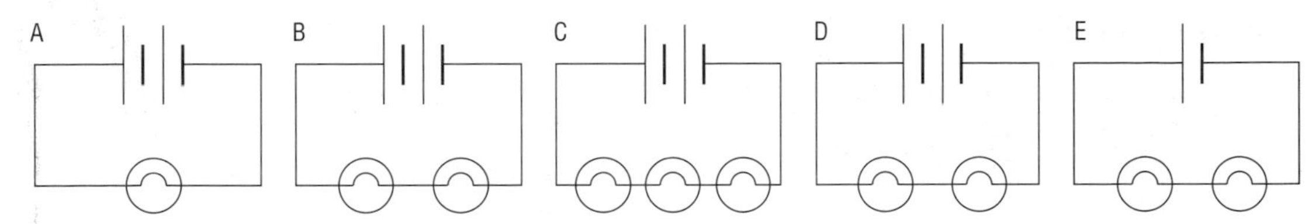

1 Which circuit has the brightest bulb?

2 Which circuit has the dimmest bulb?

3 Which two circuits have bulbs with the same brightness?

4 Which circuit contains the most resistance?

Questions 5 to 9

Look carefully at the circuit diagrams in which all the bulbs are identical.

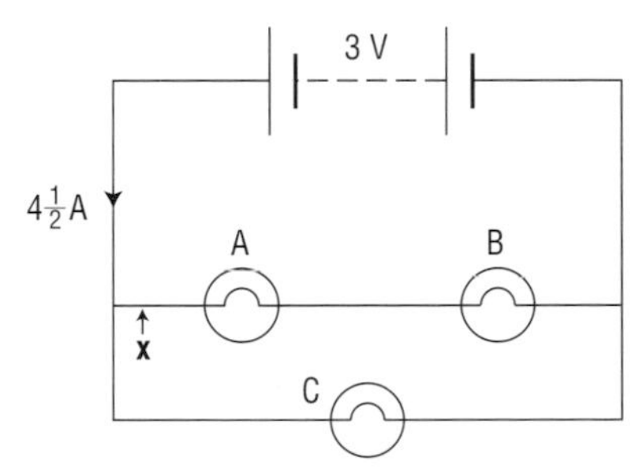

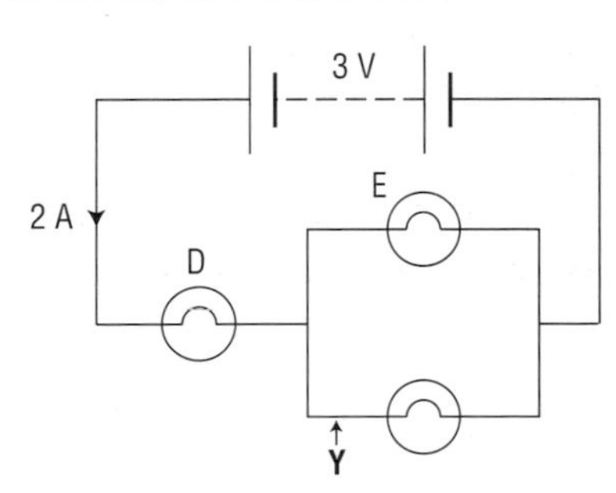

5 What is the current at point X in the left hand circuit?

6 What is the current at point Y in the right hand circuit?

7 Which of the bulbs labelled A to E is the brightest?

8 Which of the bulbs labelled A to E is the dimmest?

9 Which of the bulbs labelled A to E are equally bright?

Questions 10 to 16

In questions 10 to 16, choose the correct answer from A – stays the same; B – gets bright; C – goes dimmer; D – goes out.

What would you expect to happen to the bulb in the circuit below if:

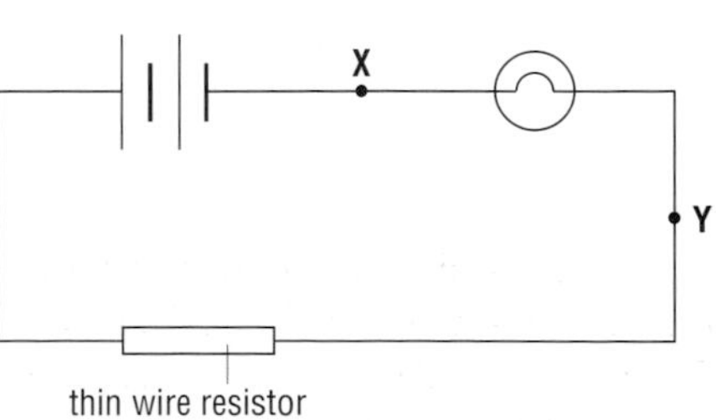

10 the resistor was replaced by another resistor of lower resistance?

11 a piece of thick copper wire was connected between X and Y?

12 another similar bulb was connected between X and Y?

13 one of the cells was turned round?

14 the resistor was replaced by another with thinner wire of the same length and same material?

15 another similar bulb was put in the circuit in series with the present one?

16 the thin wire resistor was turned round?

Short questions

17 a) Draw a circuit diagram with two lamp bulbs in which a variable resistor dims only one of them. *2 marks*

b) Suppose you were asked to wire a bell that could be rung at either the back door or the front door of a house. You are provided with a bell, a battery of three cells, two push switches and some connecting wire. Draw a circuit diagram showing how you would wire them up. (For the bell, just draw a box with two terminals on it.) *4 marks*

18 What energy changes take place when a cell is connected to a small light bulb? *4 marks*

19 a) What is the current flowing through
(i) the 6 ohm resistor in the circuit?
(ii) the 2 ohm resistor in the circuit?
(iii) the battery in the circuit? *4 marks*

b) What single resistor would have the same resistance as the 6 Ω and 2 Ω resistors in parallel? *2 marks*

6 V
6 Ω
2 Ω

20 The graph shows values of voltage against current for a light bulb.

a) Use the graph to estimate the current when the voltage across the bulb is 1 V. *1 mark*

b) What is the resistance of the bulb when the voltage across it is 1 V? *1 mark*

c) Use the graph to estimate the current when the voltage across the bulb is 4 V. *1 mark*

d) What is the resistance of the bulb when the voltage across it is 4 V? *1 mark*

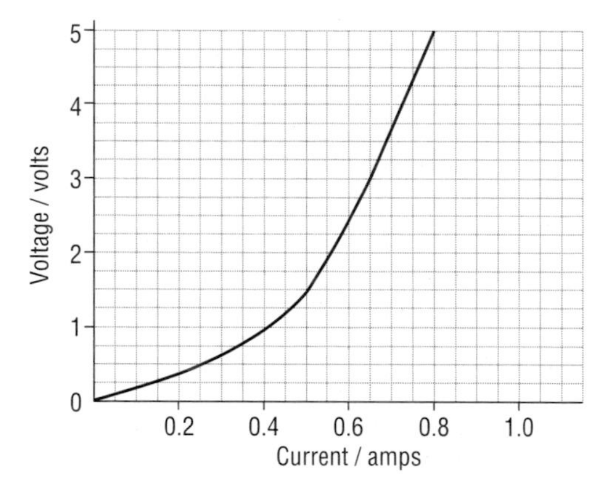

21 Look at the graph in Question 20 above.

When the current is 0.2 A, the resistance of the light bulb filament is 2 Ω.
When the current is 0.5 A, the resistance of the light bulb filament is 3 Ω.
When the current is 0.8 A, the resistance of the light bulb filament is 6.25 Ω.

Why does the resistance of the light bulb increase as the current increases? *4 marks*

Further examination questions

22 The graph shows V/I graphs for three electrical components – A, B and C.

a) Which component obeys Ohm's Law? *1 mark*

b) Which component could be a thermistor? *1 mark*

c) At what voltage does the same current pass through B and C? *1 mark*

d) Suppose the three components are put in series and the current flowing through them is 0.4 A.

(i) What is the voltage across each component?

(ii) What is the resistance across each component?

(iii) What is the total resistance of the three components in series? *7 marks*

23 a) Name the components labelled X, Y. *1 mark*

b) Current flows from the cell in the direction shown when the switch is closed.

(i) State whether lamps L_1 and L_2 are ON or OFF. *1 mark*

(ii) Give reasons for your answer to b) (i). *2 marks*

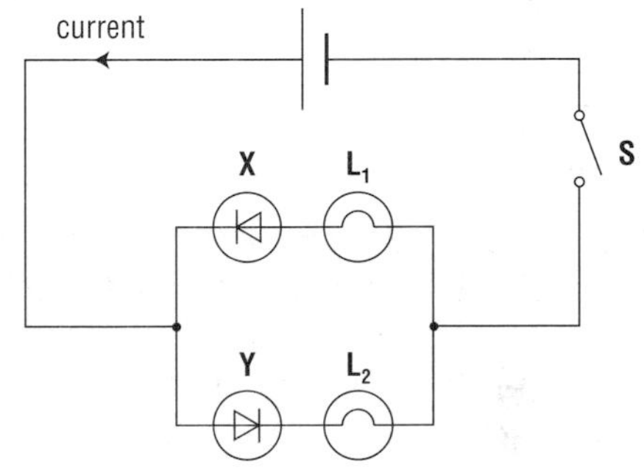

WJEC

24 The diagram shows a simple circuit.

a) (i) Redraw the diagram and add a voltmeter so that it can measure the p.d. across the lamp.

(ii) Explain how you would calculate the resistance of the lamp from the two meter readings. *2 marks*

b) A pupil measures the p.d. across the lamp to be 6 V when the current through it is 2 A. Calculate the resistance of the lamp. *2 marks*

c) The pupil finds that when the p.d. across the lamp increases, so does its resistance. Explain why this happens. *2 marks*

d) Describe how the resistance of a LDR can be varied. *2 marks*

WJEC

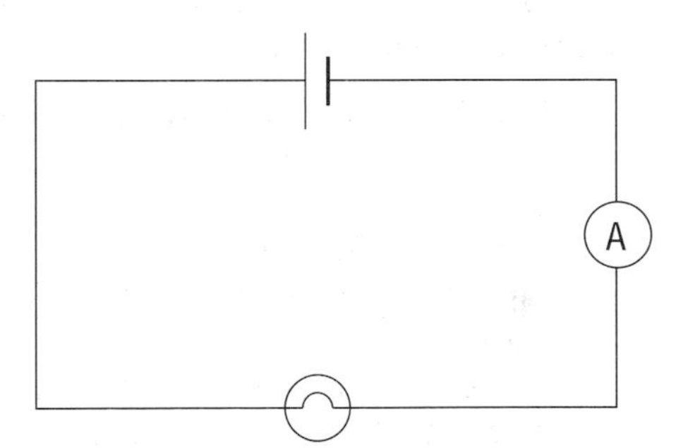

25 Some students want to find out how the current through component X changes with the voltage they use. The diagram shows the students' circuit. The graph shows the students' results.

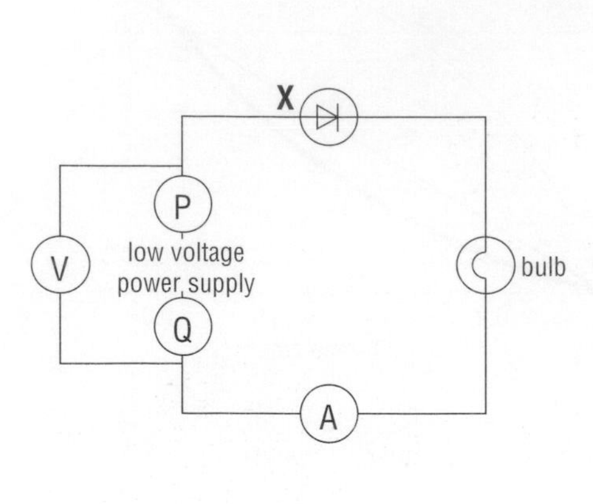

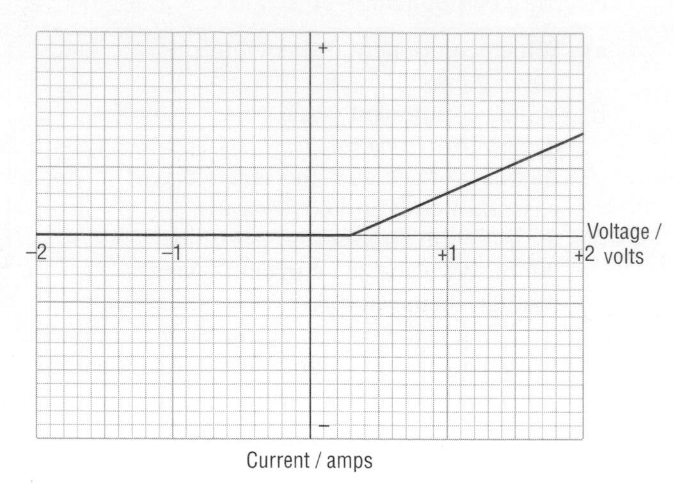

a) Describe, as fully as you can, what happens to the current through component X as the students both vary and reverse the voltage. *5 marks*

b) The low voltage supply is changed to a.c. The voltage between terminals P and Q of the supply then constantly changes as shown on graph A. Redraw graph B and show what would then happen to the current through component X. *2 marks*

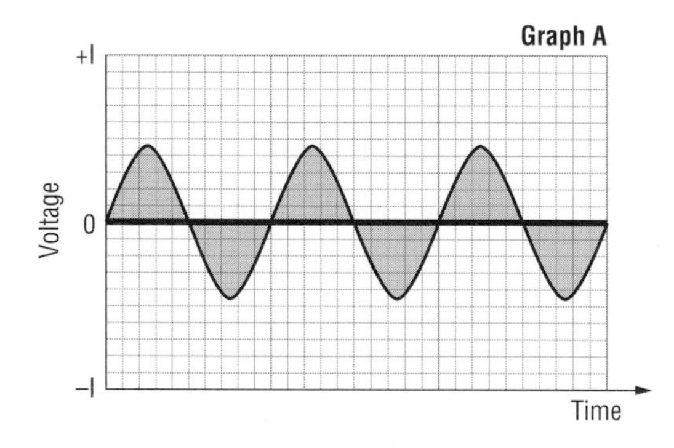

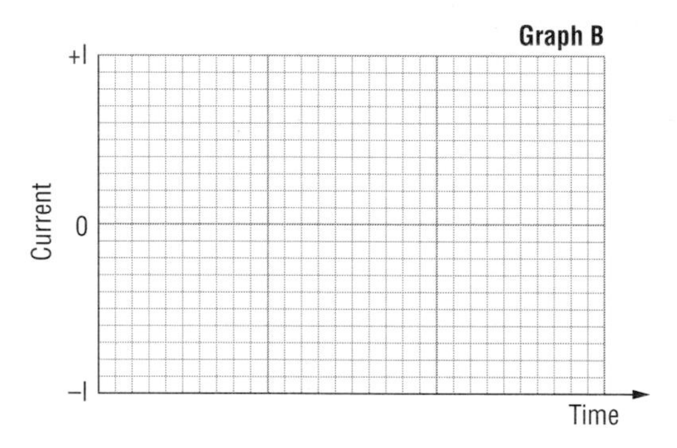

AQA

CHAPTER

25

Using electricity

SUMMARY

1 **Electric charges** (electrostatic charges) can be obtained by rubbing two non-conducting objects together causing friction.

Static charges can readily be produced by:

- combing your hair with a plastic comb
- rubbing a plastic ruler with a duster
- dragging your jumper over your hair or over your shirt.

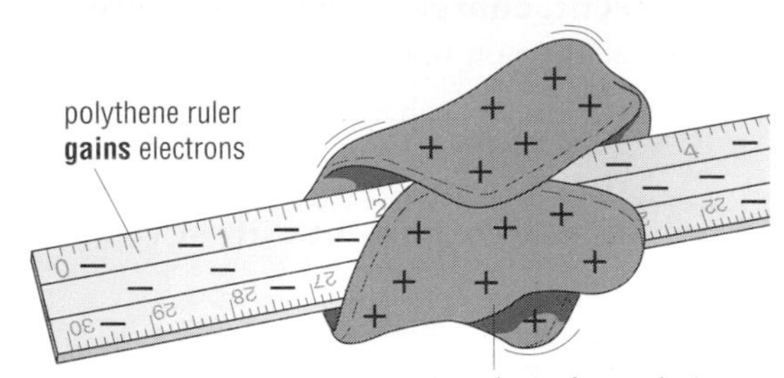

polythene ruler
gains electrons
duster **loses** electrons

Rubbing transfers electrons from the duster to the ruler

The static charges are produced by rubbing negatively-charged electrons off one material onto the other.

Remember that positive charges (protons) *never* move when static charge is generated in this way.

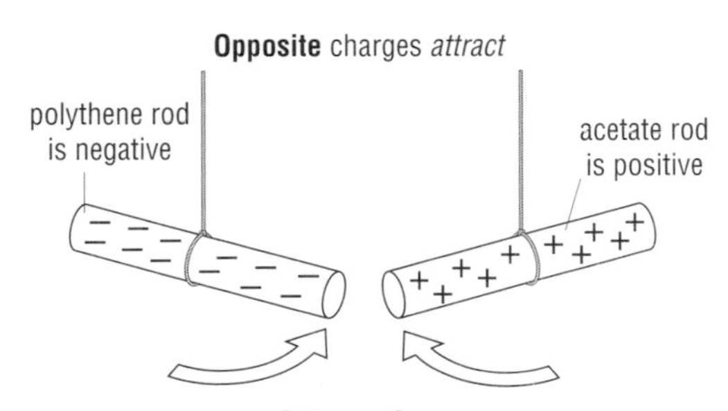

Opposite charges *attract*
polythene rod is negative
acetate rod is positive
Attraction
rods move towards each other

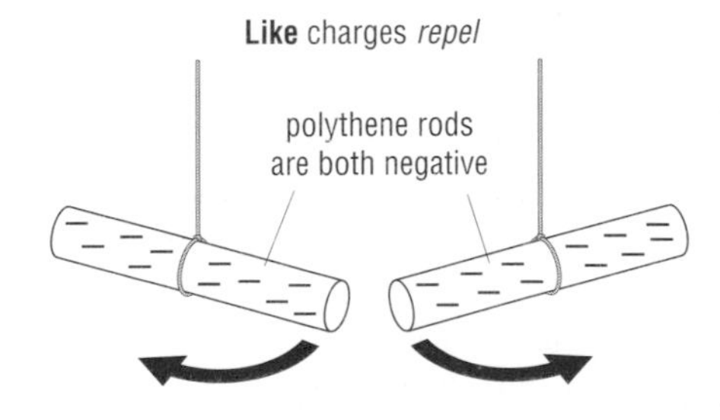

Like charges *repel*
polythene rods are both negative
Repulsion
rods move away from each other

2 **The dangers and uses of static charges**

- Lightning is caused by static charges. The bottom of dark thunder clouds build up a negative charge while the Earth below becomes positive. This negative charge builds up and eventually – Bang! Millions and millions of electrons rush to Earth as a flash of lightning.

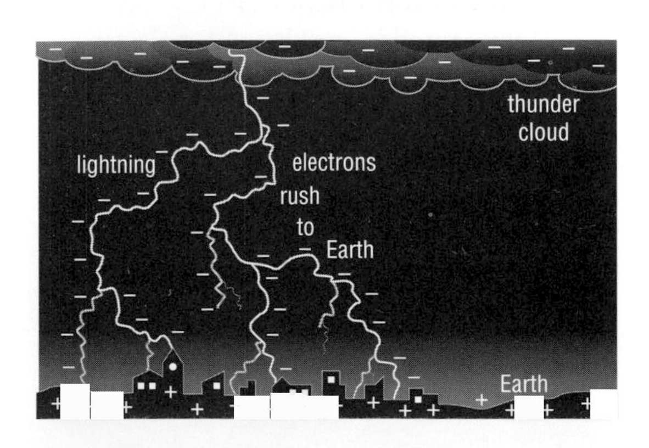

thunder cloud
lightning
electrons rush to Earth
Earth

- Air rushing past a moving car drags electrons off the car and leaves it positive. This can cause travel sickness to some people. Conducting tails allow the car to pick up electrons from the road surface and lose its charge.
- During photocopying, a roller becomes positively charged from the image to be copied. This attracts negative dark toner which is then transferred to paper.

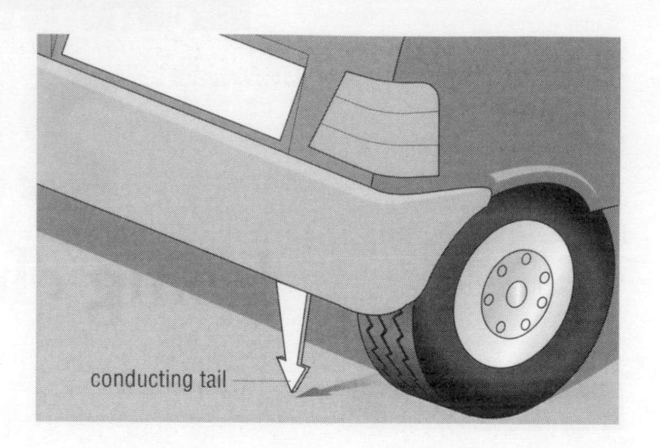

3 Current, charge, energy and power

Make sure you understand and can use the following equations.

$$\text{current} = \frac{\text{charge}}{\text{time}} \qquad I = \frac{Q}{t}$$

➡ **charge = current × time** $\qquad \mathbf{Q = I \times t}$

$$\text{energy} = \text{charge} \times \text{voltage} \qquad E = Q \times V$$

➡ **energy = voltage × current × time** $\qquad \mathbf{E = V \times I \times t}$

$$\text{power} = \frac{\text{energy transferred}}{\text{time taken}} \qquad P = \frac{V \times I \times \not{t}}{\not{t}} = V \times I$$

➡ **power = voltage × current**

➡ **energy = power × time**

4 Safety in electrical circuits

- Electrical circuits have fuses, earth wires, insulation and circuit breakers for safety.
- **Fuses** have a thin wire which melts if the current is too large. This prevents:
  - electrical wiring getting too hot which could cause a fire
  - an electric shock to the user of an electrical appliance
  - damage to the electrical appliance itself.

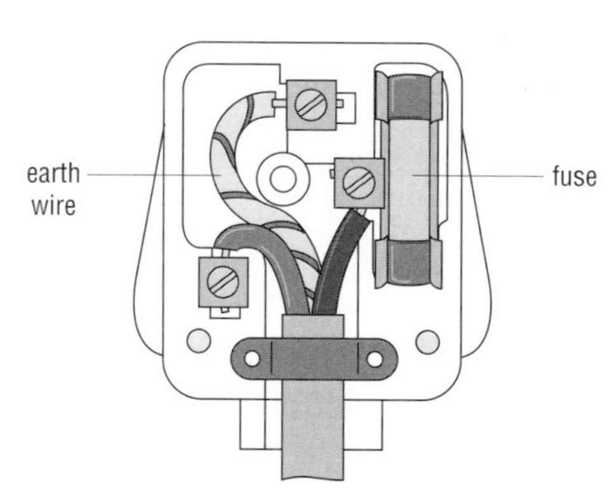

- **Earth wires** conduct currents to Earth and prevent electric shocks.
- **Insulation** – plastic (PVC) covering on the live wire, neutral wire and earth wire provides insulation for electrical wiring and appliances.
- **Circuit breakers** – when the current gets too large, contacts separate and the circuit is broken, protecting the users of electrical appliances.

STUDY QUESTIONS

Objective questions

Questions 1 to 10

In questions 1 to 10, choose from A, B, C or D which is the correct answer.

1 An ammeter measures how much
- A charge passes through the meter.
- B charge passes through the meter per second.
- C energy passes through the meter.
- D energy passes through the meter per second.

2 A plastic comb is given a negative charge by rubbing it with a dry cloth. If the dry cloth is now tested, it will have
- A a negative charge equal to that on the comb.
- B a negative charge less than that on the comb.
- C a positive charge equal to that on the comb.
- D a positive charge greater than that on the comb.

3 A car headlamp has a rating of 36 W and 12 V. Suppose the headlamp is lit by connection to a 12 V battery of negligible resistance. What is the current through the headlamp?

A 36 A B 12 A C 3 A D 1 A

4 A two-bar (2 kW) electric fire is used for 4 hours. What is the cost of this if electricity costs 7 p per kWh?

A 8 p B 14 p C 28 p D 56 p

5 The outer casing of an electric kettle is connected to earth in order to prevent
- A an electric shock to the user.
- B damage to the kettle.
- C damage to the fuse.
- D damage to the wiring.

6 Which of the following is a unit of power?

A coulomb B joule C newton D watt

7 During a flash of lightning, 10 coulombs of charge travel from the bottom of a thundercloud to the Earth in 0.1 seconds. The electric current which flows during the flash is

A 0.1 A B 1 A C 10 A D 100 A

8 A negatively-charged rod is brought close to a metal ball. Which *one* of the diagrams shows the charge distribution on the ball?

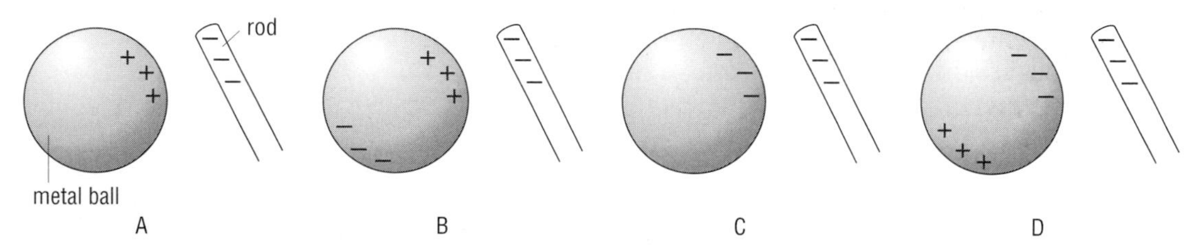

9 The rate at which electric charge passes from a lightning cloud to Earth is a measure of

A current B energy C power D work.

10 If an electrical appliance has double insulation, it means that the appliance
- A has a fuse and an earth wire.
- B has two layers of PVC on its wiring.
- C has an insulated outer casing.
- D has two fuses.

Short questions

11 A positively-charged acetate rod is held close to one of two metal balls, A and B. The balls are suspended on insulating string and are touching each other. While the rod is held in the position shown, B is moved away from A using the insulating string.

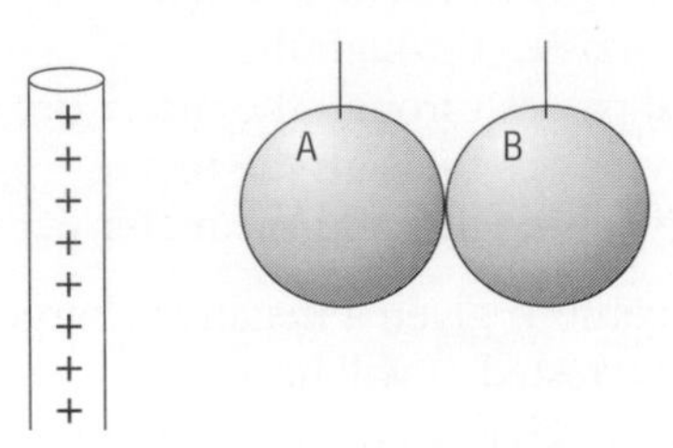

a) What is the final charge on B? *1 mark*
b) Explain your answer to part a). *3 marks*

12 a) What is the purpose of a fuse in an electrical circuit? *2 marks*
b) How does a fuse work? *3 marks*

13 In the circuit shown below, 2 units of charge pass bulb A every second.

a) How many units of charge pass bulb A in 4 seconds? *1 mark*
b) How many units of charge pass bulb B in 4 seconds? Explain your answer. *2 marks*
c) If bulb B is removed, how many units of charge pass bulb A in 4 seconds? Explain your answer. *2 marks*

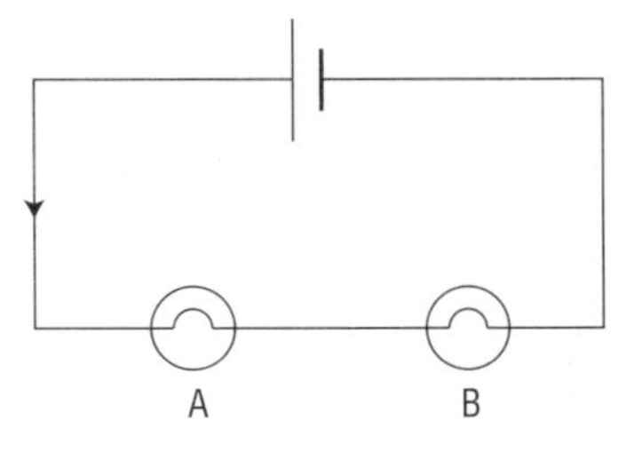

14 Balloons A, B, C, D are hung near each other by nylon threads.

a) Balloons A, B, C are given some charges and hang as in the diagram. Balloon D is given a negative charge by being touched with a wire from the negative terminal of a 5000 volt supply.
What are the charges on A, B and C? *2 marks*
b) Suppose the balloons hang as in this diagram with negative charge on D. What are the charges on balloons A, B and C now? *2 marks*
c) Cleaning a nylon carpet is a problem because it becomes charged and pieces of fluff stick to it.
Why does the charged carpet attract the fluff? *3 marks*

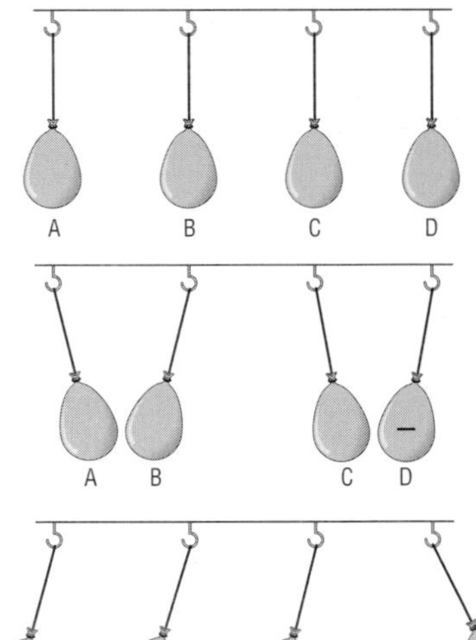

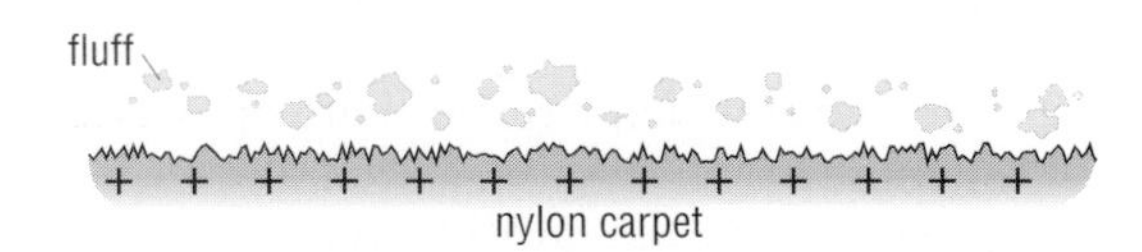

15 Some modern bedside lights are double insulated and do not have an earth connection.

a) Why do these lights not require an earth connection? *1 mark*
b) A bedside light uses an 80 W bulb connected to the 240 V mains supply.
(i) What is the equation connecting current, power and voltage? *1 mark*
(ii) What is the current in the bulb? *2 marks*
(iii) Would you use a 3 A or 13 A fuse with the light? *1 mark*

Further examination questions

16 The diagram shows a soldering iron which is marked 240 V 25 W. It is used to melt solder at 220°C.

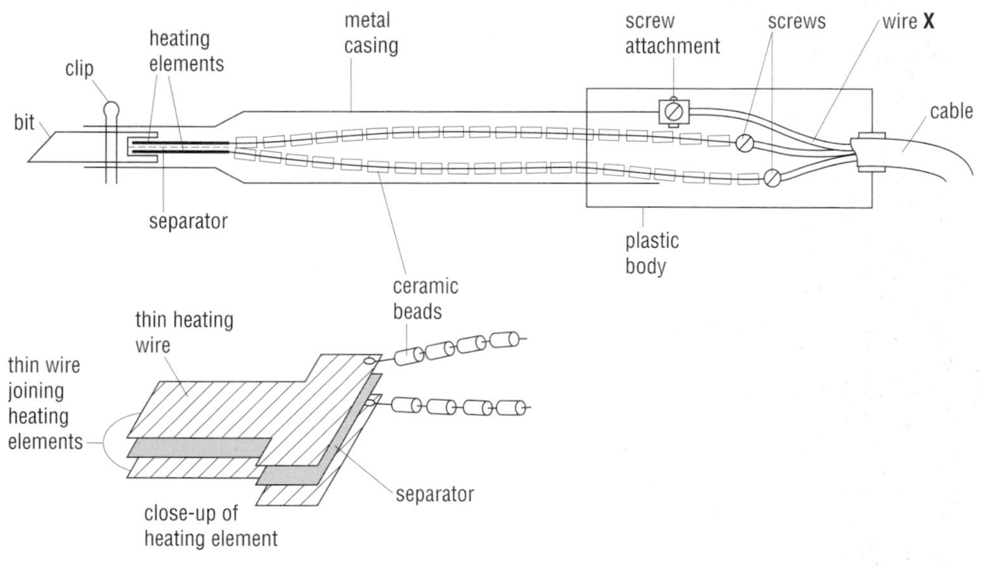

a) Calculate the current in the soldering iron when it is switched on. *3 marks*
b) It takes 100 seconds for the soldering iron to reach its working temperature. Calculate how much electrical energy is supplied in this time. *3 marks*
c) What is the function of wire X? *1 mark*
d) Explain why the wires in the heating element get hot but those in the cable do not. *3 marks*

EDEXCEL

17 The diagram shows the electrical connections from a mains supply to an electric fire. S is a switch.

a) Explain why the wires A, B, C are covered in plastic. *1 mark*
b) State which wire, A, B or C, is
(i) the live wire *1 mark*
(ii) the earth wire. *1 mark*
c) (i) Which wire is connected to the fuse? *1 mark*
(ii) Explain the purpose of the fuse. *2 marks*

WJEC

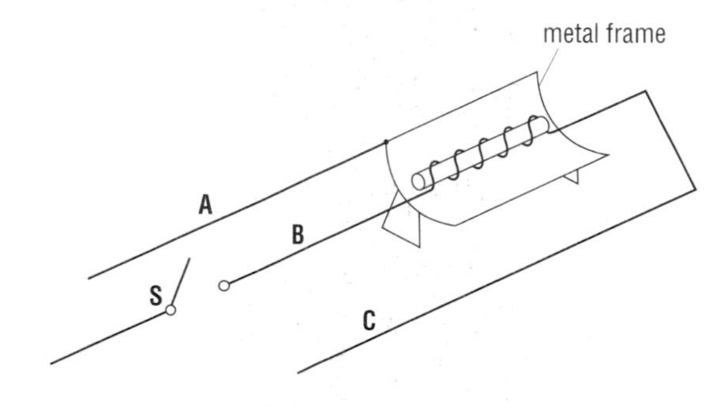

18 a) When a balloon is rubbed in your hair, the balloon becomes negatively charged.
(i) Explain how the balloon becomes negatively charged. *2 marks*
(ii) State what you know about the size and the sign of the charge left on your hair. *2 marks*
b) The negatively-charged balloon is brought up to the surface of the ceiling. The balloon sticks to the ceiling. Explain how and why this happens. *3 marks*

WJEC

19 The diagram shows an electric kettle and the label on the bottom of this kettle. The water at the bottom of the kettle will heat up first. This is because the heating element is near the bottom of the kettle. Convection currents will then cause the rest of the water in the kettle to be heated.

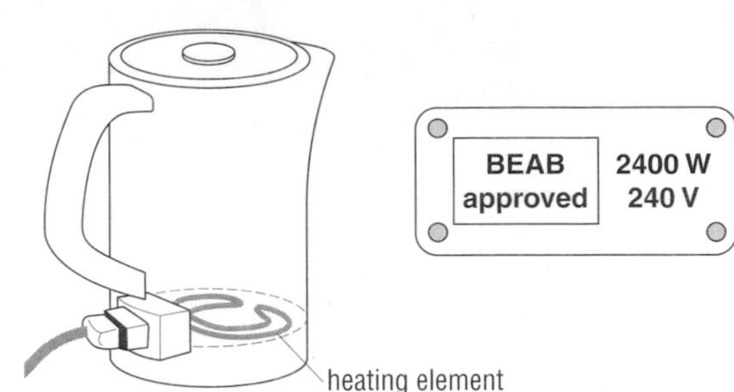

a) (i) What are convection currents? *1 mark*
(ii) Explain how convection currents are produced. (Your answer should refer to density and temperature.) *4 marks*
b) (i) Calculate the current in the heating element when the kettle is first switched on. *4 marks*
(ii) For safety reasons, the plug of the kettle is fitted with a fuse. 3 A, 5 A, 10 A and 13 A fuses are available. Which one should be used? *1 mark*
c) Calculate the electrical resistance of the heating element when the kettle is first switched on. *4 marks*
d) The temperature of 2 kg of water in the kettle is 20°C. The kettle is switched on. 588 000 joules of thermal energy is transferred to the water. (The specific heat capacity of water is 4200 J/kg°C.) Calculate the final temperature of the water. *5 marks*

AQA

20 The graph shows how the current in a filament lamp changes when the potential difference across it is increased.

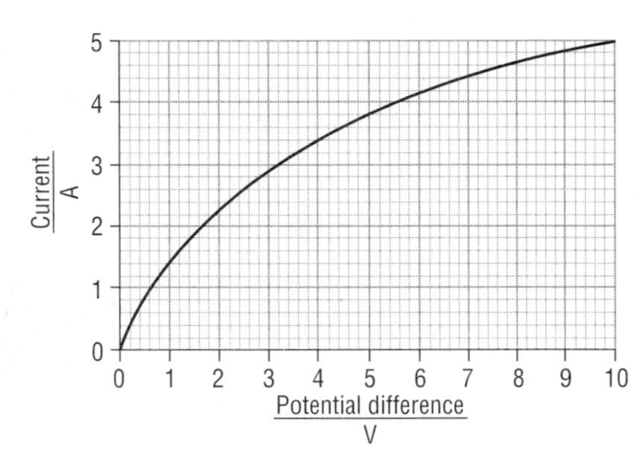

a) (i) Calculate the resistance of the filament when the potential difference is 5.0 V. *4 marks*
(ii) Describe how the resistance changes when the potential difference is increased from zero to 10.0 V. You should justify your answer by referring to data from the graph. *2 marks*
b) (i) Calculate the power of the lamp when the potential difference is 10.0 V. *3 marks*
(ii) Is it true to say that 'when the voltage is doubled, the power is doubled'? You should justify your answer by referring to data from the graph. *2 marks*

OCR

CHAPTER

26

Electromagnetism

SUMMARY

1 Magnets, magnetic poles and magnetic fields

- The ends (parts) of a magnet where the magnetic effects are the strongest are called **poles**.
- Magnets will only attract unmagnetised metals and alloys containing iron, cobalt or nickel.
- The end (pole) of a magnet which points north is called the **north-seeking pole** or just **north pole**. The pole which points south is called the **south-seeking pole** or **south pole**.
- **Like poles repel, unlike poles attract** one another.
- The region around a magnet (where its magnetic force acts) is called a **magnetic field**.
- The field around a magnet can be shown as a series of field lines called **lines of magnetic force** which always point **away from a north pole** and **towards a south pole**.

A wire with a current in it produces a magnetic field. Use the right-hand grip rule to predict the direction of the field.

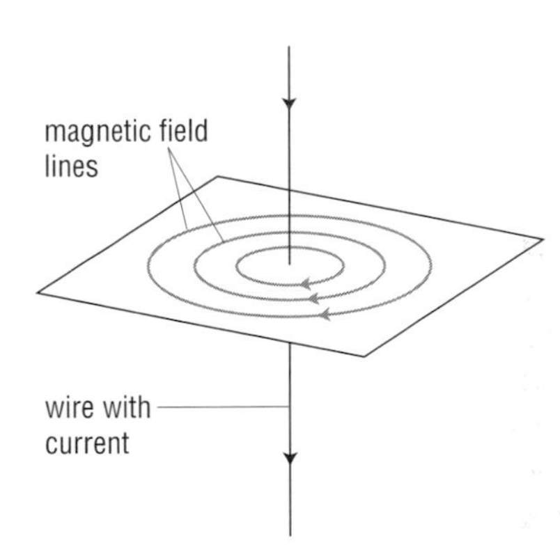

A solenoid is a long coil of wire carrying a current. Its magnetic field is like a bar magnet.

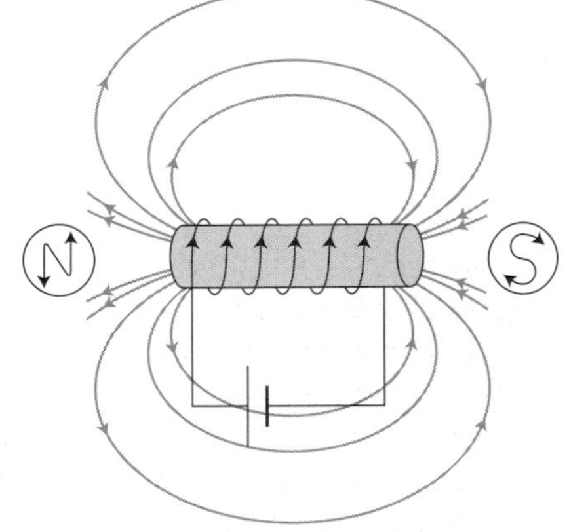

An electromagnet is usually a coil through which a current can flow with a core of soft iron

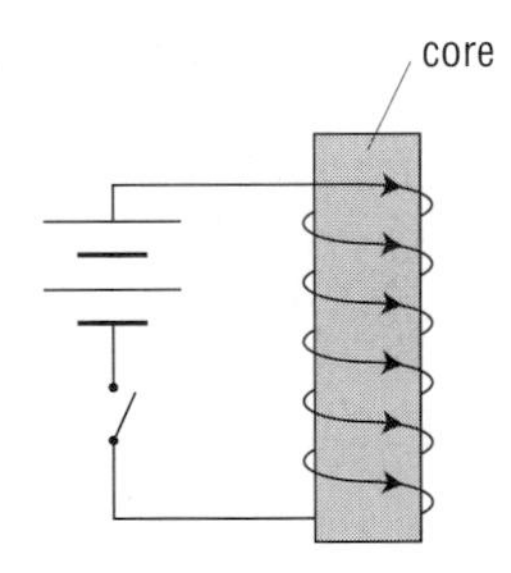

A soft iron core is needed so that its magnetism is lost when the current is switched off. If steel was used, the magnetism would stay even when the current was switched off.

2 Electromagnets

- The strength of an electromagnet can be increased by having (i) a larger current, (ii) more turns in the coil, (iii) a soft iron core.
- A **relay** is an electromagnetic switch. A small current in one circuit is used to switch on a much larger current in another circuit.

3 Motors

- In a motor: electricity + magnetism → movement (the motor effect)
  electrical energy → kinetic energy
- Use the left-hand motor rule to predict the direction of movement.
  - **Left hand** – first finger, second finger and thumb at right angles to each other.
  - **F**irst finger – **F**ield, se**C**ond finger – **C**urrent, thu**M**b – direction of **M**ovement

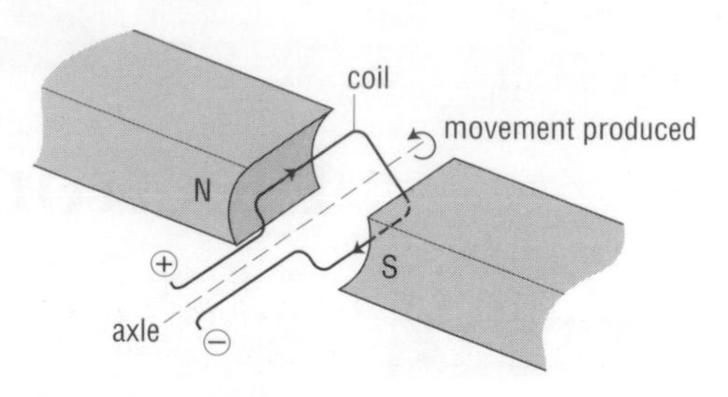

A very simple motor

4 Generators (dynamos)

- In a generator (dynamo): movement + magnetism → electricity
  kinetic energy → electrical energy
- Electric currents generated in this way are called **induced currents** and the effect is known as **electromagnetic induction**.
- The size of the induced current is proportional to the rate at which the conductor cuts through lines of magnetic force. This is **Faraday's Law of electromagnetic induction**.

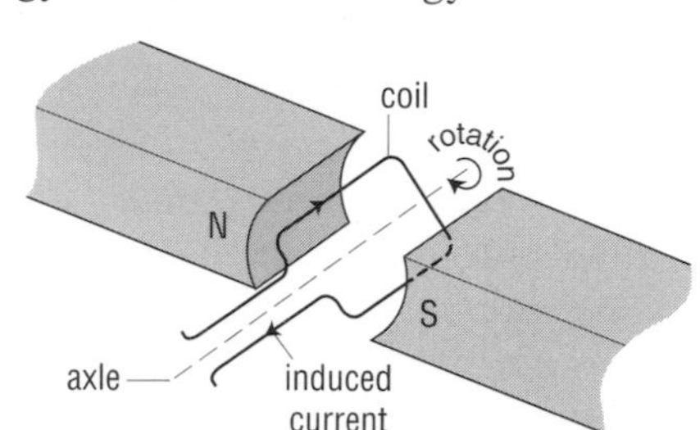

- Use the **right-hand generator (dynamo) rule** to predict the direction of the induced current.
  - **Right hand** – first finger, second finger and thumb at right angles to each other.
  - **F**irst finger – **F**ield, thu**M**b – **M**ovement direction, se**C**ond finger – **C**urrent direction

5 Transformers

An induced current is obtained in the **secondary coil** (circuit) if an **alternating current** passes through a **nearby primary coil**. This is the basis for transformers.
Equations for transformers:

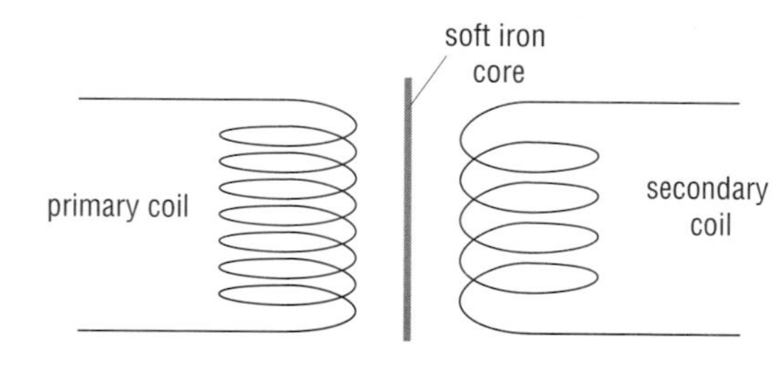

n = number of turns, V = voltage across the coil, I = current in coil, p = primary, s = secondary

Voltage induced in a coil ∝ number of turns

$$\Rightarrow \frac{V_s}{V_p} = \frac{n_s}{n_p}$$

If a transformer is 100% efficient,

$$\text{Energy input} = \text{Energy output}$$
$$V_p \times I_p \times \cancel{t} = V_s \times I_s \times \cancel{t}$$
$$\Rightarrow V_p \times I_p = V_s \times I_s$$

STUDY QUESTIONS

Objective questions

Questions 1 to 3

The figure shows the north poles of two magnets. Redraw the diagram and show:

1 the direction of the lines of force around the north pole on the left.

2 the direction of the lines of force around the north pole on the right.

3 the position of a neutral point marked by X.

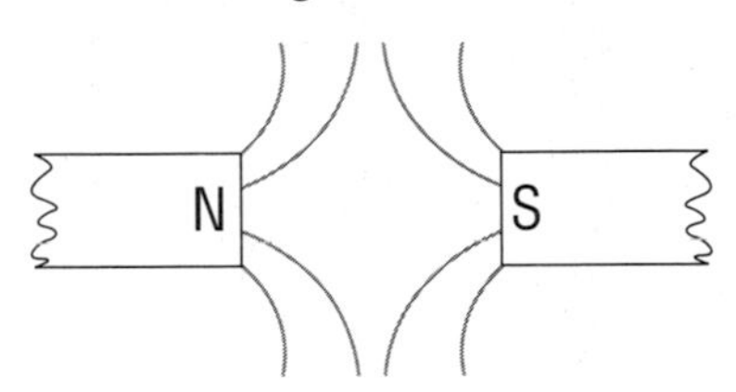

Questions 4 to 6

The diagram shows an investigation of the field around a straight wire carrying an electric current. Redraw the diagram to show:

4 the positive and negative ends of the low voltage d.c. supply.

5 the direction of the conventional current in XY.

6 the direction of the magnetic field in the lines of force.

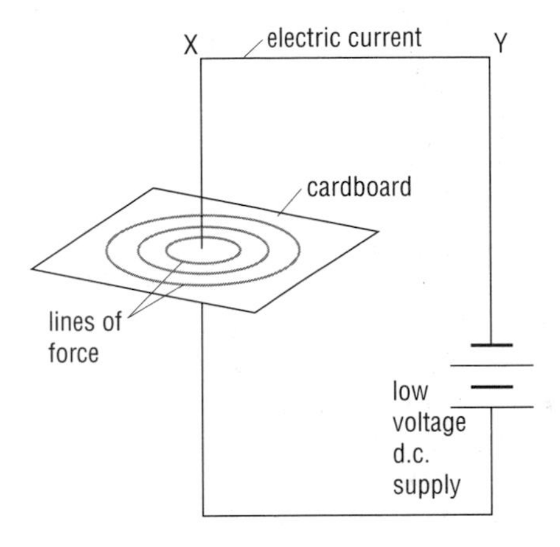

In questions 7 to 14, choose from A, B, C or D which is the correct answer.

Questions 7 to 10

A simple experiment was set up as shown in the diagram. The single coil of wire was connected to a battery and the arrows show the direction of the electric current through the coil.

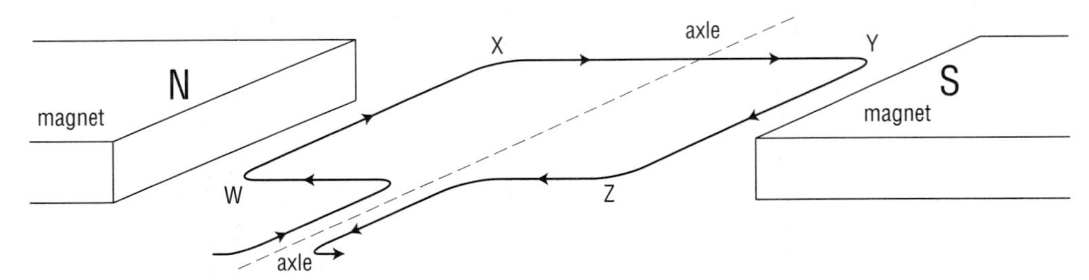

7 The experiment shows a simple
A dynamo B electromagnet
C generator D motor

8 The direction of the force on side WX of the coil is
A along WX B downwards
C towards the N pole D upwards

9 The direction of the force on side YZ of the coil is
A along YZ B towards the N pole
C towards the S pole D upwards

10 The overall effect of the forces on the coil
A keep it stationary.
B make it rotate.
C move it along the axle.
D move it downwards.

Questions 11 to 14 concern transformers

11 When a transformer is in use, the voltage in the secondary coil is said to be
A generated B induced C reduced D transformed

12 When a transformer is in use, the primary current flows
A forwards and backwards in the primary coil.
B from the primary coil to the secondary coil.
C from the primary coil and into the iron core.
D from the primary to the secondary and back.

13 Which one of the following materials would be the most useful for the core of a transformer.
A brass B copper C nickel D zinc

14 The primary coil of a transformer carries a current of 2 amperes at a voltage of 240 V. If the transformer is 100% efficient and the secondary voltage is stepped down to 24 V, the current in the secondary coil is
A 48 amps B 20 amps C 2 amps D $\frac{2}{10}$ amps

Short questions

15 The diagram shows a relay circuit used to start a car engine. Describe what happens when the key is turned and the ignition switch closes. *6 marks*

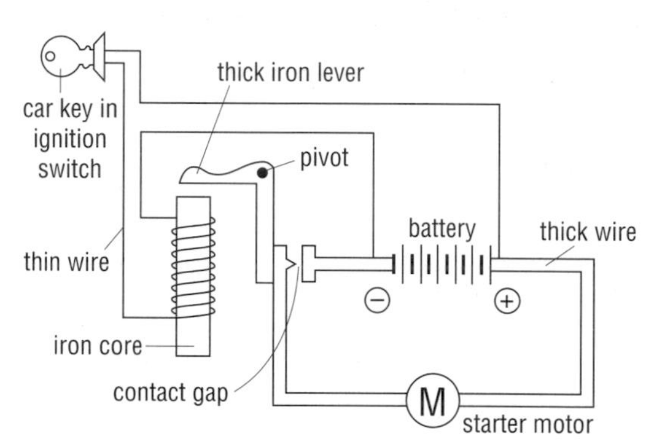

16 a) Generators at power stations produce alternating current for the National Grid with a frequency of 50 Hertz. Explain the meaning of the underlined terms. *2 marks*

b) Look at the diagram of a simple generator consisting of a single coil of wire which was made to rotate in a magnetic field. The graph shows the output voltage produced as the coil rotates from the position shown.

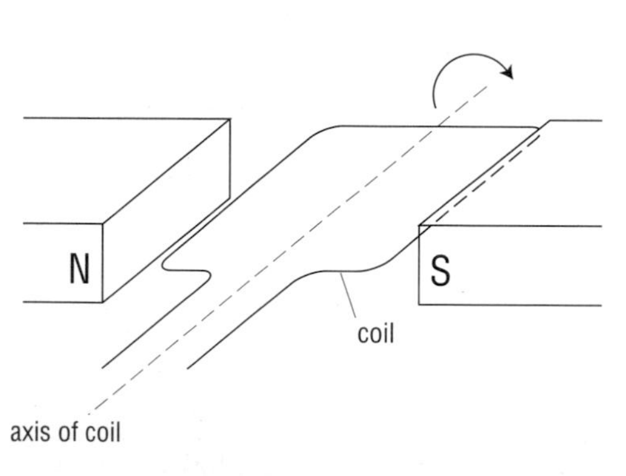

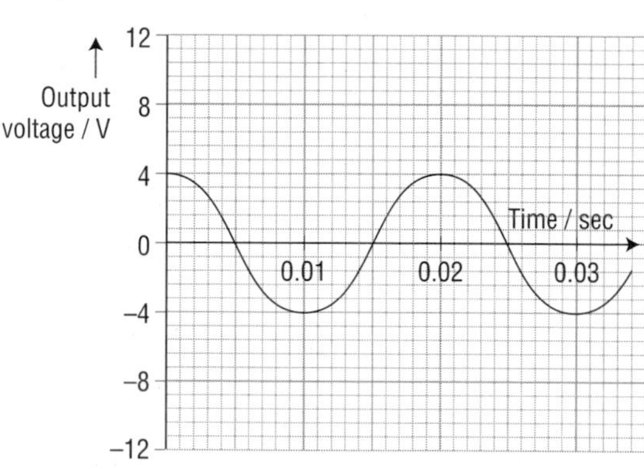

(i) Redraw the graph and show the effect on the output voltage of having two similar turns in the coil. *2 marks*

(ii) Redraw the graph again and show the effect on the output voltage of doubling the speed of rotation. *2 marks*

17 a) In which direction will the bare wire XY move when the switch is closed? *1 mark*

b) Use the motor rule to explain the movement of XY. *4 marks*

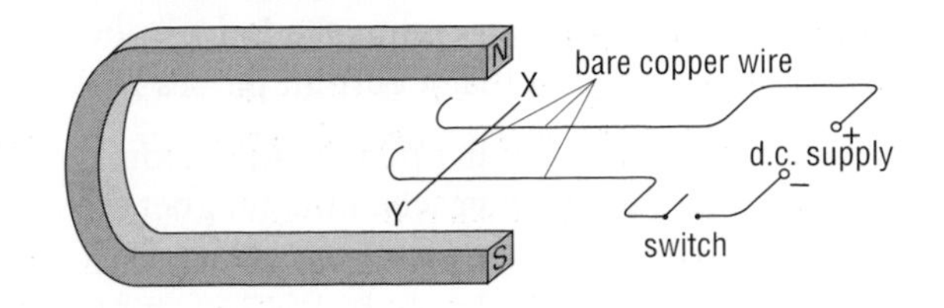

18 A transformer is used to step down an input voltage of 12 000 V to an output of 240 V to supply a remote cottage with electricity.

a) What is the name for all the power lines and transformers which connect power stations to consumers of electricity? *1 mark*

b) Why was the 12 000 V supply stepped down to 240 V for use in the cottage? *2 marks*

c) There are 1000 turns on the output (secondary) coil of the transformer at the cottage. How many turns are there on the input (primary) coil? Show your working. *3 marks*

Further examination questions

19 The diagram shows part of a burglar alarm system.

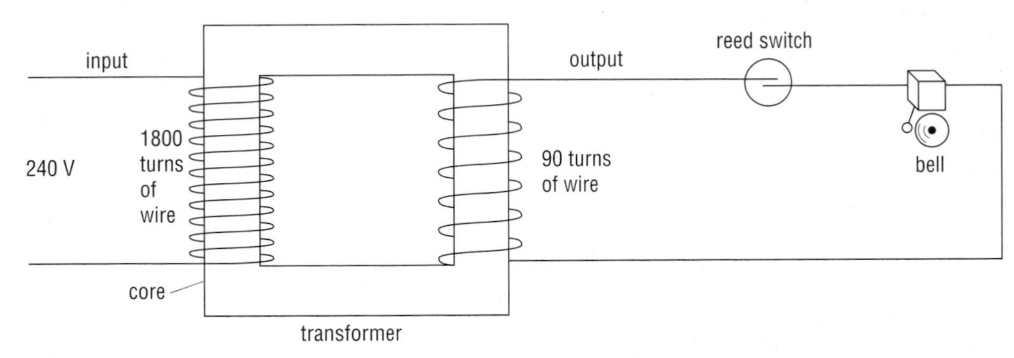

a) What would be a suitable material for the core of the transformer? *1 mark*

b) Use information on the diagram to calculate the maximum output voltage (potential difference). *3 marks*

c) Explain how a voltage is produced in the secondary coil. *4 marks*

d) Step-up transformers are used to transmit electricity at high voltages over long distances. Explain how this reduces energy losses. *2 marks*

EDEXCEL

20 The diagram shows part of an a.c. generator. As the coil rotates in the direction shown, an alternating voltage is produced across the ends of the coil, X, Y.

a) What is an alternating voltage? *1 mark*

b) (i) Explain why an alternating voltage is produced. *2 marks*

(ii) Explain why the voltage changes in size as the coil rotates. *2 marks*

c) State *three* ways of increasing the size of the alternating voltage. *3 marks*

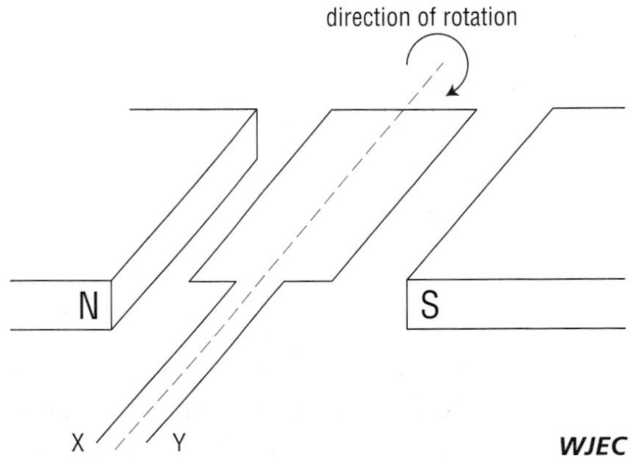

WJEC

21 The diagram shows a magnet being moved into a coil of wire. A sensitive ammeter shows that a current passes in the coil.

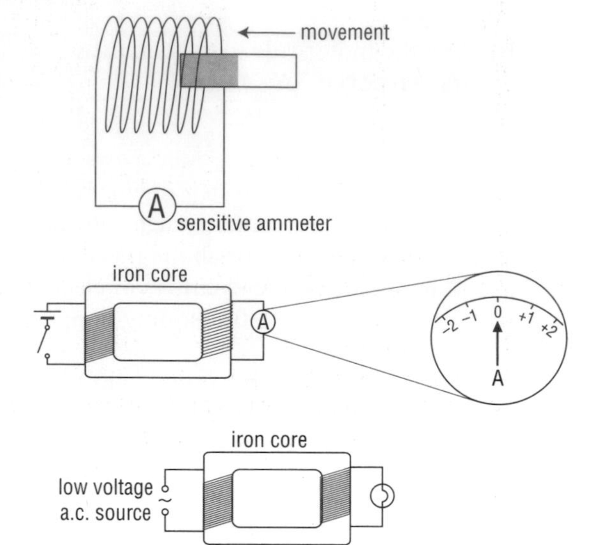

a) Explain what happens to the reading on the ammeter when the magnet is held stationary inside the coil. *2 marks*

b) Electromagnets can be used to cause a current in a coil of wire. The diagram shows a suitable arrangement. The pointer on the ammeter is in the centre of the scale when no current passes.
State and explain what happens to the reading on the ammeter:
   (i) as the current in the left hand coil is switched on. *2 marks*
   (ii) when the current in the left hand coil remains on. *2 marks*
   (iii) as the current in the left hand coil is switched off. *2 marks*

c) A transformer can be made using two coils of wire and an iron core. The diagram above shows a transformer being used to light a lamp.
Explain how the transformer uses electromagnetism to transfer energy from the low voltage source to the lamp. *2 marks*

OCR

22 The diagram shows how mains electricity is transmitted over a long distance.

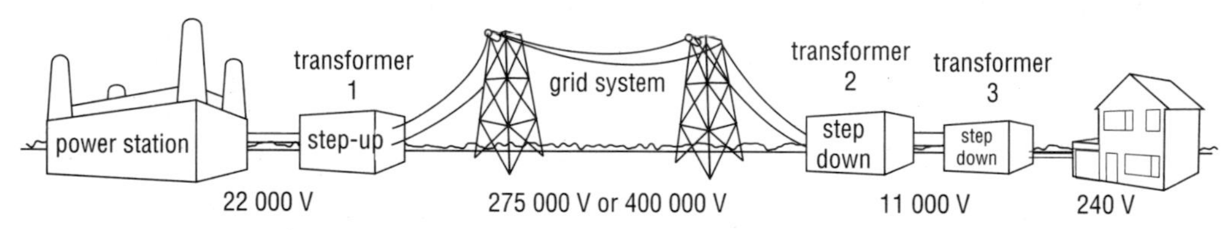

a) (i) Explain why the voltage is stepped up by transformer 1 before entering the grid system. *2 marks*
   (ii) Give one reason why the voltage is stepped down by transformers 2 and 3 before entering the houses. *1 mark*
   (iii) Explain one reason why mains electricity is transmitted as a.c. rather than as d.c. *2 marks*

b) This is a simple transformer.

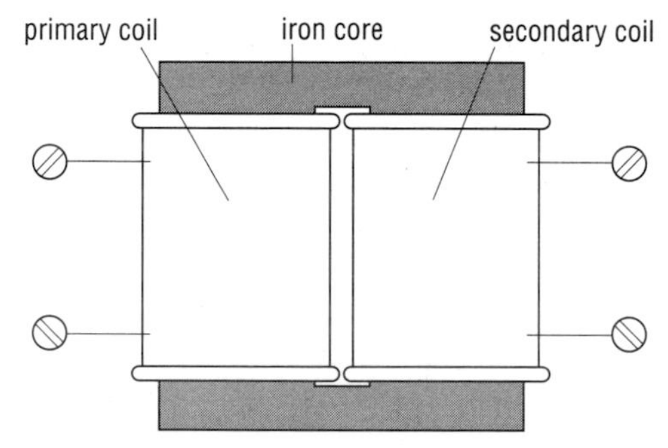

   (i) Explain the purpose of the iron core. *2 marks*
   (ii) Large transformers are often cooled by oil. Suggest one reason why oil is a suitable liquid to use as a coolant. *1 mark*
   (iii) A transformer is used to step-down a voltage from 11 000 V to 240 V. It has 3000 turns on the primary coil. How many turns will it have on the secondary coil? Show how you obtained your answer. *2 marks*

AQA

CHAPTER

27

The properties of waves

SUMMARY

1 **Light and sound** are **wave motions.** Light consists of vibrating electromagnetic waves. Sound is produced when materials vibrate. When light rays and sound waves hit materials and surfaces, three things can happen.

- The light and sound can be **transmitted** e.g. light and sound through air.
- The light and sound can be **absorbed** e.g. light by black surfaces, sound by soft furnishings.
- The light and sound can be **reflected** e.g. light by mirrors, sound by large hard surfaces, such as walls, creating echoes.

Sound waves are **longitudinal waves** – the material vibrates to and fro in the same direction as the wave.
Ultrasounds are sound waves with such high frequencies of vibration that they cannot be detected by the human ear.
Light waves and all other **electromagnetic waves** are **transverse waves** – the electromagnetic vibrations are at right angles to the direction of the wave.

2 **Images in mirrors** are

- the same size as the object.
- the same distance behind the mirror as the object is in front.
- **virtual** (light does not really go to them).
- laterally inverted.

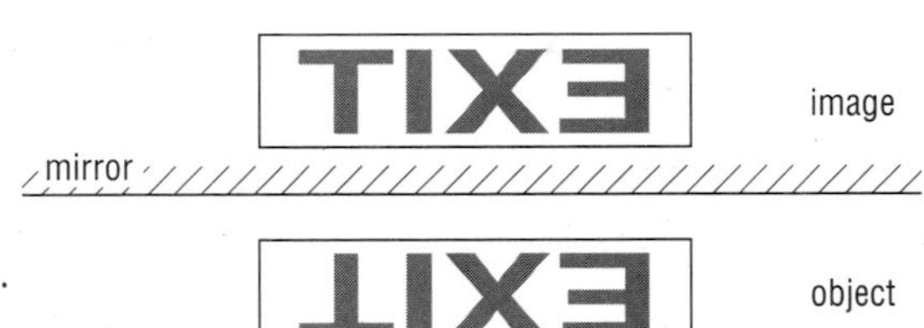

The image of an EXIT sign in a mirror

3 **Measuring waves**

- **Wavelength**, λ, is the distance from start to finish of one complete cycle.
- **Amplitude** is the maximum displacement of the vibrating material from its undisturbed (normal) position. As amplitude increases, the energy carried by the wave increases (e.g. louder sounds).
- **Frequency**, f, is the number of wave cycles per second. High pitched notes have a greater frequency than low pitched notes.

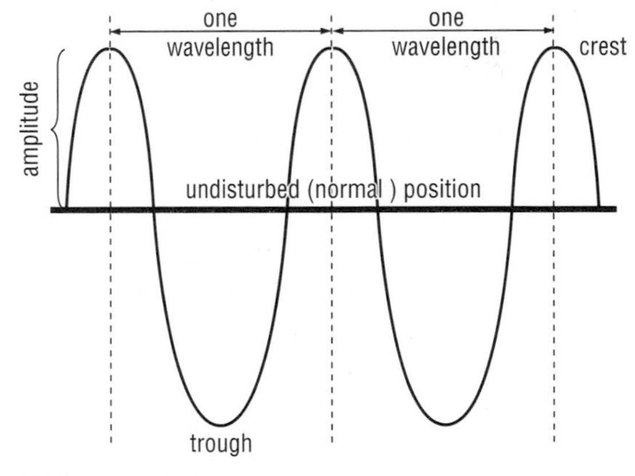

Wave properties

- **The wave equation:** wave speed = wavelength × frequency

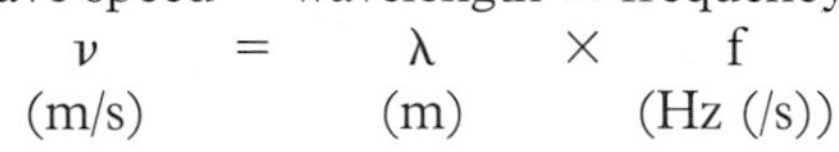

$$v = \lambda \times f$$

(m/s) (m) (Hz (/s))

4 Reflection and refraction of waves

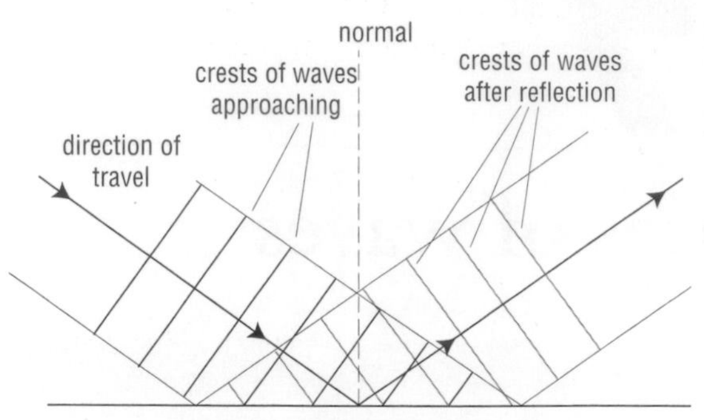

Waves reflected from a flat surface (e.g. water waves at a barrier, light waves at a mirror)

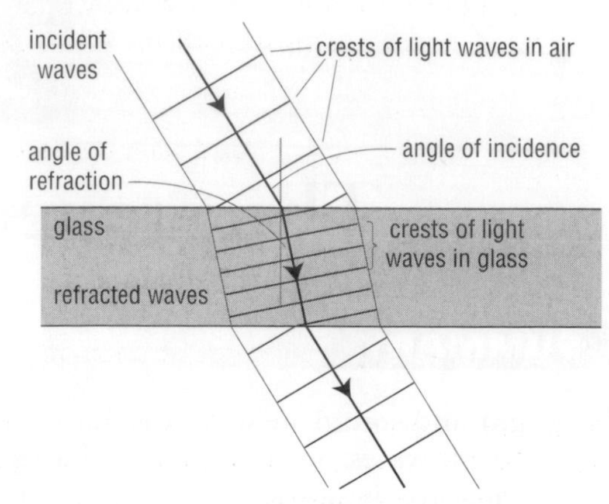

Waves (light) refracted as they pass from one material (air) into another (glass). Refraction occurs because the wavelength, and hence the speed, changes in the second material

5 Total internal reflection

When light passes from a denser to a less dense material (e.g. from water or glass into air), the angle of refraction is greater than the angle of incidence. (See lower surface of glass in the last diagram.)

When the angle of incidence reaches 42° for glass to air or 49° for water to air, the refracted ray travels along the surface of the denser material and the angle of refraction is 90°. When this position is reached, the angle of incidence is called the **critical angle**, c.

When the angle of incidence is greater than the critical angle, there is no refracted ray and all the light is reflected internally. This is known as **total internal reflection** which has an important use in bicycle reflectors, periscopes, binoculars and optical fibres.

STUDY QUESTIONS

Objective questions

Questions 1 to 3

Photographers often use large mirrors to light up a person's face when taking portrait photos.

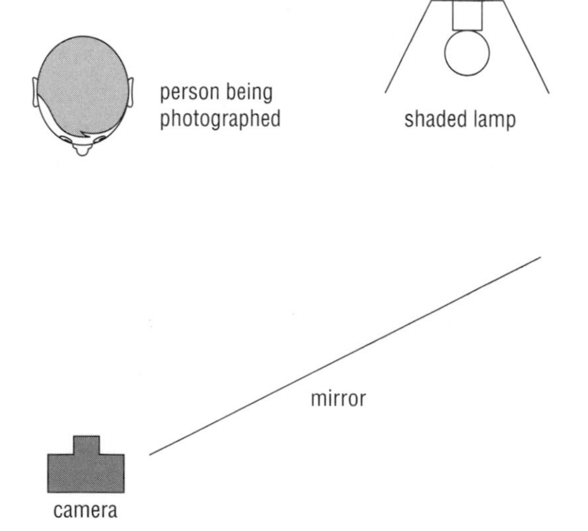

1 The person being photographed will see an image of himself in the mirror. How far is the image of his nose behind the mirror?

2 Which four words or phrases below describe the image in the mirror?

diminished, laterally inverted, magnified, real, same size, upright, upside-down, virtual.

3 In the diagram, light from the person being photographed passes through the camera lens into the film. Use two words or phrases in the list above to show how the image formed by the camera lens is different from that formed by the mirror.

Questions 4 to 6

Questions 4 to 6 concern sound waves A, B, C and D.

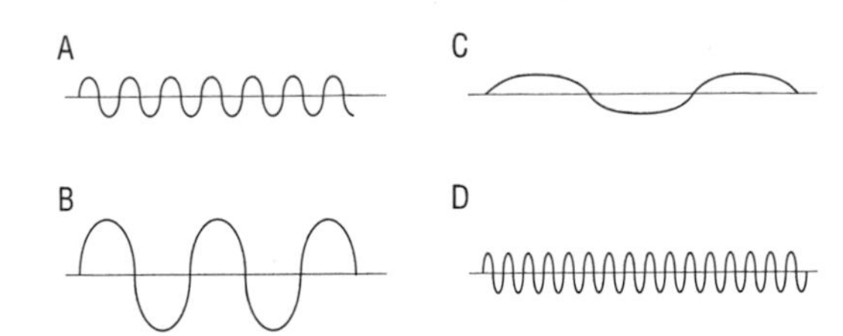

4 Which wave has the highest frequency?

5 Which wave is the loudest?

6 Which wave has the largest wavelength?

Questions 7 to 12

In questions 7 to 12, choose from A, B, C or D which is the correct answer.

7 A microphone changes
- A electrical signals into sound waves.
- B light waves into sound waves.
- C sound waves into electrical signals.
- D sound waves into light waves.

8 Which one of the following does not change when light travels from air into water?
A direction B frequency C velocity D wavelength

9 The colour of light is determined by its
A amplitude B brightness C frequency D speed

10 A high pitched sound has a large
A amplitude B frequency C speed D wavelength

11 An observer hears the thunder from lightning 5 seconds after seeing the flash. If the speed of sound is 330 m/s, how far is the observer from the lightning?
A 16.50 m B 330 m C 66 m D 33 m

12 If you sing in the bath, your voice seems louder than in other rooms because
- A water vapour in the air amplifies the sound.
- B water in the bath refracts the sound.
- C tiles on the wall reflect the sound.
- D the small bathroom creates more echoes.

Short questions

13 Radio 1 has a frequency of 1089 kHz and a wavelength of 275 m.
a) What is the formula linking frequency, wavelength and speed? *1 mark*
b) Calculate the speed of waves from Radio 1. *3 marks*
c) What would you expect the speed of waves from Radio 2 to be? Explain your answer. *2 marks*

14 Redraw and complete the diagrams a) and b) below to show plane waves crossing a straight boundary and passing into a medium in which their speed is lower.

a)

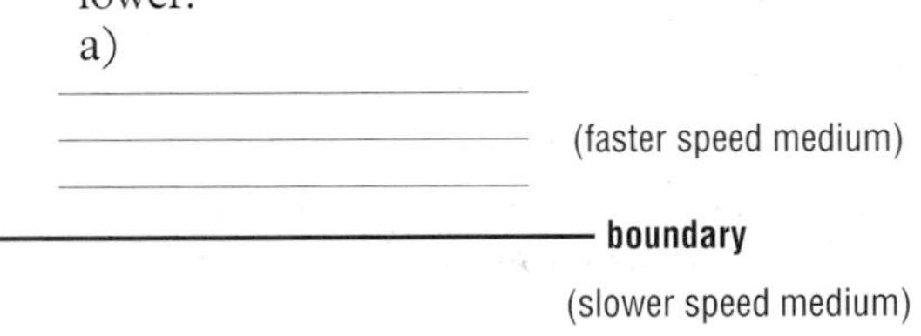

Wavelengths parallel to boundary
2 marks

b)

(faster speed medium)
boundary
(slower speed medium)

Wavelength at an angle to the boundary
2 marks

c) What changes, if any, occur in the frequency and wavelength as a result of refraction when waves pass from a faster speed medium into a slower speed medium. *2 marks*

15 Copy the diagrams and complete them, to show how the light rays behave.

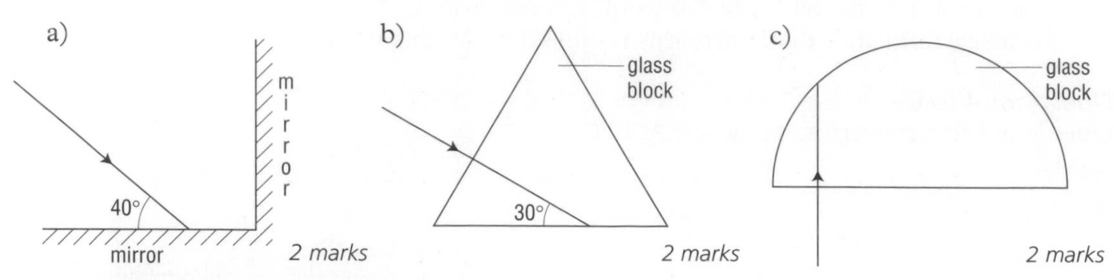

16 An electronic synthesiser produces two pure notes A and Y as in the diagram.

a) A third note, Z, is produced by adding the two wave forms together. Copy the diagram and then add the two waveforms to produce Z. *2 marks*

b) Will Z be louder or softer than X or Y? Explain your answer. *2 marks*

c) How does the frequency of Z compare with that of X and Y? *2 marks*

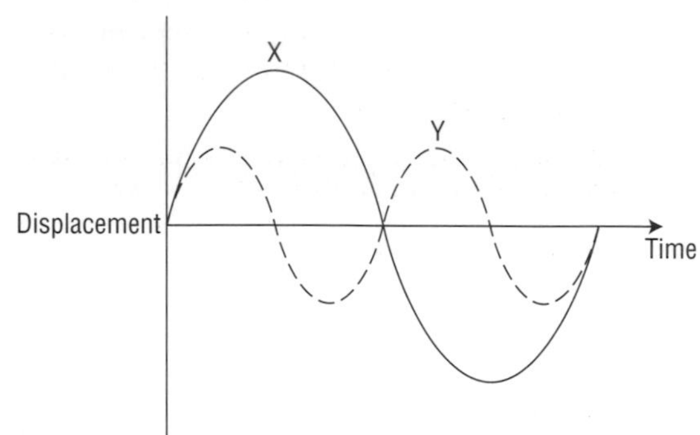

Further examination questions

17 When bats fly, they make high frequency squeaks. These are reflected from objects as echoes. The echoes give bats information about what is around them.

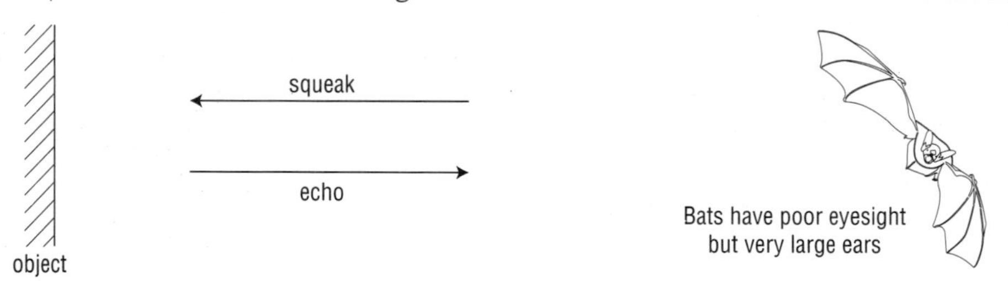

a) The diagram below shows the oscilloscope traces of the original squeak and the echo.

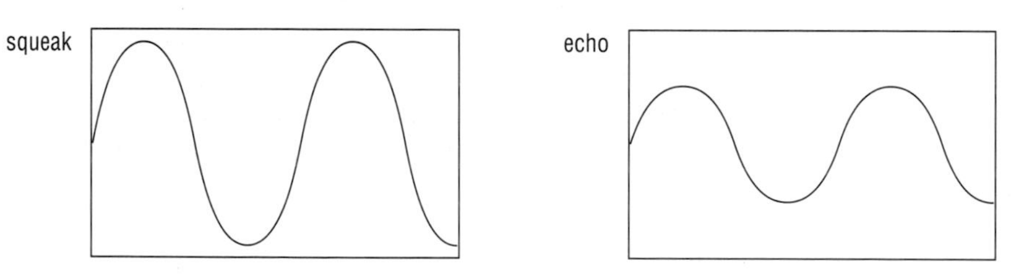

Compare these two traces and state a property that has
(i) changed, (ii) remained the same. *2 marks*

b) The bat squeaks. 0.3 s later it hears the echo from an object. The speed of sound in air is 330 m/s.

(i) Use the equation: distance = speed × time taken, to calculate how far the sound travelled in this time. *2 marks*

(ii) How far is the bat from the object? *1 mark*

c) A bat transmits a sound of frequency 33 000 Hz. The velocity of sound in air is 330 m/s.
   (i) Write down the equation which relates the velocity (v) of a soundwave to its frequency (f) and wavelength (λ). *1 mark*
   (ii) Calculate the wavelength. State the unit of your answer. *3 marks*

EDEXCEL

18 a) A pregnant woman has come into hospital for an ultrasound scan of her baby.
   (i) What is ultrasound? *2 marks*
   Ultrasound can be used to produce images of the baby. The sound waves travel through different materials at different speeds.

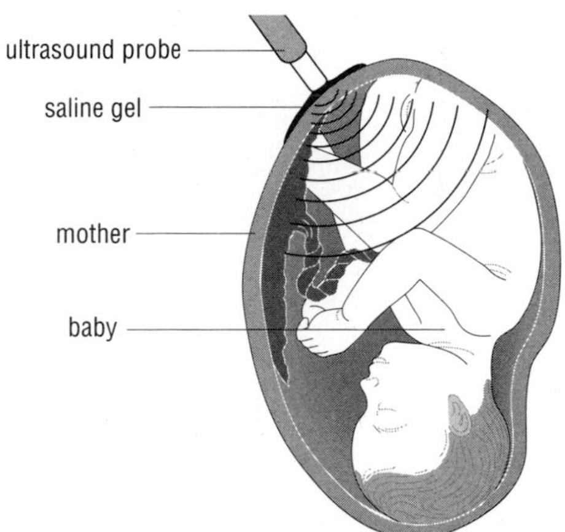

Material	Speed of sound/m/s
air	332
bone	3360
fat	1476
muscle	1540
saline gel	1515

Sound waves are reflected as they pass from one material to another if the speed of sound in the materials is different. The greater the difference in speed the greater the reflection.
   (ii) Between which two materials in the table would the reflection be:
   A greatest, B least? *2 marks*

b) A new student carried out the examination and at first she forgot to place saline gel on the mother's abdomen. Explain why the student only got a good picture of the baby when she used the saline gel. *4 marks*

EDEXCEL

19 Sound at a very high frequency of 200 000 Hz is called ultrasound. It can be used for cleaning up chemical spillage.

a) The people who operate the equipment cannot hear the ultrasound. Give a reason why ultrasound cannot be heard by humans. *1 mark*
b) Ultrasound is transmitted through water as a longitudinal wave. Describe the movement of water particles when they transmit the ultrasound. *2 marks*
c) Heavy oil spills that have seeped into the ground can be broken up by ultrasound and washed away with water. Suggest one advantage of using ultrasound instead of detergents to clean up ground that has been soaked in oil. *1 mark*
d) Ultrasound is produced by a crystal. The speed of the sound in the crystal is 4000 m/s. Calculate the wavelength of ultrasound of frequency 200 000 Hz in the crystal. *3 marks*

OCR

20 a) When light enters a block of glass from air it may change direction. What is the name of the process which occurs as the light changes direction? *1 mark*

b) Light travels as a wave. The wavefronts are at right angles to the direction of the light ray. Some of the wavefronts are shown on the diagram. Redraw the diagram and explain why the light changes direction as it enters the glass. You should add to the diagram to help you to give your answer.

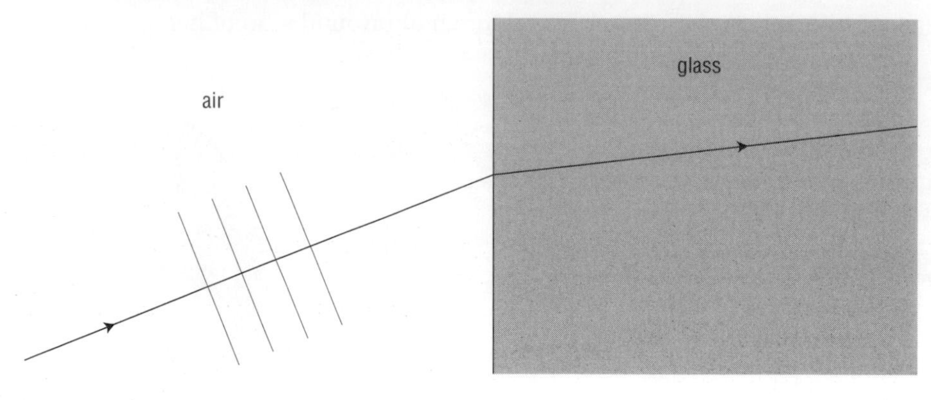

5 marks

c) The critical angle for light going from glass into air is 42°. Copy out and complete each of the diagrams in parts (i) and (ii) to show the direction of most of the light as it leaves point P.

(i) The angle **x** is less than 42°

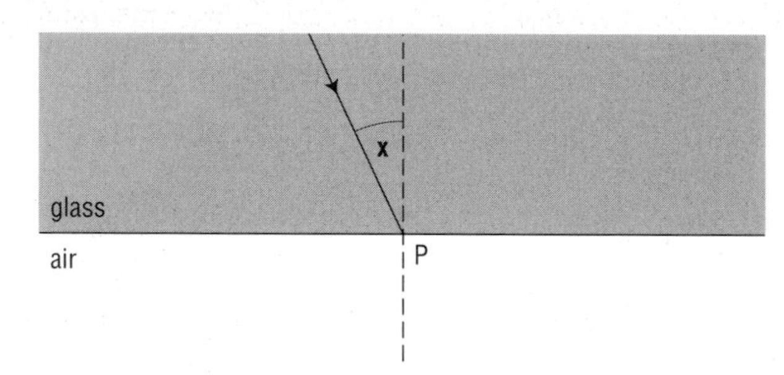

2 marks

(ii) The angle **y** is more than 42°

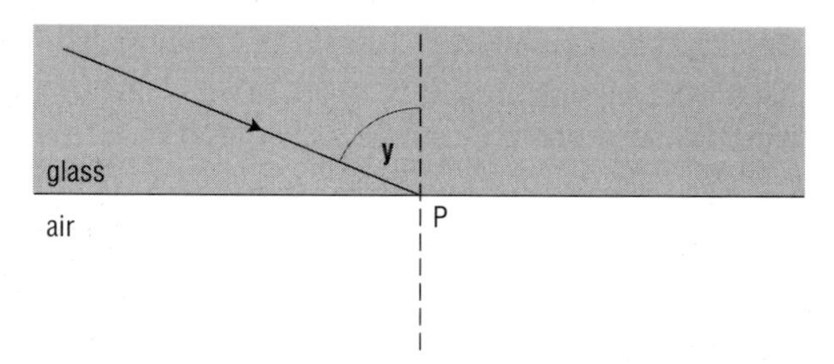

2 marks

(iii) What is the name of the dotted line shown in parts c) (i) and c) (ii)? *1 mark*

(iv) What is the name of the process which happens in part c) (ii)? *1 mark*

AQA

CHAPTER

28

Using waves

SUMMARY

1 The electromagnetic spectrum

Gamma rays, X-rays, UV rays, visible light, IR rays, microwaves and radiowaves are all members of the **electromagnetic spectrum**.

Electromagnetic waves have enormous differences in wavelength and frequency but they:

- are all produced when atoms or electrons lose energy.
- are all transverse waves.
- all transfer energy as vibrating electric and magnetic fields.
- can all be reflected, refracted and diffracted.
- all travel in a vacuum at a speed of 300 million m/s (3×10^8 m/s)

2 Properties and uses of different waves

Type of wave and wavelength	Properties	Uses	
gamma rays less than 10^{-12} m	• emitted by radioactive substances • the most penetrating e/m waves • can destroy cells and cause mutations	• treatment of cancer • sterilisation of medical instruments and food	gamma rays cancer tumour
X-rays 10^{-12} to 10^{-8} m	• penetration of matter depends on relative atomic mass of constituent atoms • absorbed very effectively by lead (Pb = 207) • can damage cells – dangerous in high doses • absorbed by bones and teeth but pass through flesh	• to diagnose broken bones and diseased teeth • treatment of some cancers	An X-ray showing fractured radius and ulna bones of the lower arm

Type of wave and wavelength	Properties	Uses
ultraviolet rays 10^{-8} to 4×10^{-7} m	• emitted by the Sun and other white hot objects • pass through air • absorbed by solids • increase formation of vitamin D and melanin (brown pigment) in skin • over-exposure burns the skin and this may lead to skin cancer	• energy efficient lamps • UV lamps and sunbeds 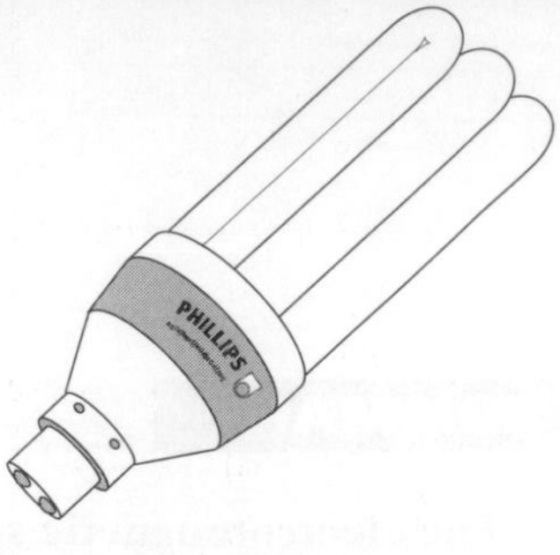 An energy efficient lamp
visible light 4×10^{-7} to 7.5×10^{-7} m	• light of different wavelength has a different colour e.g. a rainbow (4×10^{-7} violet, 7×10^{-7} red)	• electric lighting • photography  A camera consists of a lens and a light-sensitive film mounted in a light-tight box
infrared rays 7.5×10^{-7} to 10^{-4} m	• emitted by all hot and warm objects • detected by special photographic film and heat sensitive equipment	• radiant heaters • red-hot grills for cooking • detection of warm objects e.g. nocturnal animals, people buried under rubble
microwaves 10^{-4} to 1 m	• absorbed by water • reflected by metals • transmitted through glass, paper and plastic	• cooking – microwave ovens • satellite communication 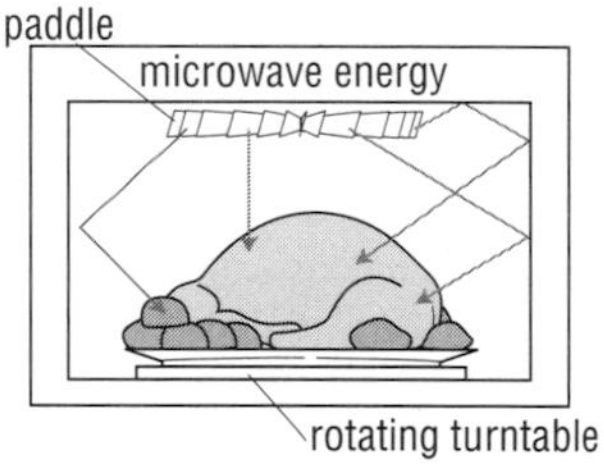 paddle microwave energy rotating turntable A microwave oven
radio waves 1 to 10^4 m	• radiowaves with $\lambda > 10$ m are transmitted around the Earth by reflection from the ionosphere • radiowaves with $\lambda < 10$ m pass through the ionosphere	• local and national radio, TV and telephone • international radio, TV and telephone communications

STUDY QUESTIONS

Objective questions

Questions 1 to 4

Name the electromagnetic radiations described below.

1 This radiation cannot be detected by the eye, but is emitted by hot objects.

2 These are the shortest radio waves used in communications and to produce heat.

3 This radiation has a short wavelength and is emitted by unstable nuclei.

4 This radiation is used for the remote control of televisions.

Questions 5 to 8

Diagrams A to E show five waveforms of sound on an oscilloscope screen when all the oscilloscope controls stayed the same.

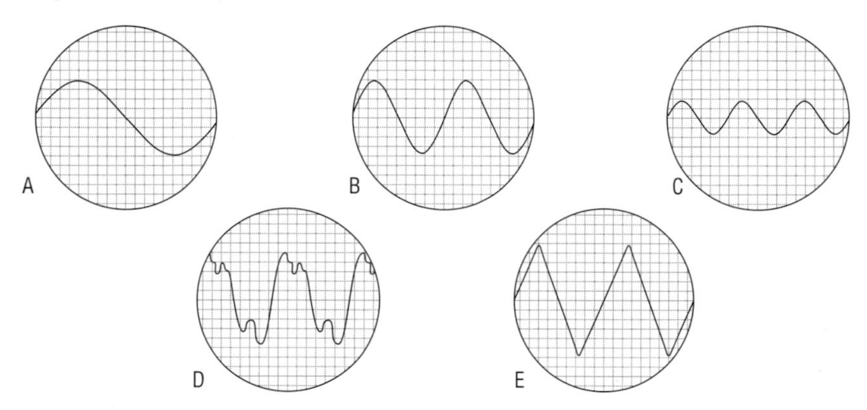

5 Which diagram shows the sound which is not a pure note?

6 Which diagram shows the sound with the longest wavelength?

7 Which diagram shows the sound with the highest pitch?

8 Which diagram shows the sound with the largest amplitude?

Questions 9 to 12

In questions 9 to 12 choose from A, B, C or D which is the correct answer.

9 Diffraction occurs when
A people are talking on the other side of a doorway.
B a surgeon uses an optical fibre endoscope.
C light passes through a triangular glass prism.
D X-rays are used to photograph fractured limbs.

10 Infrared radiation travels through a vacuum by
A conduction only B convection only C dispersion only D radiation only

11 Which one of the following will determine the colour of light?
A amplitude B brightness C luminosity D wavelength

12 Which one of the following properties always remains unchanged when light passes from glass into air?
A direction B frequency C speed D wavelength

Short questions

13 The diagram shows the different types of radiation in the electromagnetic spectrum.

a) State three properties in which these radiations are similar. *3 marks*

b) Which of these radiations is used
(i) to obtain colour photographs, (ii) to cook food in a conventional oven,
(iii) to treat certain cancers? *3 marks*

Radio waves	Microwaves	Infrared waves	Visible light	Ultraviolet light	X-rays	Gamma rays

14 The diagram shows the wavefronts of an electromagnetic wave moving towards a barrier in which there is a gap.

barrier

a) Redraw the diagram and show four wavefronts further to the right. *2 marks*

b) What is the name of the process which occurs as the wavefronts pass through the gap? *1 mark*

c) This process is very noticeable in the example shown in the diagram. Why? *1 mark*

d) Does the wavelength increase, decrease or stay the same as the wave passes through the gap? *1 mark*

15 Shock waves from earthquakes can be recorded on a seismograph similar to this diagram. The time divisions on the recording are shown every 10 seconds.

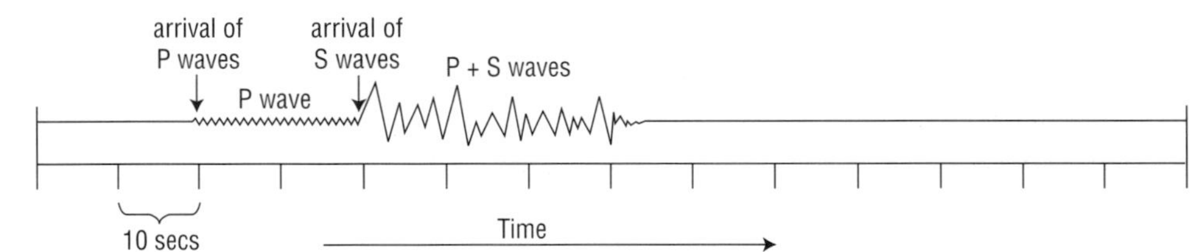

P waves travel faster than S waves so they arrive earlier at the recording station. The difference between the time of arrival is called the time lag. The time lag can be used with the graph to calculate the distance from the epicentre of the earthquake to the recording station.

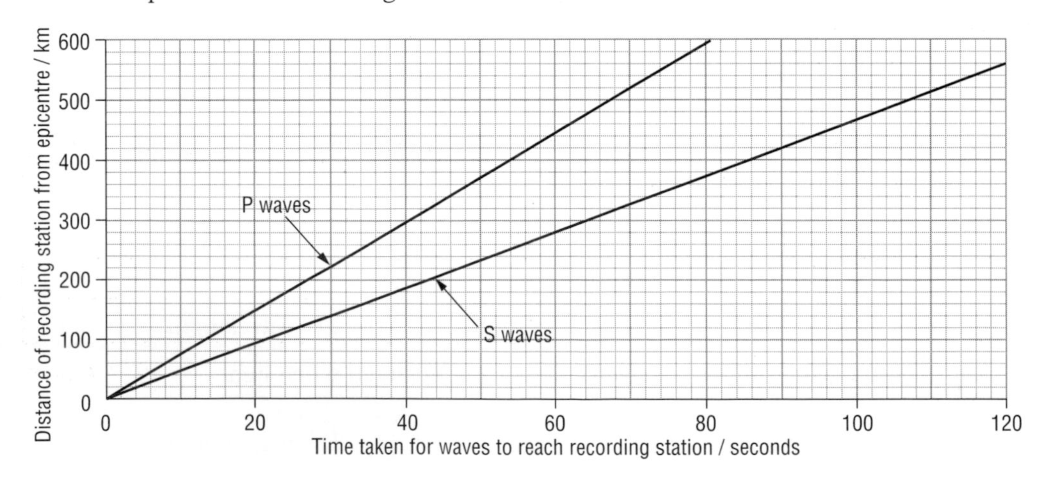

a) Use the graph to work out the actual speed of P waves. Show your working. *3 marks*
b) Use the diagram to obtain the time lag. *1 mark*
c) Use the graph to find the distance of the recording station from the epicentre of the earthquake. *1 mark*

16 An earthquake produces seismic waves which travel around the Earth's surface at a speed of 5 km/s. The graph shows how the ground moves near the epicentre of the quake as the waves pass.

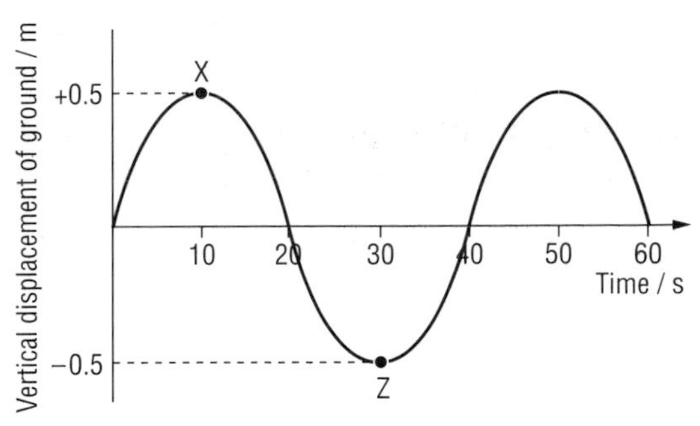

a) What is the time for one period of the waves? *1 mark*
b) What is the frequency of the waves? *1 mark*
c) Calculate the wavelength of the seismic waves. *2 marks*
d) At which of the points X, Y or Z is
 (i) the ground moving most rapidly?
 (ii) the ground accelerating at its greatest value? *2 marks*
e) Estimate the vertical speed of the ground at the time marked Y. *2 marks*

Further examination questions

17

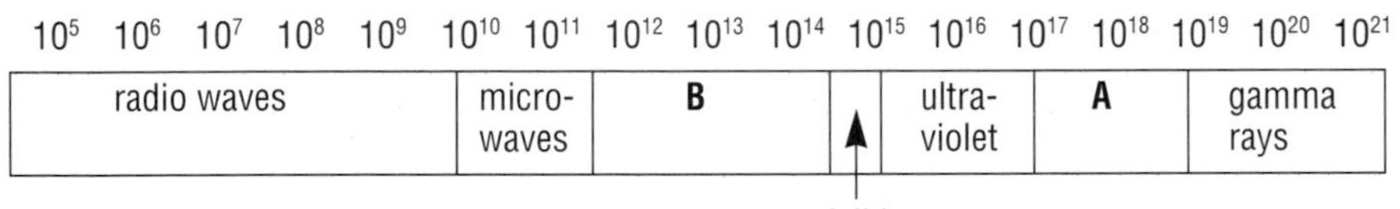

The diagram shows the main regions of the electromagnetic spectrum. The numbers show the frequencies of the waves measured in Hertz (Hz).

a) Name the regions A and B. *2 marks*
b) (i) Write down, in words, the equation connecting wave speed, wavelength and wave frequency. *1 mark*
 (ii) Calculate the frequency of the radiation with a wavelength of 0.001 m (10^{-3}m), given that all electromagnetic waves travel at a speed of 300 000 000 m/s (3×10^8 m/s) in space. *2 marks*
 (iii) State to which part of the electromagnetic spectrum the radiation in part (ii) belongs. *1 mark*

WJEC

18 Radio waves, ultraviolet, visible light and X-rays are all types of electromagnetic radiation.

a) Choose wavelengths from the list to complete the table.

3×10^{-8} m, 1×10^{-11} m,
5×10^{-7} m, 1500 m

Type of radiation	Wavelength
radio waves	
ultraviolet	
visible light	
X-rays	

4 marks

b) Microwaves are another type of electromagnetic radiation. Calculate the frequency of microwaves of wavelength 3 cm. (The velocity of electromagnetic waves is 3×10^8 m/s.) *4 marks*

c) Which type of electromagnetic radiation is used:
   (i) to send information to and from satellites, (ii) in sunbeds, (iii) to kill harmful bacteria in foods? *3 marks*

d) Electromagnetic waves may be diffracted. Explain the meaning of diffraction. *3 marks*

AQA

19 The diagram shows some cooking instructions from a packet of rice.

Cooking instructions for 100 g rice	
Microwave method	**Hot plate method**
1) add 400 cm^3 cold water and salt to taste.	1) add 500 cm^3 cold water and salt to taste.
2) microwave on full power for 10 mins.	2) bring to the boil and simmer for 15 mins.
3) strain and serve.	3) strain and serve.

a) Give one similarity and one difference between infrared and microwave radiation. *2 marks*

b) Microwave radiation is absorbed by water molecules.
   (i) Explain how this cooks the rice. *2 marks*
   (ii) A warning with the microwave stated that all food should be cooked in a container with a loose fitting lid. Explain why. *2 marks*

a) An electric hot plate produces infrared radiation. Describe how the energy from the hot plate reaches the rice. *3 marks*

OCR

20 a) A ray of light enters a glass block as shown.
   (i) Complete the path of the ray of light through the glass block and emerging into the air. *2 marks*
   (ii) What happens to the speed of light as it enters glass? *1 mark*

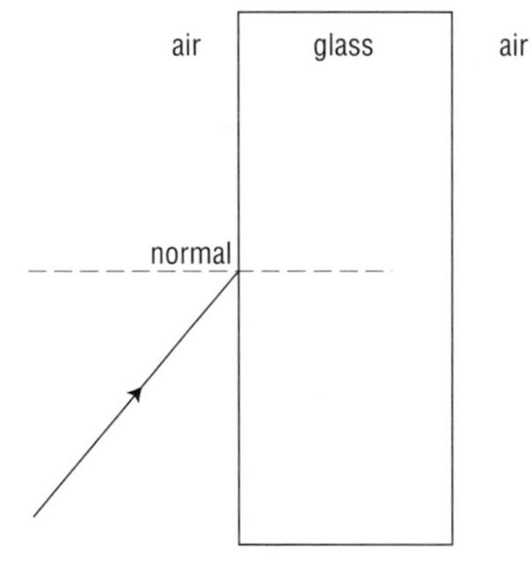

b) A beam of white light from a hot source strikes a triangular glass prism. The diagram on the next page shows the path of a ray of red light passing through and emerging from the glass prism.
   (i) Redraw the diagram and draw the path of a ray of blue light passing through and emerging from the prism. *2 marks*
   (ii) What is the name of this process of splitting white light into different colours? *1 mark*
   (iii) Use the letters UV to mark the region on the screen in your diagram where ultraviolet radiation could be detected. *1 mark*

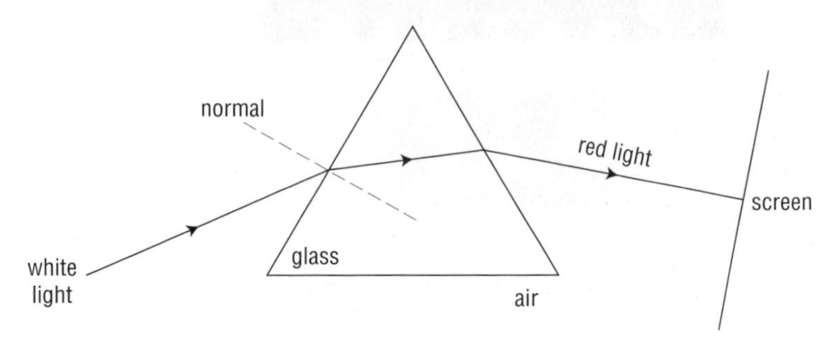

c) The manager of a holiday camp suggests installing ultraviolet lamps in the region of the swimming pool. Some of the holiday-makers object.
   (i) State a reason for installing the ultraviolet lamps. *1 mark*
   (ii) State a reason against installing the ultraviolet lamps. *1 mark*

d) Ultraviolet radiation is not suitable for finding earthquake victims buried by rubble. Suggest a type of radiation which would be suitable for this purpose. *1 mark*

e) Another type of radiation is used by a security firm to check luggage at an airport.
   (i) What type of radiation is used to detect metal objects in the luggage? *1 mark*
   (ii) Why is it unsafe for passengers to be exposed to this radiation? *1 mark*

f) State one property which is common to all parts of the electromagnetic spectrum. *1 mark*

NICCEA

CHAPTER

29

Radioactivity

SUMMARY

1 **Radioactivity (radioactive decay)**

Radioactivity (radioactive decay) is the spontaneous breakdown of unstable atoms. These breakdowns involve the nuclei of the unstable atoms. They are called **nuclear reactions**.

Nuclear reactions involve nuclei in the centre of atoms. Chemical reactions involve electrons in the outer parts of atoms.

There are three kinds of radiation emitted during radioactive decays:

Ionising Power | Penetrating Power

- **α-particles** (helium nuclei, ${}^{4}_{2}He^{2+}$)

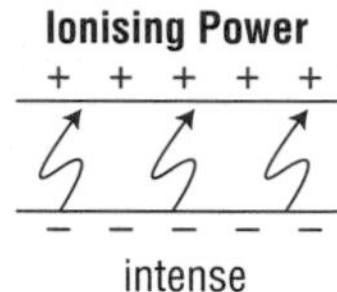

paper

- **β-particles** (electrons, ${}^{0}_{-1}e$)

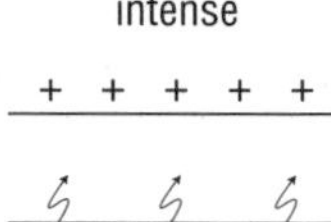

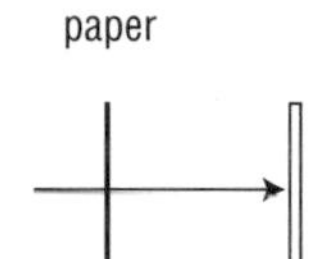

- **ψ-rays** (electromagnetic rays)

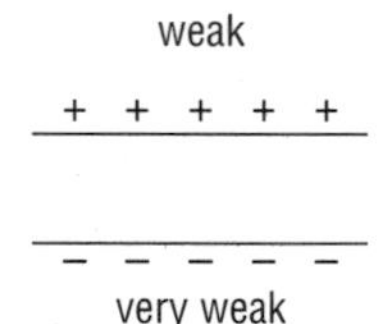

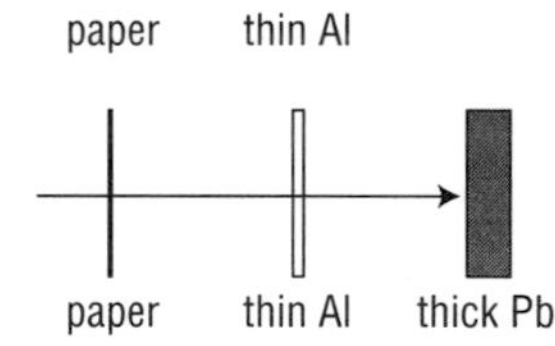

2 **Background radiation** is the natural radiation to which we are exposed. It comes from:

- various *rocks* in the Earth, especially granite
- *the Sun* which emits gamma radiation
- traces of *radioactive materials* in bricks and other building materials

Background radiation is normally very low and causes no risk to our health.

3 **Half-life** is the time it takes for the activity or the amount of a radioactive isotope to halve.

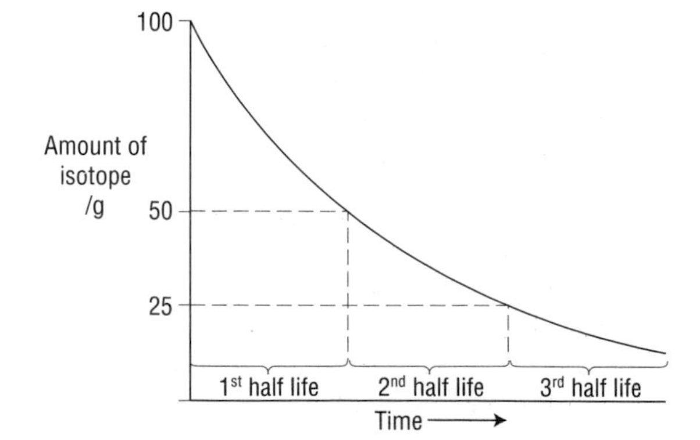

4 Radiation can be:

Harmful		Beneficial
causing • nausea • skin burns • loss of hair • sterility • cancer • death	increasing harm ↓	uses include • treatment of cancers • sterilising medical instruments and dressings • tracking leaks • thickness and level gauges • dating archaeological remains and rocks

STUDY QUESTIONS

Objective questions

Questions 1 to 4

Each line in the table shows the number of neutrons (n), protons (p) and electrons (e) in two particles.

	First particle			Second particle		
A	13n	12p	12e	12n	12p	10e
B	10n	9p	9e	10n	10p	10e
C	12n	10p	10e	10n	10p	10e
D	12n	11p	11e	12n	11p	10e
E	10n	9p	9e	9n	9p	10e

Which line shows

1 two particles with the same mass number?

2 an atom X and a negative ion X^-?

3 two atoms with different atomic numbers?

4 an atom Y and a positive ion Y^{2+}?

Questions 5 and 6

In questions 5 and 6, choose from A, B, C or D which is the correct answer.

5 An element is best defined as a substance which contains
A atoms which are identical.
B atoms with the same atomic number.
C different kinds of atom.
D protons, neutrons and electrons.

6 Natural background radiation comes mainly from
A granite rock in the area.
B holes in the ozone layer.
C nearby nuclear power stations.
D the Sun's activity.

Questions 7 and 8

The paths of four different types of radiation are shown here.

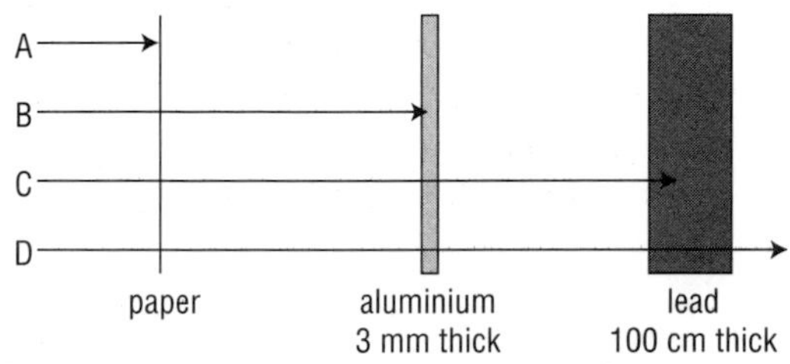

7 Which path is probably that of a β-particle?

8 Which path is probably that of a γ-ray?

Questions 8 to 14

In questions 8 to 14, choose from A, B, C or D which is the correct answer.

9 Carbon-14 is radioactive with a half-life of about 5500 years. The radioactivity of carbon-14 in trees growing today is 100 units. The radioactivity in a prehistoric fossilised tree is 25 units. How old is it?
A 5500 years B (5500 × 2) years
C (5500 × 4) years
D (5500 × 75) years

10 Which one of the following isotopes has 15 protons and 18 neutrons?
A $^{18}_{15}Z$ B $^{33}_{15}Z$ C $^{33}_{18}Z$ D $^{36}_{15}Z$

11 Radioactive substances must be used with care because they give off
A electrically-charged particles.
B protons, neutrons and electrons.
C rays which make substances radioactive.
D rays which damage living cells.

12 Barium sulphate is used in radiography. Which *one* of the following properties of barium sulphate is most important for this use?

A It contains dense Ba^{2+} ions.
B It is a white solid.
C It is an insoluble solid.
D It does not react with acids in the stomach.

Questions 13 and 14

The isotope $^{232}_{90}Th$ loses an alpha particle and then a beta particle.

13 The atomic number of the final isotope is
A 89 B 88 C 87 D 86

14 The mass number of the final isotope is
A 231 B 230 C 229 D 228

Short questions

15 The radioactive isotope plutonium-241 can be used to check whether glass bottles are full.

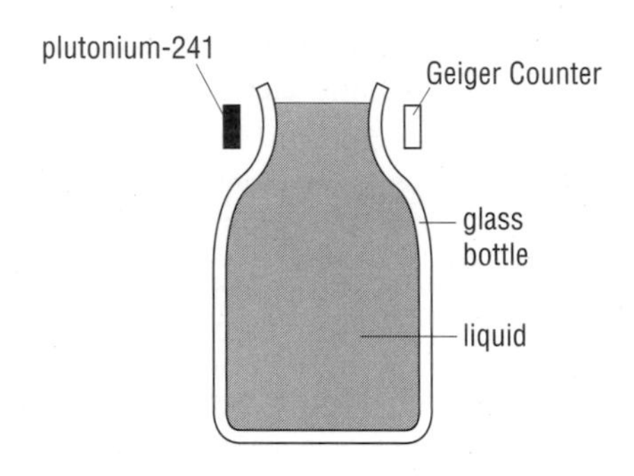

a) What type of radiation is needed for this use? *1 mark*
b) Why is this type of radiation suitable for this use? *1 mark*
c) Explain why the plutonium-241 can be used to check whether the bottles are full. *2 marks*

16 a Name *two* devices for detecting radiation. *2 marks*
b) Say what happens to each device when it is exposed to radiation. *2 marks*
c) Radiation can be dangerous, but it also has some beneficial uses. State *one* harmful effect and *one* beneficial use of radiation. *2 marks*

17 a) Potassium-40 decays by emitting β-particles (electrons).

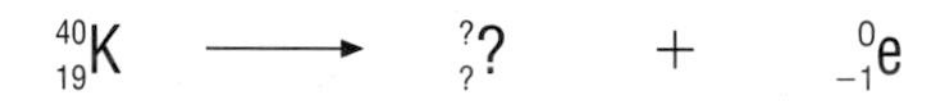

$$^{40}_{19}K \longrightarrow \ ^{?}_{?}? + \ ^{0}_{-1}e$$

Write the symbol and numbers for the particle shown by question marks in the decay equation. You may need to look at a periodic table.
b) A piece of rock contains potassium-40 which has a half-life of 1.3×10^9 years. If 20g of potassium-40 has decayed to 2.5 g, how old is the rock? Explain your working. *3 marks*

18 a) Why does the decay of radioactive isotopes inside the Earth produce heat? *1 mark*
b) Attempts have been made to transfer energy from hot rocks in the Earth's crust to produce hot water. Describe briefly how this might be done. *2 marks*
c) Suggest *one* advantage and *one* disadvantage of using energy from hot rocks in this way. *2 marks*

Further examination questions

19 $^{226}_{88}Ra$ and $^{210}_{82}Pb$ are both formed when $^{238}_{92}U$ undergoes radioactive decay. $^{226}_{88}Ra$ decays by emitting an alpha particle, whilst $^{210}_{82}Pb$ decays by emitting a beta particle.

a) Use the periodic table to state what isotope will be formed by each decay. *2 marks*

b) An extended investigation was carried out to find the half-life of $^{210}_{82}Pb$. The count rate of a sample was measured at various times. The results (corrected for background radiation) are shown in the table.

Time/years	0	10	20	30	40	50
Count rate/Bq	160	112	84	63	44	33

(i) What is meant by count rate, and what instrument is used to measure it? *2 marks*

(ii) Draw a graph to show these results. Plot 'Count rate (Bq)' vertical and 'Time (years)' horizontal. *2 marks*

(iii) From your graph, find the half-life of $^{210}_{82}Pb$. *2 marks*

c) Yellow lead chromate is insoluble in water, and is used as a pigment in some yellow paints. It is made by the reaction of aqueous solutions of lead nitrate and sodium chromate. The formula of each of the ions is shown below.

lead Pb^{2+}, nitrate NO_3^-, sodium Na^+, chromate CrO_4^{2-}

Write an ionic equation including state symbols to show the formation of lead chromate by this method. *2 marks*

d) An experiment was carried out on a sample of lead chromate paint from a painting to find out whether it was an old master, dating back to 1650, or a forgery by a 20th century artist. Lead chromate paint contains small amounts of $^{210}_{82}Pb$. The original count rate of this is 160 Bq per kg of lead paint. The count rate of $^{210}_{82}Pb$ in the sample of lead chromate paint was 20 Bq per kg of paint. Use this, and your value for the half-life of $^{210}_{82}Pb$, to deduce whether the painting was genuine or a forgery. Give reasons for your answer. *3 marks*

EDEXCEL

20 a) The radioactive isotope $^{24}_{11}Na$ (sodium-24) is used as a tracer in some medical procedures.

(i) Copy and complete the table for the nucleus of an atom of sodium-24. *2 marks*

Number of neutrons	
Number of protons	

(ii) Sometimes people who have been badly burned have new skin grafted onto the damaged skin. It is important that blood flows to the new skin or it will die. Suggest how a solution of sodium-24 chloride could be used to check if blood is flowing to the new skin. *2 marks*

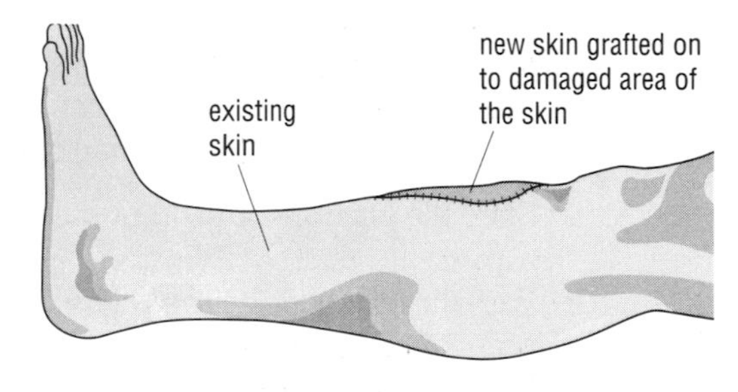

b) (i) The graph shows how the activity of a sample of sodium-24 changes with time.
Use the graph to estimate the half-life of sodium-24 to the nearest hour.
You should show your method on the graph. *2 marks*

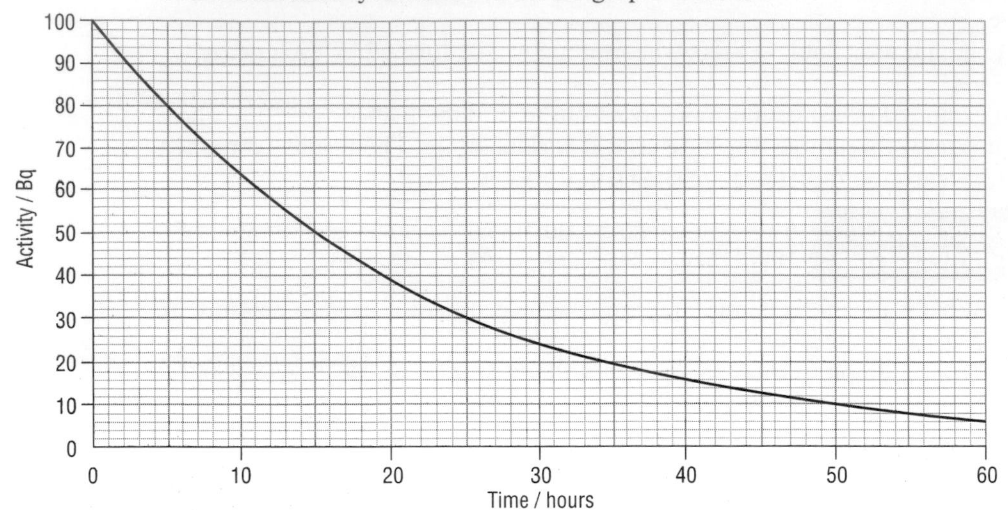

(ii) The following equation shows what happens when a nucleus of sodium-24 decays.
$^{24}_{11}Na \rightarrow ^{X}_{Y}Mg + ^{0}_{-1}e$

What type of ionising radiation is produced? Find values for X and Y.
3 marks

EDEXCEL

21 a) Rewrite and complete the table about atomic particles.

Atomic particle	Relative mass	Relative charge
proton		+1
neutron	1	0
electron	negligible	

2 marks

b) Use the periodic table to help you to answer some parts of this question. Read the following passage about potassium.

Potassium is a metallic element in Group I of the periodic table. It has a proton (atomic) number of 19. Its most common isotope is potassium-39, ($^{39}_{19}K$). Another isotope, potassium-40, ($^{40}_{19}K$), is a radioisotope.

(i) State the number of protons, neutrons and electrons in potassium-39.
2 marks

(ii) Explain why potassium-40 has a different mass number from potassium-39. *1 mark*

(iii) What is meant by a radioisotope? *1 mark*

(iv) Atoms of potassium-40 change into atoms of a different element. This element has a proton (atomic) number of 20 and a mass number of 40. Name, or give a symbol of, this new element. *1 mark*

(v) Explain in terms of atomic structure, why potassium-39 and potassium-40 have the same chemical reactions. *1 mark*

c)

Type of radiation	Change in proton (atomic) number	Change in mass number
alpha α	goes down by 2	goes down by 4
beta β	goes up by 1	no change
gamma γ	no change	no change

Use the table together with the periodic table to help you to answer the following questions.

(i) Name the type of radiation given out in part b)(iv). *1 mark*

(ii) Give the name, or symbol, of the element formed when an atom of sodium-24 (proton number = 11) emits gamma radiation. *1 mark*

(iii) State the group number of the element formed when an atom of radon-220 (proton number = 86) emits an alpha particle. *1 mark*

d) (i) Name a suitable detector that could be used to show that potassium-40 gives out radiation. *1 mark*

(ii) Name a disease which can be caused by too much exposure to a radioactive substance such as potassium-40. *1 mark*

AQA

22 a) A Geiger–Müller tube detects radiation even when there is no radioactive source near it. This radiation is called background radiation.

(i) What is background radiation? *1 mark*

(ii) Give *one* source of background radiation. *1 mark*

b) Americium-241 is a radioactive isotope. Its atomic number is 95.

(i) Work out the number of protons, neutrons and electrons in an atom of americium-241. *3 marks*

(ii) Another isotope of americium has a mass number of 243. What is an isotope? *2 marks*

c) The radioactive isotope radium-226 decays by emitting alpha (α) particles.

(i) What is an alpha (α) particle? *1 mark*

(ii) Explain why a different element is formed when an atom loses an alpha (α) particle. *2 marks*

d) A radioactive isotope is used in pipes so that leaks can be detected without the need for digging.

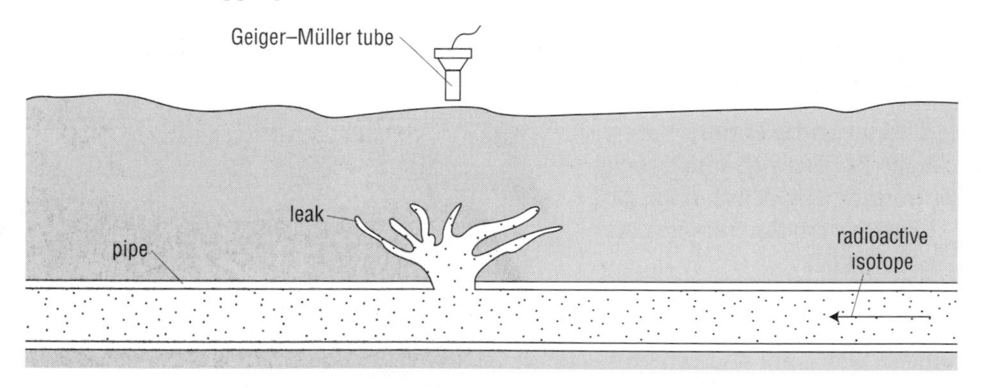

(i) What type of radiation, alpha (α), beta (β) or gamma (γ) should the radioactive isotope emit? Explain the reason for your choice. *3 marks*

(ii) The radioactive isotope should have a half-life of about 15 hours. Explain why. *2 marks*

AQA

CHAPTER

30

The Earth and beyond

SUMMARY

1 Our **Sun** is a small **star**. It is the nearest star to Earth.

2 Millions upon millions of stars cluster together to form a **galaxy** and billions of galaxies make up the whole **universe**. Most of the stars we see at night are in our own galaxy – the Milky Way.

3 The study of stars, planets and other bodies in the sky is called **astronomy**. Astronomical distances are so huge that they are measured in **light years**. A light year is the distance travelled by light in one year (10^{16} metres). The distance across our galaxy (the Milky Way) is 100 000 light years.

4 **The solar system**

The caption in the drawing will help you to remember the order of the planets from the Sun.

- Our solar system has nine **planets** orbiting the Sun. The Sun is huge compared to its planets.
- All the planets move in elliptical orbits in the same direction around the Sun and in the same plane except Pluto whose orbit is at an angle to this plane.
- The four inner planets nearest the Sun – Mercury, Venus, Earth and Mars – have hard, solid and rocky surfaces with thin, gaseous atmospheres.
- The next four planets – Jupiter, Saturn, Uranus and Neptune – are much larger planets with small, rocky cores surrounded by dense, thick gas.
- Pluto, furthest from the Sun, is a very small, rocky planet.
- Between the inner and outer planets, there is a band of rocky bodies (like tiny planets) which are also orbiting the Sun. These are called **asteroids**.

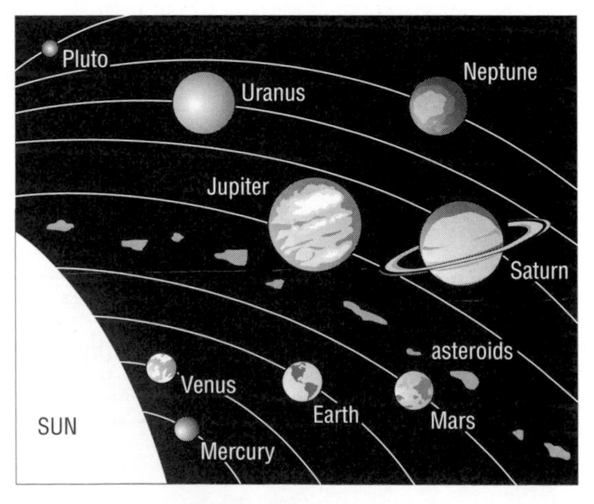

5 The **conditions on a planet** are affected by two major factors:
- the distance from the Sun, which determines the surface temperature and the evaporation of volatile substances
- the relative size of the planet, which determines the gravitational pull on its atmosphere.

6 **Gravitational forces** act between all masses. These gravitational forces increase
- as the masses increase
- as their distance apart decreases.

Gravitational forces from the Sun are very strong and the pull from the Sun's gravity keeps all the planets in their orbits. These forces hold the Moon in orbit around the Earth. Gravitational forces from the Earth cause your weight.

7 A **satellite** is something which orbits a planet. Moons are 'natural' satellites. Artificial satellites have three main uses.
- **Communication**/navigation e.g. TV, radio and telephone signals. They are usually put into orbit high above the equator so that they always remain above the same point on Earth as it spins. They are described as **geostationary**.
- **Monitoring**/observation e.g. weather forecasting, mapping, spying. They are usually put into orbit circling the poles so that they can scan the whole Earth.
- **Exploration** e.g. the Hubble telescope is a satellite orbiting the Earth way above our atmosphere.

8 **The evolution of stars** is shown in the diagram below.

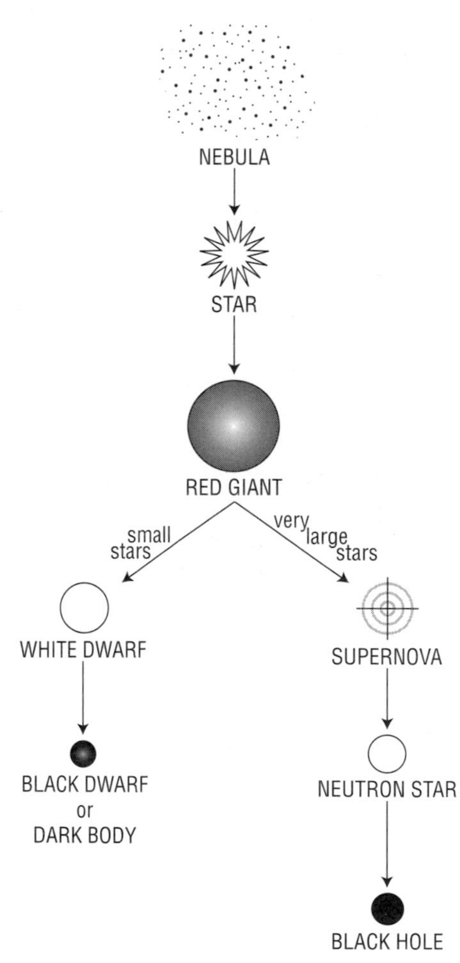

9 **Evolution of the universe**
Edwin Hubble was the first astronomer to investigate other galaxies. Hubble discovered that
- the light from other galaxies had longer wavelengths than expected. Hubble called this the **red shift** because red light has longer wavelengths than other colours. Hubble explained this observation by suggesting that other galaxies are moving away from us.
- the further a galaxy is from Earth, the greater is its red shift. This is called **Hubble's Law**.

Hubble explained this second observation by suggesting that other galaxies further away are moving faster.

Hubble's observations and suggestions support the **'big bang' theory** for the origin of the universe. Since the 'big bang', the universe has been expanding and cooling as galaxies move away from each other.

STUDY QUESTIONS

Objective questions

Questions 1 to 6

In questions 1 to 6, choose from A, B, C or D which is the correct answer.

1 An astronaut travels from the Earth into outer space. Which of the following describes the changes in his weight and mass?

A Mass remains constant, weight remains constant.
B Mass remains constant, weight decreases.
C Mass decreases, weight remains constant.
D Mass decreases, weight decreases.

2 A space ship is travelling in deep space where the gravitational force is zero. If the drive motors are switched off, the space ship will be

A stationary.
B decelerating.
C accelerating.
D moving at constant speed.

3 Which *one* of the following happens once in a year?

A The Earth spins on its axis.
B The Moon orbits the Earth.
C Each planet orbits the Sun.
D The Moon orbits the Sun.

4 Our Sun has a surface temperature of 5800°C. The star Antares has a surface temperature of 3100°C and a diameter 300 times greater than our Sun. Antares is best described as

A a black hole.
B a nebula.
C a red giant.
D a supernova.

5 If you look at Pluto through a telescope, it looks like a star. Shortly after Pluto was discovered in 1930, astronomers realised Pluto was a planet and not a star because

A it moves around the sky.
B it did not produce light.
C it was too close to the Sun.
D it was much too small.

6 The galaxies in Ursa Major, which are 900 light years away, are moving away from us at 15 000 km/s. The galaxies in Corona Borealis which are 1200 light years away are moving away from us at 20 000 km/s. These figures suggest that the speed of a galaxy

A decreases as it moves away.
B is proportional to its distance away.
C depends on its size.
D is less the further it is away.

Questions 7 to 10

Stars pass through different stages of development during their lifetime. Complete the sentences below by choosing the correct words from this list.

nebula neutron star white dwarf
red giant black dwarf

7 About 10 million years ago our Sun began to form by accumulating dust and gas as a __________ .

8 A star that no longer shines is called a __________ .

9 A very dense star with a hot core is a __________ .

10 As hydrogen is used up, the mass of a star gets less, gravitational forces decrease and the star expands to a __________ .

Short questions

11 The diagram shows the orbit of a satellite around the Earth.
The satellite transmits television programmes from the USA to Europe.
This satellite is in geosynchronous (geostationary) orbit.

a) What is meant by geosynchronous orbit? *2 marks*

b) Explain why this is the most suitable type of orbit for a communications satellite. *1 mark*

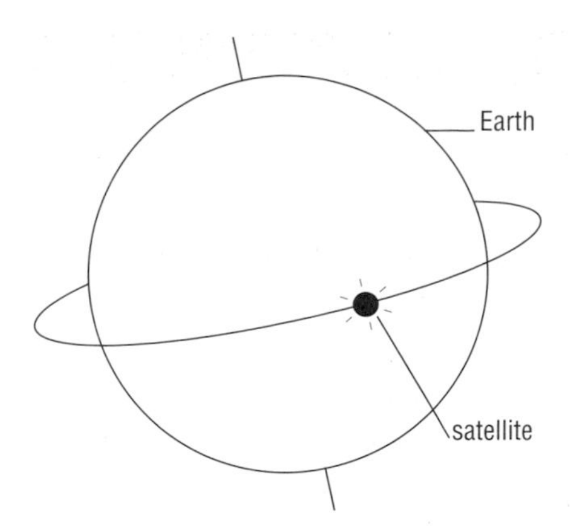

12 A comet moves around the Sun in an elliptical orbit.
Describe and explain changes in speed as the comet moves in its orbit from:

a) A to B *3 marks*

b) B to C *2 marks*

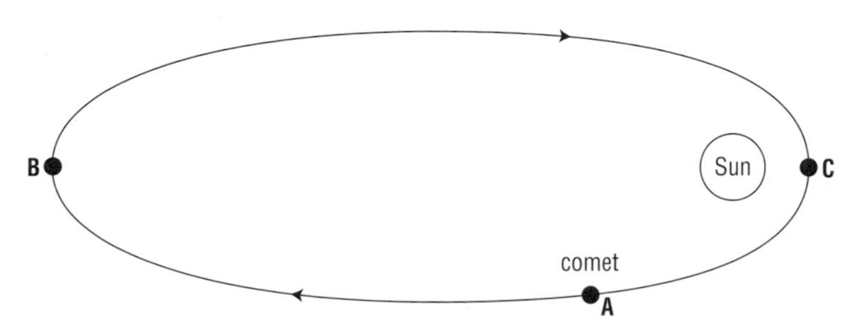

13 The table lists information about seven of the planets.

Planet	Distance from Sun /million km	Radius /km	Average density /g/cm^3
Earth	149	6350	5.52
Jupiter	773	70 960	1.33
Mars	227	3360	3.94
Mercury	58	2400	5.43
Uranus	2886	25 275	1.30
Venus	108	6025	5.24
Neptune	4469	25 200	1.76

Use the information in the table to answer the following questions.

a) On which of these planets will the surface temperature probably be the lowest? *1 mark*

b) Which of the planets will come closest to the Earth? *1 mark*

c) What pattern is there between the size and density of the planets? *1 mark*

d) What pattern is there between the density of the planets and their position in the solar system? *1 mark*

14 a) Explain how stars form from large gas clouds. *5 marks*

b) Use words from the following list to show our Sun's likely evolution.

black body   black hole   neutron star   red giant   supernova   white dwarf

Sun → __________ → __________ → __________ *3 marks*

15 The diagram shows an astronaut wearing a space suit.

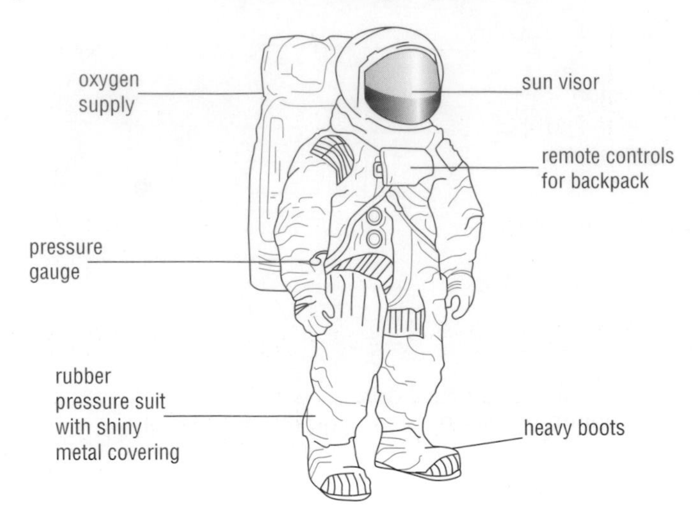

Explain why such a space suit is needed on the Moon. *4 marks*

Further examination questions

16 The table shows the period of orbit for Earth satellites which have different radii of orbit.

Radius of orbit (×1000 km)	Period of orbit (×1000 s)
10	10
20	28
30	50
40	78

a) Plot a graph of period of orbit against radius of orbit. *3 marks*
b) Describe how the period of orbit varies with radius of orbit. *1 mark*
c) (i) What is the radius of orbit for a satellite with a period of orbit of 24 hours? Explain how you have arrived at your answer. *2 marks*
(ii) State and explain a use for such a satellite. *2 marks*
d) How does the speed of an orbiting satellite vary with the radius of its orbit? Explain how you arrived at your answer. *2 marks*
e) The diagram shows an astronaut in an orbiting spacecraft. Explain why the astronaut appears to have no weight even though gravity is keeping the spacecraft in orbit. *3 marks*
f) Training astronauts is an expensive procedure. To simulate weightlessness, an aeroplane may be flown in a special path. Describe the motion that the aeroplane must follow so that the passengers may appear weightless. Explain your reasoning. *3 marks*

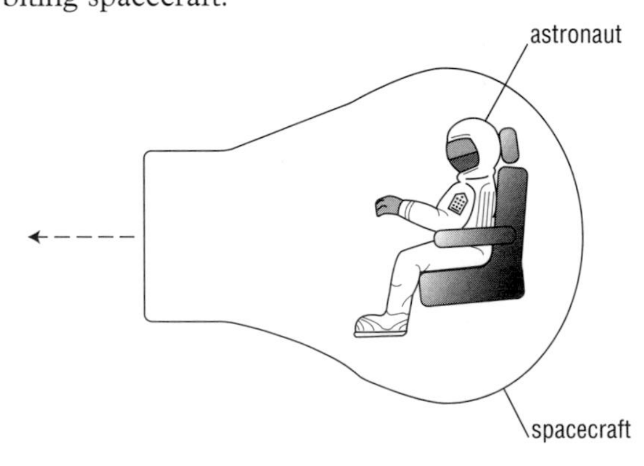

EDEXCEL

17 a) Describe as fully as possible the type of nuclear reaction which explains how the Sun releases its energy. *4 marks*

b) Describe the stages that a star may go through at the end of its stable period. *4 marks*

c) State the evidence which suggests that the solar system was formed from material produced when earlier stars had exploded. *2 marks*

WJEC

18 Scientists are hoping to send a space probe to Mercury. Mercury is a small planet, slightly larger than the Earth's moon. The table compares some features of Mercury and the Earth's moon.

Feature	Mercury	Earth's Moon
atmosphere	none	none
surface temperature	varies between −180°C and +430°C	varies between −170°C and +120°C
density	5.4 g/cm^3	3.3 g/cm^3
magnetic field	strong	very weak
gravitational pull on each kg of mass near the surface	3.7 N	1.6 N

a) Before landing, the space probe would orbit Mercury several times in order to transmit photographs back to Earth. The diagram shows a probe in orbit around Mercury.
The arrow represents the force acting on the probe. Write a description of this force. *2 marks*

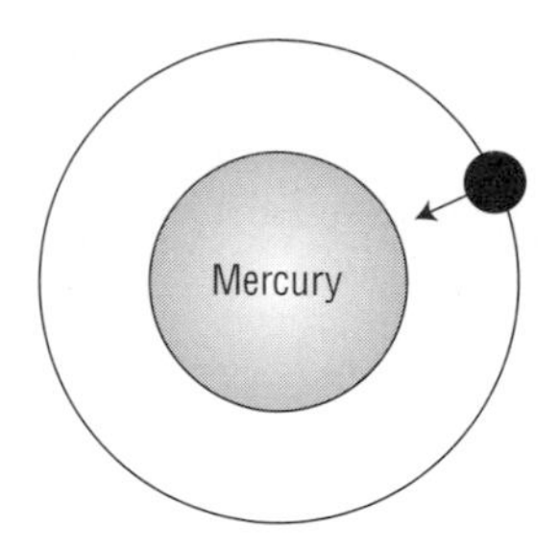

b) A space probe would be travelling at 4000 m/s as it reached the surface of Mercury. Explain why parachutes would not be effective in slowing down the space probe. *1 mark*

c) The space probe would have rocket engines which could be rotated to fire in any direction. Explain how it could use these to slow down as it approached the surface of Mercury. *2 marks*

d) Scientists suspect that Mercury, unlike the Earth's Moon, has an iron core. Give *two* pieces of evidence that support the idea that Mercury might have an iron core. *2 marks*

e) Both Mercury and the Earth's Moon have surfaces covered in impact craters. Many of these craters are millions of years old. Explain how the lack of atmosphere has preserved these craters. *2 marks*

f) Explain why the variation in surface temperature on Mercury is greater than that on the Earth's Moon. *2 marks*

OCR

19 The table gives information about some planets in our solar system.

Planet	Surface type	Average surface temperature /°C	Atmosphere	Average distance from Sun/ km × 10^6
Mercury	rocky, covered in craters	350	none	58.0
Venus	rocky craters, volcanic mountains	480	sulphuric acid, carbon dioxide	108.2
Earth	solid, two thirds covered in water	26	oxygen, nitrogen, carbon dioxide	149.7
Mars	rocky craters and volcanoes	−30	carbon dioxide	228.0

a) (i) Use the information in the table to explain why the surface temperatures on Mercury are different from that on Earth. *1 mark*
(ii) Use the information to explain why the surface temperatures on Venus are different from that on Earth. *2 marks*

b) Between Mars and Jupiter there are several thousand lumps of rock called asteroids. Use the information in the table to help you estimate the following:
(i) the average surface temperature of the asteroids. *1 mark*
(ii) the average distance from the Sun of the asteroids. *1 mark*

c) Deep inside the core of the Sun, hydrogen is being converted to helium. What is the name given to this process? *1 mark*

d) The diagram shows the value of the gravitational force on an object of mass 1 kg at several positions between the Earth and the Moon.

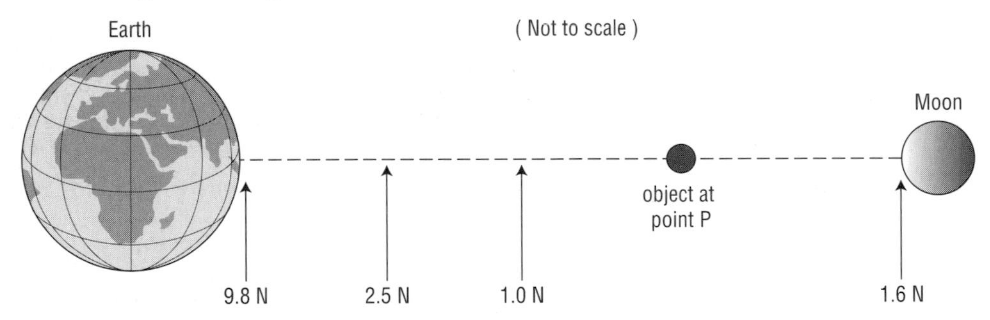

(i) Why is the value of the gravitational force at the surface of the Earth greater than the value at the surface of the Moon? *1 mark*
(ii) Why does the value of the gravitational force acting on the object
- decrease as it moves from the Earth to point P *1 mark*
- increase as it gets close to the Moon? *1 mark*

e) Depending on its mass, a protostar can develop into other types of star. The sentences below describe the formation of a neutron star. Complete the sentences by choosing the correct words from the following.

black dwarf   nova   red giant   red supergiant   supernova   white dwarf

A massive bright blue sun type expands to become a __________ . Next, a giant explosion called a __________ blows most of the outer layers off. The part that remains begins to shrink and becomes a neutron star or black hole. *2 marks*

AQA

Answers

Answers to objective questions (Chapter 1)

1 Cell wall
2 Photosynthesis
3 D
4 E
5 C
6 Day 5
7 7 days
8 Day 7
9 4 days
10 B
11 C
12 A

Answers to short questions (Chapter 1)

13 a) Nutrition *[1]*
b) The release of waste products *[1]*
c) Sensitivity *[1]*
d) Movement, growth, respiration (any two for *[2]*)

14 a) Cell membrane *[1]*
b) Nucleus *[1]*
c) Cytoplasm *[1]*
d) B *[1]*
e) They have a long tail to swim and find an egg/the head contains enzymes to digest the egg membrane so that the nuclei can join. (any one for *[1]*)

15 a) Nucleus
b) Chromosomes
c) Genes
d) Tissues
e) Organs
f) Organisms *[6]*

16 Acts as a boundary/gives the cell shape. *[1]* It controls movement of water and food molecules into and out of the cell. *[1]*

Answers to further examination questions (Chapter 1)

17 a) Both have a cell membrane, cytoplasm and a nucleus. (any two for *[2]*)
b) Cell wall, vacuole or chloroplast (any one for *[1]*)

18 a) 73 beats per minute *[1]*
b) (i) His pulse rate would increase. (ii) His pulse rate would begin to fall back to the resting rate. *[2]*

19 a) Diffusion *[1]*
b) Water concentration is higher outside the cells than inside. *[1]* Therefore, water passes through the partially permeable cell membrane into the root hair cells *[1]* by osmosis. *[1]*
c) Water concentration, in this case, is higher inside the root hair cells of plants than in the sea water which has flooded them. *[1]* This causes a net movement of water molecules out of the root cells, causing the plants to wilt. *[1]*

20 Movement/contact of water animal *[1]* sets off the trigger, releasing the lid. *[1]* Tube pierces skin of water animal and injects poison. *[1]* Barbs can hold the prey while poison incapacitates it. *[1]* (any three for *[3]*)

21 Rubber plant can respire/reproduce/respond to stimuli/grow/photosynthesise. (any three for *[3]*)

22 a) Control all chemical reactions inside the cell *[1]* and play important part in cell division/reproduction. *[1]*
b) Energy producing reactions occur in mitochondria. *[1]*

23 a) Drawing of muscle cell(s) – long, thin and striated, nucleus. *[2]* Groups of muscle cells can relax and elongate. *[1]* Then, they contract causing movement. *[1]*
b) Drawing of nerve cell showing branches, nucleus, long and thin fibre. *[2]* Branches pick up stimuli *[1]* and transfer messages along the long thin fibre. *[1]*

Answers to objective questions (Chapter 2)

1 C
2 E
3 A
4 D
5 D
6 E
7 A
8 C
9 Mouth and small intestine *[2]*
10 Small intestine *[1]*
11 Liver *[1]*
12 Bile *[1]*
13 Salivary glands, stomach, small intestine, pancreas (any three for *[3]* – deduct a mark for each wrong suggestion)
14 Increases surface area for enzyme action. *[1]*
15 Oesophagus *[1]*
16 Roughage/semi-solid indigestible matter *[1]*

Answers to short questions (Chapter 2)

17 a) They provide vitamins A and C. *[1]*
b) Rice from which the husks are removed *[1]*
c) The husks of whole-grain rice contain vitamin B. *[1]*
d) Citrus fruits contain vitamin C, preventing scurvy. *[1]*
e) Aerobic *[1]* respiration *[1]*
f) Smoke pollution prevented Sun shining through, so children deprived of a source of vitamin D. *[2]*

18 a) Each food bar identifiable, *[1]* energy axis labelled *[1]* with kJ units, *[1]* correct plotting of food bars (not points) *[1]*
b) 4500 kJ *[1]*
c) Provides energy/provides insulation/protects vital organs (any two for *[2]*)

d) Vitamin C or minerals [1]

19 a) Axes labelled, [1] scales on axes, [1] points plotted accurately [1]
b) Line of best fit [1]
c) Points accurately plotted, [1] line of best fit [1]
d) The % of starch remaining falls faster at 20°C than 10°C [1]
∴ the rate of amylase activity increases from 10°C to 20°C [1]

Answers to further examination questions (Chapter 2)

20 a) The breakdown of large molecules in food (carbohydrates, fats and proteins) [1] into smaller, soluble molecules [1] to pass into the bloodstream or be stored [1] (any two for [2])
b) The enzyme pepsin in the stomach [1] has begun to break down the protein to peptides [1] leaving fibrous strands of indigestible material. [1] (any two for [2])
c) Enzymes catalyse the chemical reactions involved in digestion. [1]

21 a) Enzymes catalyse the chemical reactions needed to break down food during digestion. [1]
b) It must diffuse [1] through the walls of the small intestine [1] and into the blood vessels serving all the body cells. [1]
c) Glucose is used to produce energy. [1] It combines with oxygen in cellular respiration [1] producing carbon dioxide, water and energy. [1] (any two for [2])

22 a) Vitamins, minerals, fibre (any two for [2])
b) Fresh fruit is rich in vitamins and provides roughage (fibre). [1]
c) Alcohol in excess acts as a poison [1] drugging the brain and hindering digestion. [1]
d) A diet high in animal fat may cause fatty deposits to appear in the lining of blood vessels. [1] This will impede the flow of blood [1] and may eventually cause high blood pressure/heart disease. [1]
e) The fatty food is taken into the mouth and is broken up by chewing. [1] It goes down the oesophagus, into the stomach and then the small intestine. Bile from the liver emulsifies the fat, [1] separating it into smaller droplets. Lipase enzymes in pancreatic juice [1] catalyse the breakdown of fats to fatty acids and glycerol. [1] The fatty acids and glycerol pass through the villi in the small intestine to the bloodstream. [1] Fatty acids and glycerol not needed for energy reform fat [1] which is stored under the skin, providing insulation. (any four for [4])

23 a) (i) Food is broken down by chewing. [1] Saliva containing the enzyme amylase breaks down starch to maltose. [1] (ii) Muscles in the oesophagus push food and liquid down towards the stomach. [1]
b) Sugar : energy, butter : insulation, white bread rolls : carbohydrate, onion : fibre, beefburger : protein (all five correct [3], four correct [2], three correct [1])
c) (i) The fats in butter and cheese [1] can lead to fatty deposits in blood vessels [1]. This can impede the flow of blood [1] resulting in higher blood pressure [1] and heart disease. (any three for [3]) (ii) Vitamins, minerals, fibre [3]

24 a) (i) Yellow-brown [1] (ii) brick red [1]
b) (i) Black (ii) blue [1]
c) Boiling the amylase destroys its enzyme action.[1]
d) An enzyme is a catalyst [1] for a biological reaction. [1]
e) The pH or temperature at which the enzyme works most effectively. [1]

Answers to objective questions (Chapter 3)

1 E
2 C
3 A
4 A
5 B
6 A
7 C
8 D

Answers to short questions (Chapter 3)

9 a) 15 times the normal risk [1]
b) The risk increases [1] in proportion [1] to the number of cigarettes smoked per day. It is highly likely that a person smoking 50 cigarettes a day will die from lung cancer. [1]
c) Sketch graph with axes labelled and scale, [1] gradient of graph half that of graph shown [1]

10 a) Mechanical energy is needed for all muscular movement. [1]
b) Thermal energy/heat [1]
c) When we cannot breathe fast enough/obtain enough oxygen [1] to provide all the energy required by the body [1]

11 a) $C_6H_{12}O_6 \rightarrow 2C_3H_6O_3$ [1]
glucose → lactic acid
b) The reaction is exothermic. [1] 120 kJ of energy [1] are produced when 1 mole (1 formula mass) of $C_6H_{12}O_6$ reacts. [1]

12 a) Without air (oxygen) [1]
b) Aerobic [1]
c) It enables them to survive/obtain energy [1] when the availability of oxygen is limited/absent. [1]

13 a) Yeast metabolises sugar/glucose [1] in the absence of oxygen [1] producing ethanol [1] and carbon dioxide. [1] The gas, CO_2, makes the bread dough rise. [1] (any four for [4])

b) The heat kills the yeast. *[1]*
c) The alcohol evaporates when the bread is baked. *[1]*

Answers to further examination questions (Chapter 3)

14 a) Their cell membrane is only one cell thick. *[1]* Their total number and surface area in contact with capillaries is very large. *[1]* They are clothed in blood capillaries allowing oxygen to diffuse into the bloodstream. *[1]* Their surface is moist which aids the diffusion of oxygen. *[1]* (any three for *[3]*)
b) (i) In the body cells – often muscle cells, *[1]* (ii) cellular respiration *[1]*
c) The brain, *[1]* breathing, *[1]* fall, *[1]* homeostasis *[1]*

15 a) Organs of respiration/the lungs *[1]*
b) The TB bacteria invade body cells and multiply *[1]* if the white blood cells are not effective against the infection. *[1]*
c) Viruses *[1]*
d) The disease is caused by droplet infection (i.e. drops of moisture from a cough or sneeze) containing bacteria *[1]* so TB will spread in crowded conditions. *[1]*
e) Phagocytes in the blood *[1]* engulf the bacteria *[1]* and secrete an enzyme to kill them. *[1]* OR Lymphocytes in the blood *[1]* produce chemicals called antibodies, *[1]* killing the bacteria and neutralising their harmful effects. *[1]*
f) Lung cancer is not caused by microbes (bacteria). *[1]* It is due to abnormal cell division which can be due to the carcinogen, nicotine, in cigarettes. *[1]*

Answers to objective questions (Chapter 4)

1	Arteries	**7**	B
2	Veins	**8**	D
3	Muscle	**9**	A
4	Atria	**10**	B
5	Ventricles	**11**	B
6	A	**12**	C

Answers to short questions (Chapter 4)

13 a) Phagocyte *[1]*
b) The phagocyte approaches the bacteria, engulfs them *[1]* and secretes an enzyme which kills them. *[1]*
c) Antibodies *[1]*

14 a) Digestive system – reduced
skin – increased
brain – unchanged
arteries of heart – increased
muscles of skeleton – increased
bone – reduced
(All six correct *[4]*, five correct *[3]*, four correct *[2]*, three correct *[1]*)
b) During exercise, there is a large overall increase in supply of blood to the body *[1]*, and especially for muscles. *[1]*

15 a) An artery *[1]*
b) A vein *[1]*
c) Oxygen and nutrients *[2]*
d) The lungs cannot be used as the foetus is bathed in fluid inside the uterus. *[1]*
e) If the blood supply is reduced, the foetus will not grow so well. *[1]* If it is cut off, the foetus will die and the cow will abort. *[1]*

16 a) 625 ± 10 *[1]*
b) Steady increase in successful operations from 1983 to 1989, *[1]* then a jump to approximate constant successes from 1990 to 1993. *[1]*
c) 1990 *[1]*

Answers to further examination questions (Chapter 4)

17 a) Phagocyte *[1]*
b) Phagocyte cell passes through the cell wall of the blood vessel *[1]* to engulf bacteria, *[1]* secreting an enzyme to kill them. *[1]* (any two for *[2]*)
c) Transport oxygen *[1]* plasma *[1]*

18 a) Radioactive solution has been diluted $\frac{7350}{15}$ times *[1]* = 490 times *[1]*
$\therefore$ 10 cm^3 of the radioactive solution has been diluted to 490×10 cm^3 *[1]*
$\therefore$ volume of patient's blood = 4900 cm^3 *[1]*
b) (i) Radioactive substance may be retained by the cells of different organs/radioactive substance is decaying/some of the radioactive substance may have been excreted in the urine/the radioactive substance may not be spread evenly throughout the blood. (any three for *[3]*)
(ii) Due to retention by cells, excretion and decay of the radioactive substance, the count from the blood sample will be too low. *[1]* $\therefore$ The dilution factor will be too large and the blood volume calculation will be too high. *[1]*
c) (i) A radioactive substance circulating in the blood is harmful to cells, *[1]* so its use in diagnosis must be as short as possible. *[1]*
(ii) Choose a substance which is not chemically toxic/choose a substance which is readily excreted/use a relatively low dosage. (any two for *[2]*)

19 a) Arteries and veins in centre of the foot are wider to enable a good flow of blood. *[1]* Veins closer to the bear's base of foot are narrow to reduce heat loss. *[1]*
b) The surface blood vessels dilate *[1]* to allow heat to be lost by conduction. *[1]*

c) By keeping still, they lose no energy in moving *[1]* so more energy goes into growing and keeping warm *[1]*/fat-rich milk enables a layer of fat to be laid down under their skin *[1]* for insulation *[1]*/by huddling together they decrease their total surface area *[1]* so less heat is lost from their bodies. *[1]* (any two for *[4]*)

20 a) The heart muscle wastes away. *[1]*
b) A heart attack/coronary thrombosis *[1]*
c) Have a low fat diet *[1]*
d) Infection requires more white blood cells. *[1]* Body processes speed up to produce them *[1]* and so the heart needs to beat faster. *[2]*
e) Exercise, *[1]* fright, *[1]* anxiety, *[1]* (any two for *[2]*)

21 a) White blood cells – phagocytes and lymphocytes *[1]*
b) Phagocytes engulf microbes *[1]*/phagocytes secrete enzymes which kill microbes. *[1]*/Lymphocytes produce antibodies which kill microbes. *[1]*/Lymphocytes neutralise poisonous chemicals. *[1]* (any three for *[3]*)
c) By having the disease itself (such as measles), the person becomes immune to further attacks. *[1]*/By having an injection of a weakened or similar virus, antibodies are produced against a disease. *[1]*
d) (i) The first infection caused severe illness because Susan had no antibodies. *[1]* (ii) Antibodies remained in Susan's blood after the first infection, even after 79 days, and were able to reduce the symptoms. *[1]* (iii) The third infection did not cause illness because the antibodies were much more concentrated. *[1]*

Answers to objective questions (Chapter 5)

1 C
2 B
3 E
4 A
5 Light (sunlight)
6 Carbon dioxide and water vapour
7 Oxygen
8 Starch
9 Water
10 Glucose or sugars

Answers to short questions (Chapter 5)

11 a) (i) June *[1]* (ii) Total light falling on the wood is greatest in June. *[1]*
b) (i) April *[1]* (ii) Light reaching ground level is greatest in April. *[1]*

12 a) Rate increases from W to X, *[1]* because CO_2 is a reactant in photosynthesis *[1]* and its concentration increases from W to X. *[1]*
b) The plants are using CO_2 as fast as possible *[1]* and some other factor (e.g. temperature, light, water) is limiting the rate of photosynthesis. *[1]*
c) It tells them the optimum concentration of CO_2 *[1]* above which photosynthesis will not increase *[1]* given the other conditions in the greenhouse.

13 Make sure the greenhouse is well lit (maybe night and day), *[1]* provide the plants with plentiful water, *[1]* increase the CO_2 concentration in the greenhouse, *[1]* provide the plants with other key nutrients (i.e. N, P and K). *[1]* (any three for *[3]*)

14 a) Axes labelled, *[1]* scales on axes, *[1]* points plotted accurately, *[1]* smooth curve of best fit. *[1]*
b) Dusk marked at 18 ± 1 hours on the graph. *[1]*

15 a) Either kills cells or softens the leaf *[1]*
b) Removes chlorophyll *[1]* which is soluble in ethanol
c) Washes/cleans the leaf *[1]*
d) Starch in the leaf produces a dark blue/black colour *[1]* with iodine.

Answers to further examination questions (Chapter 5)

16 a) (i) Side of ridge *[1]* (ii) Side of ridge *[1]*
b) Plants on the side of the ridge were at higher temperatures (e.g. >16°C) for the longest part of the day. Those on the top of the ridge were at higher tempcratures for the next longest time and plants on flat soil were at higher temperatures for the shortest time. *[1]* So plants on the side of the ridge grew fastest, then those on top of ridge and poorest growth on flat soil. *[1]*
c) The light (sunlight) intensity/the water provided/the kind of soil/any nutrients added to the soil. (any two for *[2]*)
d) To avoid poor growing conditions early or late in the season/to have an early crop to sell at a higher price/to be able to use the land for another crop sooner than before. (any two for *[2]*)
e) C *[1]*
f) Type 2 corn plants contain less chlorophyll. *[1]* Chlorophyll absorbs sunlight and is needed for photosynthesis to occur. *[1]* $\therefore$ Photosynthesis is slower in Type 2 plants.

17 a) Darkness prevents further starch being made because photosynthesis stops. *[1]* During the dark period, respiration will occur and use up any starch which is already there. *[1]*
b) (i) Chlorophyll *[1]* (ii) oxygen *[1]*
c) Disc Q_3 had no CO_2 available for photosynthesis, thus no starch will form in Q_3. *[1]* Disc Q_2 was between the split bung and effectively in darkness. $\therefore$ Photosynthesis could

not occur there [1] and no starch was produced.
Discs P_2 and Q_2 were able to photosynthesise and produce starch. [1]

18 a) A condition which will affect (and limit) the rate of photosynthesis [1]
b) Temperature/light intensity/pressure of air bubbled in or volume of air bubbled in (any two for [2])
c) (i) Gas A – oxygen, [1] gas B – carbon dioxide [1]
(ii) From midnight until dawn (say 6 a.m.) respiration occurs but not photosynthesis. ∴ O_2 is used up and CO_2 is produced. [1]
From about 6 a.m. and until dusk (say 6 p.m.) both respiration and photosynthesis will occur, [1] so O_2 will go through a minimum and CO_2 will go through a maximum. [1]
Between dusk (say 6 p.m.) and midnight, respiration continues but photosynthesis stops. [1] So O_2 gets to a maximum and CO_2 gets to a minimum. (any four for [4])
(iii) Poor light intensity (cloudy day) [1]
(iv) The concentrations of both CO_2 and O_2 would remain constant. [1]

Answers to objective questions (Chapter 6)

1 F
2 B
3 C
4 D
5 A
6 E
7 Nitrate ion or ammonium ion
8 Root hair cells
9 Xylem
10 Active transport
11 Amino acids or proteins or nucleic acids

Answers to short questions (Chapter 6)

12 a) Phloem and xylem [2]
b) Rate of water transport will be reduced, [1] water uptake occurs via root hairs – severing the roots will therefore reduce the uptake, [1] and water may be lost from the cuts. [1]
c) (i) It may diffuse out of the root cells. [1]
(ii) Keep the roots damp and cover with soil. [1]
d) Roots in an air space will die back. The tree itself will not be properly anchored in the soil bed. [1]

13 a) Water is required for all processes that occur in the plant. [1] The water stream carries dissolved mineral nutrients to all parts of the plant. [1] Water is needed to give plants support by turgor. [1] (any two for [2])
b) Water can pass into the root hair cells of a plant through the partially permeable cell membrane. [1] Soil water outside the roots contains lower concentrations of dissolved solutes than the cytoplasm inside the root hair cells. [1] These dissolved solutes hamper the passage of water molecules through the partially permeable membrane, [1] so more water molecules pass into the hair cells than out. [1]

14 a) (i) Y [1] (ii) Water is lost from leaves via the stomata and there are more stomata on the lower surface. [1] If vaseline covers the lower surface, it will reduce water loss. [1]
b) Loss of water vapour to the air was equal to gain of water vapour from air. [1]
c) Loss from untreated leaf
= loss from Y + loss from Z [1]
= 0.01 + 0.025 = 0.035 g [1]

Answers to further examination questions (Chapter 6)

15 a) Because of their uptake by the tomato plants [1]
b) (i) As the concentration is increased, the amount of nutrients absorbed increases rapidly to a maximum value. [1] Further increases in oxygen concentration do not increase the amount of nutrients absorbed. [1]
(ii) Small concentrations of oxygen cause a larger absorption of nutrients by rice (approx. four times the effect). [1] The uptake by rice does not reach a plateau like lettuce after about 5% oxygen, but rises at a slower rate. [1]
(iii) Oxygen is required for respiration and growth. [1] As oxygen concentration increases, the amount of nutrients will increase. [1]
(iv) Via transport through the phloem tissue after absorption by the leaves [1]
(v) Temperature affects the rates of reactions and hence the rate of photosynthesis/growth in plants. [1] At lower winter temperatures, the plants grow slower and so the nutrient uptake is reduced. [1]

16 a) Support, [1] carrying/transporting water (and minerals) from the roots [1]
b) Food is transported from the leaves to the roots for growth. [1] The grooves cause a partial blockage to the food going further down, so a build up of food just above the groove causes extra growth. [1]
c) It has become thicker above the groove than below it. [1]
d) (i) Bark cells
(ii) Xylem cells, [1] they are long and thin like capillary tubes, [1] and they have extra thick walls. [1]
(iii) Phloem cells, [1] they contain a nucleus plus cytoplasm. [1]

(iv) Xylem *[1]*

17 a) (i) Axes labelled – Time (horizontally), Mass of plant/g (vertically) *[1]* scales on axes, *[1]* points plotted accurately, *[1]* smooth curve through points *[1]*
(ii) 355.5 ± 0.5 g
(iii) Mass lost between 12 noon and 4 p.m. = 358 − 347 = 11 g ∴ Average rate of water loss = $\frac{11\text{ g}}{4\text{ hr}}$ = 2.75 g/hour
b) (i) All plant processes (e.g. photosynthesis, transport) either require water or aqueous conditions. *[1]*
(ii) Either reduce rate of photosynthesis *[1]*/or reduce transpiration by closing stomata *[1]*/or growing a thicker, more waxy epidermis *[1]* (any one for *[1]*)

Answers to objective questions (Chapter 7)

1 D
2 C
3 A
4 E
5 F
6 Lens
7 Iris
8 & 9 A and F
10 B
11 H
12 & 13 E and G

Answers to short questions (Chapter 7)

14 a)Motor neurone, b) the central nervous system, c) to stretch from parts of a limb to the spinal cord, d) to insulate the neurone (nerve cell) from other neurones

15 a) Driver sees dog (stimulus) *[1]*/receptors in her eyes detect stimulus *[1]*/impulse passes along sensory neurone (optic nerve) to brain *[1]*/impulse passes via relay neurone and/or pyramidal neurones to motor neurone *[1]*/motor neurone carries impulse to effector (muscle) *[1]*/effector (muscle) in foot slams on brake (response). *[1]* (any five for *[5]*)

16 a) Temperature, *[1]* pressure (touch) *[1]*
b) Reflex action *[1]*
c) They are instantaneous/very rapid *[1]*
d) Our fingers need to be much more sensitive *[1]* to different kinds of stimuli. *[1]* Evolution has caused more nerve endings to be developed in the tips of our fingers. *[1]* (any two for *[2]*)

17 a) Crying *[1]*, b) hunger/thirst *[1]*, c) cooing or smiling *[1]*, d) talk/sing/wave, jingle a toy (any two for *[2]*)

Answers to further examination questions (Chapter 7)

18 a) C, A, F, E, B, D *[6]* if all in correct order; one letter out of order *[5]*; two letters out of order *[3]*
b) Nerve endings (receptors) are sensitive to stimuli. *[1]* (e.g. receptors in the retina of the eye are sensitive to light) *[1]* The stimuli create impulses (like minute electric currents) *[1]* which pass along the neurone. *[1]* (any three for *[3]*)

19 a) Receptors *[1]*, b) tiny electrical *[1]* impulses (currents) *[1]*, c) muscles in the leg *[1]*, d) bright light (stimulus) *[1]* → receptors in the eye (retina) *[1]* → sensory neurone carries impulse to brain *[1]* → relay neurones/ pyramidal neurones in brain (co-ordinator) *[1]* → motor neurone carries impulse to muscle (effector) *[1]* → muscle in eye lid makes you blink (response) *[1]* (any five for *[5]*)

20 a) A = cornea, B = optic nerve *[2]*, b) the lens *[1]*, c) Stimulus of bright light affects receptors in the retina. *[1]* Impulse passes along sensory neurones in optic nerve to brain. *[1]* Brain co-ordinates impulse and a return impulse passes along a motor neurone to muscles in the iris. *[1]* Iris extends (response) and covers more of lens. *[1]*
d) Rods – sensitive to low light intensity, *[1]* cones – provide detail and colour vision. *[1]*
e) (i) Synapse *[1]*, (ii) Impulse reaches the end of a neurone, *[1]* chemicals released into the synapse *[1]* stimulate the next neurone to produce an impulse. *[1]*

21 a) Named reflex action *[1]* e.g. withdrawing finger from hot object. Description of path should include: stimulus, *[1]* receptors, *[1]* sensory neurone carries impulse to CNS, *[1]* motor neurone carries return impulse, *[1]* to effector *[1]* which produces a response.
b) They are automatic/inborn/rapid/ uncontrollable. (any two for *[2]*)

22 a) (i) Reflex action *[1]*, (ii) It prevents further damage/discomfort. *[1]*
b) Sting from bee acts as a stimulus, *[1]* receptors in the toe detect the stimulus, *[1]* sensory neurone carries impulse to CNS, *[1]* motor neurone carries return impulse to muscle (effector), *[1]* muscle contracts as a response. *[1]* (any four for *[4]*)
c) The backbone *[1]* d) Riding a bicycle is a conscious action and requires co-ordination by the brain. *[1]* e) System of endocrine glands which produce hormones. *[1]*

Answers to objective questions (Chapter 8)

1 The endocrine system or the hormonal system
2 C
3 D
4 B
5 The pituitary gland
6 Oestrogen and progesterone
7 A

8 F
9 D
10 It secretes sweat
11 Blood
12 It gets narrower/thinner/constricts

Answers to short questions (Chapter 8)

13 a) After 17¾ days and 28¼ days
b) Progesterone
c) Ovaries
d) Menstruation commences
e) An egg has been fertilised.

14 a) Axes labelled, *[1]* scales on axes, *[1]* points plotted correctly, *[1]* line of best fit *[1]*
b) At first, the increase in auxin concentration causes large percentage increases in length. *[1]* As the auxin concentration gets larger, the % increases in length do not increase so much. *[1]*

15 a) Taking a meal *[1]*
b) (i) It falls from about 37°C to about 35.5°C. *[1]*
(ii) Metabolism slows down so respiration, which generates body heat, slows down and temperatures fall. *[1]*
c) (i) It rises from 35.5°C to 35.65°C. *[1]*
(ii) The body is less relaxed, metabolism speeds up a little and more heat is generated. *[1]*

16 a) When we get hot, sweat glands secrete sweat (mainly water) onto our skin. *[1]* The sweat (water) evaporates from our skin. *[1]* As it does so, it takes heat needed for evaporation from the skin and cools us down. *[1]*
b) When we become cold, shivering causes our muscles to contract and relax. *[1]* This activity stimulates respiration in muscle cells. *[1]* This is an exothermic process, so we warm up. *[1]*

17 a) The person has had a meal containing carbohydrate. *[1]*
b) Insulin *[1]*
c) Insulin is secreted from the pancreas into the blood. *[1]* It then causes liver cells to convert glucose into glycogen. *[1]* The glycogen is then stored in the liver. *[1]*
d) Glucagon *[1]*
e) Adrenaline increases heart beat and the rate of metabolism. *[1]* So, carbohydrates are broken down to glucose faster *[1]* and the blood carries glucose to our muscles faster. *[1]* (any two for *[2]*)

18 a) The dummy patches act as a control. *[1]*
b) More convenient/dose of testosterone spread over a period of time (either point for *[1]*)
c) (i) May lead to unwanted bone growth/muscle development/strength/sex drive (any one for *[1]*)
(ii) Excessive load on heart/lungs etc. *[1]*

Answers to further examination questions (Chapter 8)

19 a) Kidney removes waste materials from the blood. *[1]* Bladder collects urine. *[1]*
b) (i) Renal artery has thick elastic walls to withstand the pressure from the pumping action of the heart. *[1]* Renal vein does not need to withstand such pressure and is wider for easy exit of blood. *[1]*
(ii) Urea is removed, *[1]* salts are removed, *[1]* water is regulated. *[1]*
(iii) Blood enters kidneys via renal arteries which branch into a network of capillaries. *[1]* These capillaries are wrapped around thousands of tiny looped tubules. *[1]* As blood flows in the capillaries over the tubules, urea and other waste substances diffuse into the tubules, forming urine. *[1]*
c) (i) The membrane acts as a partially permeable membrane, *[1]* allowing particles in the blood *[1]* to diffuse through the membrane into or out of the blood. *[1]*
(ii) Salts will diffuse through the membrane to make their concentration in the blood *[1]* the same as that in the dialysis fluid. *[1]*
d) The stimulus is the water content of the blood. *[1]* The receptor is the brain. *[1]* The effector is the hormone secreted to affect cells in the kidney tubule. *[1]*

20 a) At 5½ days and at 17 days
b) Menstruation *[1]*
c) (i) Oestrogen promotes repair/build up of lining of uterus (or oestrogen inhibits the production of FSH). *[1]*
(ii) Progesterone maintains the lining of the uterus.
d) The progesterone level stayed high. *[1]*
e) From day 16, graph rises until about day 22 to 24 where it is at a maximum. *[1]* It then falls sharply to the base level (6 units) at about day 28 to 30. *[1]*
f) It increases in thickness and in blood supply *[1]*/to receive the fertilised egg, provide it with plenty of nutrients *[1]*/and protect it. *[1]*/The fertilised egg attaches itself to the uterus. A placenta forms which takes over production of progesterone *[1]*/to maintain the lining of the uterus. *[1]* (any four for *[4]*)

21 a) (i) The pancreas is not secreting enough insulin *[1]* to control levels of glucose in the blood. *[1]*
(ii) Insulin causes the liver to remove sugar (glucose) from the blood *[1]* and convert it to glycogen *[1]* which is then stored in the liver. Too much insulin will reduce sugar (glucose) to an abnormally low level. *[1]*
(iii) Exercise requires energy *[1]* which is provided by chemical respiration in our body

cells *[1]* and this process uses sugar (glucose) taken from the blood. *[1]*
b) Sugar is easily digested (converted) to glucose, *[1]* absorbed into the blood and then taken up by cells for respiration. *[1]* Starchy foods are digested and converted to glucose much more slowly. *[1]*
c) When the blood sugar level becomes too low:
• the pancreas reduces its release of insulin *[1]* which reduces the conversion of glucose (sugar) to glycogen. *[1]*
• the pancreas also produces a second hormone glucagon *[1]* which promotes the conversion of glycogen to glucose. *[1]* (any three for *[3]*)

22 a) An untreated shoot: mass when fresh = 60 g
mass after drying = 24 g
⇒ mass of water present = 36 g *[1]*

$\therefore$ % water = $\frac{36}{60} \times 100$ = 60% *[1]*

b) Growth of shoots is promoted by five times. *[1]* Excluding water, the growth is about three times. *[1]* Growth of roots is inhibited to about half. *[1]* Excluding water, the growth is unchanged. *[1]*
c) No. *[1]* The hormone has no effect on the growth of the root tissue. *[1]* It also reduces the water content of the fresh roots so the yield will have a smaller mass *[1]* and produce a lower cash return to the farmer. *[1]* (any three for *[3]*)

Answers to objective questions (Chapter 9)

1 Oak tree
2 Lacewing larvae
3 4
4 The Sun/sunlight
5 The direction/movement of food/energy
6 Lacewing larvae
7 Nitrogen fixation/nitrogen fixing bacteria
8 Feeding/eating
9 Excretion or death
10 Decomposition/decay/denitrifying bacteria
11 Temperatures will rise
12 Greenhouse effect
13 Flooding/damage in coastal areas

Answers to short questions (Chapter 9)

14 a)

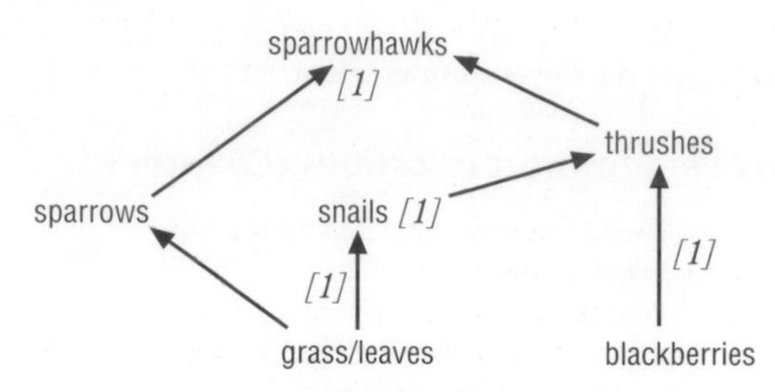

b) (i) Arrows from all other organisms to 'Decomposers' *[1]*
(ii) They decompose dead material from animals and plants *[1]* and return it to the environment as CO_2, water and nitrogen. *[1]* This decomposition is crucial for the recycling of materials. *[1]* (any two for *[2]*)

15 a) Cattle *[1]* b) It is lost to the environment. *[1]*
c) % energy transferred = $\frac{12}{5000} \times 100$ *[1]* = 0.24% *[1]*
d) Cattle in the shed don't lose heat energy to the surroundings *[1]* or use energy to move around so much. *[1]*

16 a) $\frac{900 \times 100}{1\,500\,000}$ *[1]* = 0.06% *[1]*
b) Respiration, *[1]* excretion, *[1]* decay/decomposition *[1]*
c) It may pass to decomposers *[1]* or microbes *[1]* as the decaying materials are converted to CO_2, water and nitrogen. *[1]* (any two for *[2]*)

17 a) Tunnels increase the surface area of the cow pat *[1]* where decomposers start to break down the cow pat. They also make oxygen more available for decomposition. *[1]*
b) Decomposers break down materials in the cow pat releasing CO_2, water, *[1]* nitrogen compounds and minerals. *[1]* These are precisely the materials which plants need for photosynthesis and growth. *[1]* The decomposition processes are exothermic which keeps the temperature at the edge of the cow pat higher for faster metabolism and growth. *[1]* (any three for *[3]*)

18 a) Any two of fungi/worms/bacteria/viruses/insects *[2]*
b) There would be reduced amounts of nitrites and nitrates. *[1]* Ammonium salts and dead leaves would accumulate. *[1]*
c) Increased numbers of decomposers/plentiful supply of oxygen/moist conditions/temperatures close to 37°C. (any two for *[2]*)

Answers to further examination questions (Chapter 9)

19 a) Seeds and plant shoots → voles → barn owls *[2]*
b) Fewer barns *[1]* for nest sites *[1]* OR disturbance of owl habitat by humans *[1]* hence unsuccessful nesting *[1]* OR fewer voles along roadsides *[1]* which are killed by cars/killed by pesticides. *[1]*
c) Vole numbers went up as there were fewer owls (predators). *[1]* Vole numbers levelled off as the owl numbers levelled off. *[1]*
d) Carry out the surveys more frequently. *[1]* Employ more people in their surveys. *[1]*

20 a)

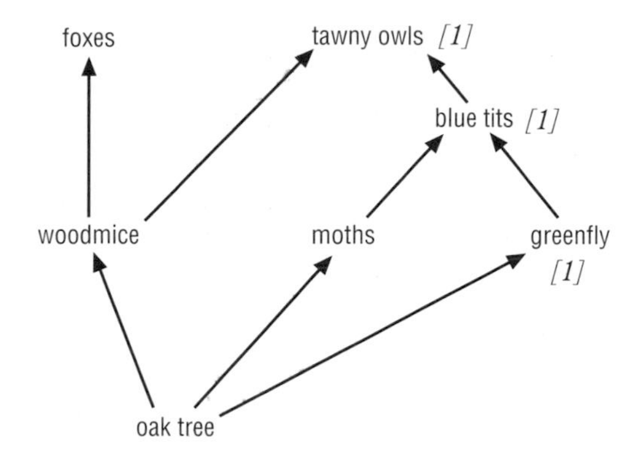

b) (i) Fewer greenfly result in far fewer blue tits, so moths are not taken so much. *[1]*
(ii) Fewer greenfly mean that blue tits take more moths. *[1]*
c)

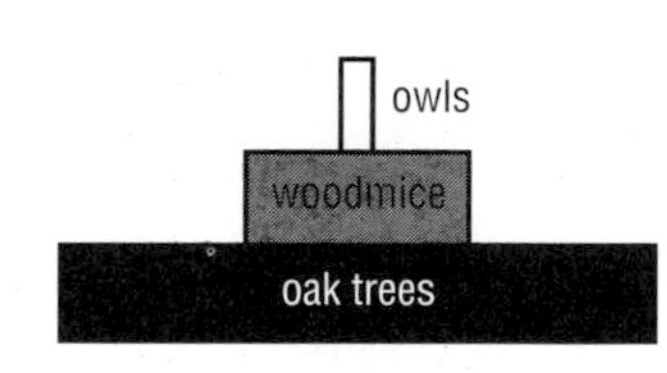

[1]

d) Large eyes improve hunting at night, *[1]* pointed, powerful beak to kill/tear up prey, *[1]* sharp, large claws to take/hold prey. *[1]*

21 a) (i) Animal dropping, *[1]* dead worm *[1]*
(ii) Aluminium can, *[1]* broken bottle *[1]*
b) (i) In a cupboard *[1]*
(ii) Temperature needs to stay near to about 35°C. *[1]* Fridge is much too cold, radiator is too hot. *[1]* Bread needs to remain soft and damp. *[1]* It would get hard and cold in the fridge and hard and hot by the radiator. *[1]* (any three for *[3]*)
(iii) Water to keep it moist. *[1]*

22 a) Respiration, *[1]* excretion, *[1]* decomposition/decay *[1]* (any two for *[2]*)
b) Primary consumers are likely to increase *[1]* as more food is available due to increased photosynthesis. *[1]*
c) Peas are leguminous plants with nodules containing nitrogen fixing bacteria *[1]* which convert nitrogen in the air into amino acids and proteins. *[1]* The roots of the pea plants are usually left in the soil to decay and leave the amino acids and proteins for the barley the next year. *[1]*
d) As dead plant material decomposes, bacteria convert it to ammonia *[1]* and ammonium salts. *[1]* Nitrifying bacteria then convert the ammonia and ammonium salts to nitrates, *[1]* $NH_3/NH_4^+ \rightarrow NO_3^-$. *[1]*

Answers to objective questions (Chapter 10)

1	D and F	**7**	D
2	A and E	**8**	B
3	B and C	**9**	E
4	B	**10**	A
5	F	**11**	C
6	C	**12**	B

Answers to short questions (Chapter 10)

13 a) (i) Mitosis *[1]* (ii) meiosis *[1]*
b) During fertilisation, egg cells containing 18 chromosomes fuse with sperm cells containing 18 chromosomes *[1]* to form a zygote containing the normal 36 chromosomes of hen cells. *[1]*

14 Fertilisation, *[1]* zygote, *[1]* DNA (deoxyribonucleic acid), *[1]* genes *[1]*

15 a) Meiosis *[1]* b) In meiosis, the homologous pairs of chromosomes separate quite randomly. *[1]* c) Suppose the homologous pairs are Aa Bb Cc. The possible gametes have either A or a plus either B or b and either C or c. *[1]* i.e. $2 \times 2 \times 2 = 8$ different gametes. *[1]*

Answers to further examination questions (Chapter 10)

16 a) To attract the female *[1]*/to facilitate release of the sperm sac *[1]*/to move the sperm sac towards the female. *[1]* (any two for *[2]*)
b) (i) Fertilisation is much more likely, or eggs and sperm are not wasted. *[1]*
(ii) Frogs or any fish *[1]*
c) (i) Growth is not restricted *[1]*/embryos are independent *[1]*/allows more offspring from the parents. *[1]* (any one for *[1]*)
(ii) Embryos are vulnerable to predators or harsh conditions. *[1]*
(iii) Human or any other mammal *[1]*
d) The cells divide again and again, *[1]* different cells develop different functions. *[1]*

17 a) They originate from the same single

embryonic cell *[1]* and so they contain exactly the same chromosomes (genes). *[1]*
b) The 'foster' eggs did not contribute any chromosomes (genetic identity) to the lambs *[1]* because they had had their original nuclei removed. *[1]*
c) It reduces the total genetic bank for future use. *[1]* Any genetic defect may affect large numbers of animals with disastrous consequences. *[1]* As species evolve over long periods, adaptation is less likely if the genetic bank is restricted. *[1]*

18 a) There are equal numbers of chromosomes in the body cells and in the fertilised egg. *[1]*
b) Chromosomes first replicate to form chromatids. *[1]* Nuclear membrane breaks and homologous pairs of chromatids come to the centre of the cell. *[1]* Homologous chromatids separate, move to opposite ends of the cell as it divides. *[1]* Chromatids move to middle of the two new cells. *[1]* Chromatids split and the separate chromosomes move to opposite ends of the cells and the second cell division occurs. *[1]* The nuclear membrane reforms around each set of chromosomes. *[1]* Each of the four new cells has only half as many chromosomes as the original parent cell. *[1]* (any four for *[4]*)

Answers to objective questions (Chapter 11)

1 D
2 A
3 C
4 B
5 AS
6 SS
7 AA or AS
8 R and S
9 Sufferer
10 Carrier or normal
11 C
12 B
13 A
14 C

Answers to short questions (Chapter 11)

15 a) (i) 4 *[1]* (ii) 8 *[1]* (iii) 8 *[1]*
b) Gametes (sperm cells and eggs) have half the number of chromosomes *[1]* so that they have the correct number when sperm and egg fuse to form a zygote. *[1]*
c) (i) Mutation *[1]* (ii) A chromosome may get broken into two *[1]* or the chemical structure of the chromosome may be changed by radiation or toxic chemicals. *[1]*

16 a) A cold pad caused the fur to become black. A warm pad caused the fur to become white. *[1]*
b) The rabbit's bulky body remains fairly warm so the body fur stays white. *[1]* The ears, feet, nose and tail at the extremities lose heat easily, become cold and their fur becomes black. *[1]*
c) Warm pads on the ears of the parent rabbits would make those parts white. However, all their offspring had black ears which suggests that the species naturally has black ears, *[1]* which are affected by environmental conditions. *[1]*

17 a) The dominant colour will tend to predominate in the species. Pale organisms can have the genotype PP or Pp. Dark organisms can only be pp. *[1]*
b) He needs to catch a much larger number of the offspring to be sure that the pale variety is dominant. *[1]* Other factors in the environment and simply chance may have influenced which colour of butterfly he caught. *[1]*
c) (i) PP or Pp *[1]* (ii) pp *[1]* (iii) Pp or pp *[1]*

18 a) Selecting animals or plants *[1]* with desirable characteristics *[1]* one generation after another *[1]* (any two for *[2]*)
b) Two examples (two for *[2]*)
c) Variation *[1]*
d) Meiosis *[1]* due to random separation of chromosomes. *[1]* Fertilisation *[1]* brings together two entirely different sets of chromosomes. *[1]* Mutation *[1]* causes a change in the structure of a gene. *[1]* (any two for *[2]*)

Answers to further examination questions (Chapter 11)

19 a)

Female parent \ Male parent	B	B
b	Bb	Bb
b	Bb	Bb

[1]

Possible genotypes of all (F1) offspring are Bb
∴ all black *[1]*

Next generation

Parents	B	b
B	BB	Bb
b	Bb	bb

[1]

Ratio of genotypes in (F2) offspring are BB : Bb : bb = 1 : 2 : 1
∴ 3 black to every 1 white (i.e. $\frac{3}{4}$ black, $\frac{1}{4}$ white)

b) (i) Genotypes of parents should be Bb and bb. *[1]*
(ii) A male from the F1 offspring with any white female or a female from the F1 offspring with any white male. *[1]*
c) (i)

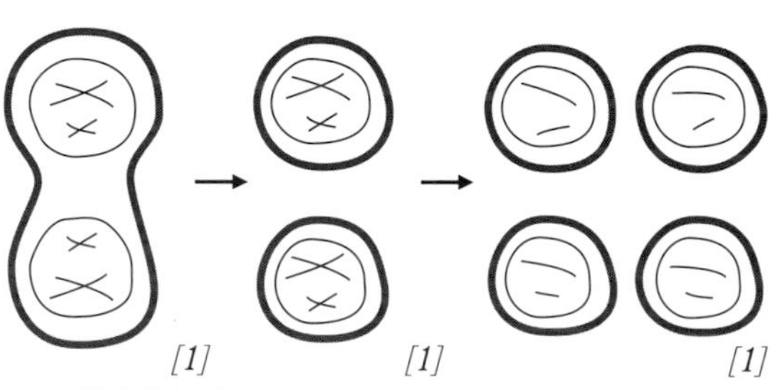

(ii) Meiosis *[1]*
(iii) Variation can occur: when the chromatid pairs separate at the first cell division, *[1]* when

the gametes produced at the second cell division fuse with gametes from the other parent on fertilisation. *[1]*
d) A change in the chemical structure of a gene *[1]*

20 a) The beak *[1]*
b) They are best adapted to obtaining different foods. *[1]*
c) Variations arise in organisms from the same species. *[1]* Some variations are inherited (genetic). *[1]* Other variations are caused by the environment. *[1]* Some of these variations help the individual organisms to survive more successfully *[1]* leading to a natural selection – 'survival of the fittest'. *[1]* The offspring from these best adapted organisms are also more likely to survive so their parents are more successful in reproducing. *[1]* (any four for *[4]*)

21 a) *[4] [2]*

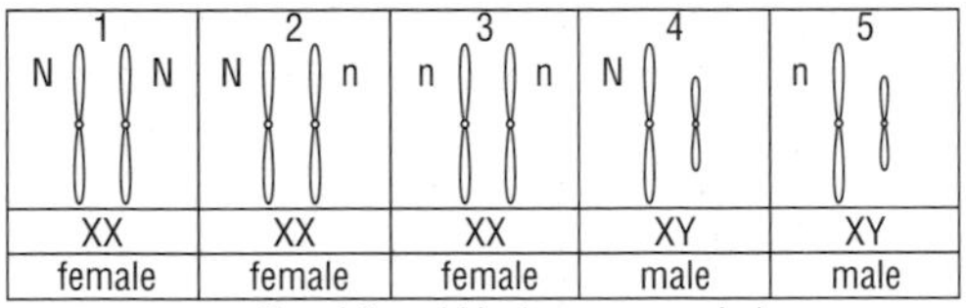

1	2	3	4	5
N N	N n	n n	N	n
XX	XX	XX	XY	XY
female	female	female	male	male

(2 and 3 could be in reverse order)

b) 22 *[1]*
c) Female haemophiliac (person 4) must have the genotype nn. *[1]* Carrier female (person 3) must have the genotype Nn. *[1]* Female 1 must be able to supply both N and n since the male Y chromosome supplies neither. Therefore, female 1 is Nn. *[1]*
d) A haemophiliac female would have to have the recessive allele (n) on both her X chromosomes. *[1]* Her father must also be a haemophiliac and her mother a carrier which is a rarity. *[1]*

22 a) (i) Chromosomes *[1]*
(ii)

Genotype	Phenotype	Homo/Hetero zygous	
bb		homozygous	*[1]*
	brown		*[1]*
	brown	heterozygous	*[1]*

b) Bb and Bb *[1]*
c)

	b	b
B	Bb	Bb
b	bb	bb

If genotype is Bb, about half offspring are white *[1]*

[1]

	b	b
B	Bb	Bb
B	Bb	Bb

If genotype is BB, then no offspring are white *[1]*

[1]

23 a) (i) rr *[1]* (ii) Rr *[1]* or RR *[1]*
b) (i) Red *[1]* (ii) Rr *[1]*
c)

	R	r
R	RR	Rr
r	Rr	rr

Genotypes of offspring are RR : Rr : rr *[1]*
in the ratio 1 : 2 : 1
number of red offspring = 75
number of white offspring = 25 *[1]*

Answers to objective questions (Chapter 12)

1 C
2 A and F
3 D
4 B
5 Madtom
6 Salamander
7 Skink
8 D
9 B
10 C
11 D
12 C

Answers to short questions (Chapter 12)

13 a) (i) Mammals *[1]* (ii) furry (hairy) or young fed on milk *[1]*
b) Supply of food plentiful/warm weather *[1]*
c) The extra food will provide a store of energy needed during hibernation. *[1]*

14 Temperature *[1]*/amount of food *[1]*/space *[1]*/type of flooring *[1]*/type of food *[1]*/frequency of feeding *[1]*/growth promoters *[1]* (any four for *[4]*)

15 a) 2500 cm^2 *[1]*
b) 0.25 m^2 *[1]*
c) $\frac{50}{6.25}$ *[1]* = 8 per m^2 *[1]*
d) Soil quality *[1]*/fertiliser used *[1]*/water supply *[1]*/thickness of grass *[1]* (any two for *[2]*)

16 a) Dandelions *[1]* more than three times as many grow 0.5 m from road compared to 2.5 m away. *[1]*
b) Possible factors – CO_2 in exhaust fumes *[1]* will improve plant growth/photosynthesis *[1]*/SO_2 or NO or NO_2 in exhaust fumes *[1]* will hinder plant growth. *[1]*/Soot from exhaust fumes *[1]* will coat plants and hinder photosynthesis and growth. *[1]*/Movement of traffic and air will help seed dispersal *[1]* and increase the number of plants. *[1]* (any two for *[2]*)
c) The combined effects which help and those which hinder plant growth give an optimum position for thistle growth at 2 m from the road. *[1]*

17 a) Temperature or sunlight. *[1]* Lower temperature of lake water in Britain, especially in winter, would not allow enzyme processes in the algae to continue. Reduced hours of sunlight in Britain, especially in winter, would not allow photosynthesis to sustain sufficient algae. (either point for *[2]*)
b) (i) Sewage in lake will discolour water, impede photosynthesis, *[1]* reduce algae and other water plants on which the flamingos, fish and other aquatic animals feed. *[1]* This would lead to a reduction in their numbers. Sewage in the lake will be decomposed by the respiration of bacteria and other decomposers. *[1]* This process will remove oxygen from the water *[1]* to the detriment of algae, fish, water plants and bacteria. *[1]* Large numbers of aquatic plants and animals may die leaving a rotting, stinking mess. *[1]* (any four for *[4]*)
(ii) The industrial expansion will provide jobs, *[1]* improve the standard of living for the local population, *[1]* lead to better community relations, *[1]* bring increased wealth to the area for improvements in housing, education, medical care, sanitation, etc. *[1]* (any three for *[3]*)

18 a) Infra-red radiation penetrates the atmosphere and reaches the Earth. *[1]* This causes the Earth and the atmosphere to warm up. *[1]* The Earth emits radiation but it is much less penetrating than that from the Sun and it cannot pass through 'greenhouse gases' like CO_2 as easily as it passes through clean air. *[1]* This causes the temperature of the Earth to rise as the concentrations of greenhouse gases increase. *[1]* (any three for *[3]*)
b) Greenhouse effect: CO_2
$= 0.035 \times 1 = 0.035$
Greenhouse effect: N_2O
$= 0.00003 \times 160 = 0.0048$
Greenhouse effect: $CH_4 = 0.0051$
Greenhouse effect: $H_2O = 0.1$
calculation of greenhouse effects *[3]*
The two gases with the most powerful greenhouse effect – CO_2 and H_2O *[1]*
c) Temperatures on the Earth will rise, *[1]* polar ice caps will melt, *[1]* leading to higher sea levels and coastal flooding. *[1]*
d) Reduce CO_2 levels in the atmosphere by burning less fossil fuels in power stations/industry, *[1]* improving the efficiency of vehicle engines, *[1]* using energy sources that do not produce CO_2, *[1]* restrict the use of CFCs. *[1]* (any two for *[2]*)

19 a) Contribution of methane is 3 parts in 20
[1] $= \frac{3 \times 100}{20} = 15\%$ *[1]*
b) Burning fossil fuels in our homes *[1]*/in industry (power stations) *[1]*/ in vehicles *[1]* (any two for *[2]*)
c) Unburnt fuel *[1]*
d) See answer to Q 19 a)

20 a) More useable land for crops or animals, *[1]* ∴ increased profits *[1]*/ larger fields mean more efficient use of machinery *[1]*/ less likelihood of weeds spreading from hedges *[1]* so better crop quality *[1]* (any two for *[2]*)
b) Removal of wind breaks/tendency towards a monoculture/fewer pollinating insects/less interesting landscape – possible adverse affect on tourism/birds and small mammals disappear creating imbalance in ecosystem. (any two for *[2]*)
c) (i) Artificial fertilisers – may change the soil pH/may harm plants and animals in the soil/allow elements not required by plants to accumulate. (any two for *[2]*)
(ii) Artificial fertilisers are washed off the land into rivers. Algae and water plants grow more rapidly, *[1]* using up oxygen in the water for respiration. *[1]* The removal of oxygen from the water causes aquatic plants and animals to die and the river becomes a stinking sewer. *[1]*

Answers to objective questions (Chapter 13)

1 D
2 C
3 D
4 D
5 C
6 A
7 B
8 A
9 C
10 D

Answers to short questions (Chapter 13)

11 a) Bubble into lime water *[1]* which goes cloudy *[1]*
b) Condense the water vapour *[1]*, add anhydrous $CuSO_4$ *[1]* which turns blue *[1]*
c) Insert a glowing splint *[1]* which is rekindled. *[1]*

12 a) It is hard/malleable/fairly cheap (any two for *[2]*)
b) It rusts/needs repainting/is not inexpensive (any two for *[2]*)
c) Plastics are easily moulded/light in weight/very cheap (any two for *[2]*)

13 a) A = oxygen, B = copper oxide, C = graphite, D = carbon dioxide, E = copper carbonate *[5]*
b) Copper + oxygen → copper oxide *[1]*
carbon (graphite) + oxygen → carbon dioxide *[1]*
copper carbonate → copper oxide + carbon dioxide *[2]*

14 Measure out 10 cm^3 of sea water using a measuring cylinder. *[1]* Evaporate very carefully/slowly in an evaporating basin. *[1]*

Be careful to avoid any spitting as the last drops of water evaporate. *[1]* Allow the hot dry residue plus evaporating basin to cool. *[1]* Weigh the dry evaporating basin + residue. *[1]* Wash the residue from the evaporating basin. *[1]* Dry the evaporating basin and reweigh. *[1]* The difference in the two masses is the mass of solid dissolved in 10 cm^3 of sea water. *[1]* (any six for *[6]*)

Answers to further examination questions (Chapter 13)

15 a) B, C, E *[3]*
b) E *[1]*, c) D *[1]*, d) F *[1]*, e) C *[1]*,
f) 1330°C *[1]*
g) A = bromine, B = cadmium, C = gold, D = krypton, E = lithium, F = sulphur *[6]*

16 a) To spread out the heat/heat more evenly *[1]*
b) To help to condense the vapours *[1]*
c) It would be more efficient *[1]* as vapours are contained/cooled more fully. *[1]*
d) Fraction 1 (20–70°C) *[1]*
e) Fraction 1 (20–70°C) *[1]*
f) Fraction 1 (20–70°C) *[1]*
g) Fraction 1 (20–70°C) *[1]*
h) Fraction 4 (170–220°C) *[1]*
i) Fraction 4 (170–220°C) *[1]*
j) It retains solids dissolved in the oil/solids which have not vaporised. *[1]*

17 a) Two different materials *[1]* which work together to produce a more suitable product than either of the separate materials *[1]*
b) (i) It is low in density/light. *[1]* It is relatively strong. *[1]*
(ii) Steel/It is stiff and low in cost. *[2]*
c) (i) Glass is cheap/relatively hard (though brittle) (any one for *[1]*)
(ii) Plastic in the laminate holds the cracked glass in place. *[1]* Plastic has the pliability to allow the laminate to change shape without shattering. *[1]*

18 a) (i) Universal indicator *[1]*, (ii) sulphur *[1]*, (iii) It is both basic and acidic *[1]*
b) X contains both carbon *[1]* and hydrogen. *[1]*

Answers to objective questions (Chapter 14)

1	E	4	D
2	C	5	A
3	B	6	C
7	C	9	D
8	A	10	B

Answers to short questions (Chapter 14)

11 a) (i) Hydrogen, oxygen, azote (any two for *[2]*)
(ii) Mercury *[1]*
(iii) Carbon, phosphorus, sulphur (any two for *[2]*)
(iv) Iron, zinc, copper, lead, silver, gold (any four for *[4]*)
b) magnesia – magnesium oxide
lime – calcium oxide
soda – sodium hydroxide
potash – potassium hydroxide
strontium – strontium sulphate
barytes – barium sulphate (any two for *[2]*)
c) Nitrogen *[1]*
d) Hydrogen = 1 *[1]*

12 a) (i) The gases will mix *[1]*
(ii) Gas particles are moving rapidly *[1]* through all the space available. *[1]*
(iii) Diffusion *[1]*
b) As the container is warmed, more energy is given to the particles. *[1]* The particles therefore move faster *[1]* hitting the walls of the container harder and more often *[1]* so the pressure rises. *[1]*

13 a) (i) An oxygen atom *[1]*
(ii) A nitrogen atom *[1]*
b) Molecules contain two or more atoms chemically joined together *[1]* and particles of nitrogen and oxygen each contain two atoms. *[1]*
c) The gas particles are moving rapidly/throughout their whole container/bombarding each other and the walls of the container. (any two for *[2]*)
d) A molecule *[1]* of nitrogen oxide *[1]*

14 a) Potassium nitrate *[1]* and lead chloride *[1]*
b) Potassium nitrate is an aqueous solution/lead chloride is a solid/2 moles of KNO_3 are produced with 1 mole of $PbCl_2$. (any two for *[2]*)

15 a) $Mg + Cl_2 \rightarrow MgCl_2$ *[1]*
b) $4Al + 3O_2 \rightarrow 2Al_2O_3$ *[2]*
c) $CuCO_3 \rightarrow CuO + CO_2$ *[1]*
d) $CaO + H_2O \rightarrow Ca(OH)_2$ *[1]*
e) $Fe + 2HCl \rightarrow FeCl_2 + H_2$ *[2]*

16 a) $2Fe + 3Cl_2 \rightarrow 2FeCl_3$ *[1]*
b) $2NH_3 + H_2SO_4 \rightarrow (NH_4)_2SO_4$ *[1]*
c) $CH_4 + 2O_2 \rightarrow CO_2 + 2H_2O$ *[1]*
d) $2K + 2H_2O \rightarrow 2KOH + H_2$ *[1]*
e) $4FeS + 7O_2 \rightarrow 2Fe_2O_3 + 4SO_2$ *[1]*

Answers to further examination questions (Chapter 14)

17 a) $M_r(NaCl) = 23.0 + 35.5 = 58.5$ *[1]*
b) $2NaCl \rightarrow 2Na + Cl_2$
117 g *[1]* 46 g *[1]* 71 g
∴ 234 g NaCl produces 92 g Na *[1]*
c) 71 g Cl_2 *[1]* are produced in the same time as 46 g Na *[1]*

$\therefore$ 142 g Cl_2 are produced whilst 92 g Na are formed *[1]*
d) At 25°C and 1 atmosphere, 1 mole of Cl_2 occupies 24 dm^3 *[1]*
number of moles of Cl_2 produced = 2 *[1]*
$\therefore$ volume of this Cl_2 at 25°C and 1 atm = 48 dm^3 *[1]*

18 a) Lead oxide + hydrogen → lead + water (hydrogen oxide) *[1]*
b) Mass of lead produced
= 54.11 − 51.00 = 3.11 g *[1]*
mass of oxygen in the lead oxide
= 54.5 − 54.11 = 0.48 g *[1]*
number of moles of Pb : O

$$= \frac{3.11}{207} : \frac{0.48}{16} \quad [1]$$

= 0.015 : 0.03 *[1]*
= 1 : 2 *[1]*
$\therefore$ formula = PbO_2 *[1]*

19 a) Exothermic *[1]*
b) (i) $CaO + H_2O \rightarrow Ca(OH)_2$
56 g *[1]* 18 g *[1]*
(ii) The heat produced caused some of the water added to vaporise. *[1]*
(iii) Wear safety spectacles/wear gloves/wear lab coat (any one for *[1]*) calcium oxide (quicklime) is very alkaline *[1]*
c) Iron is a better conductor than brick. *[1]* It therefore loses heat much more quickly than brick, so the baking process would not be completed. *[1]*

20 a) (i) $H_2SO_4 + Ca(OH)_2 \rightarrow CaSO_4$ *[1]* $+ 2H_2O$ *[1]*
(ii) Neutralisation *[1]*
b) (i) $M_r(H_2SO_4)$ =
$2 \times 1 + 32 + 4 \times 32 = 98$ *[1]*
(ii) $H_2SO_4 + Ca(OH)_2 \rightarrow \ldots$
98 g $\quad 40 + 2(16 + 1)$
= 74 g *[1]*
74 g $Ca(OH)_2$ reacts with 98 g H_2SO_4 *[1]*
$\frac{74}{98}$ kg $Ca(OH)_2$ reacts with 1 kg H_2SO_4 *[1]*
$\Rightarrow \frac{74}{\not{98}_2} \times \not{4}\not{9}^{1}00\,000$ reacts with
4 900 000 kg H_2SO_4
= 3 700 000 kg $Ca(OH)_2$
(iii) It would be very difficult to transport such a large mass. It would be very difficult to mix the calcium hydroxide throughout the lake. Insoluble $CaSO_4$ would be produced. Excess $Ca(OH)_2$ in areas of the lake could make the water alkaline. (any two for *[2]*)

Answers to objective questions (Chapter 15)

1	A	9	E
2	C	10	A
3	E	11	B
4	D	12	D
5	B	13	A
6	D	14	C
7	C	15	A
8	A	16	C

Answers to short questions (Chapter 15)

17 a) (i) The cut surface will be shiny. *[1]*
(ii) The shiny, 'silvery' surface will go dull white. *[1]* The sodium reacts with oxygen *[1]* in the air to form white sodium oxide. *[1]*
b) $4Na(s) + O_2(g)$ *[1]* $\rightarrow 2Na_2O(s)$ *[1]*

18 a) Shiny appearance when freshly cut/good conductor/solid at room temperature/malleable (any two for *[2]*)
b) Low density/low melting point/low boiling point/can be cut with a knife (any two for *[2]*)

19 a) An ore is an impure substance containing a metal *[1]* and found in rocks or in the Earth. *[1]*
b) (i) Reduction
(ii) tin(IV) oxide + carbon → tin *[1]* + carbon monoxide *[1]*
or tin(IV) oxide + carbon → tin + carbon dioxide
c) Any Group I or Group II metal or aluminium *[1]*

20 a) Sodium, magnesium, zinc, copper *[2]* (one metal out of order *[1]*)
b) (i) Magnesium + steam (water) → magnesium oxide + hydrogen *[2]*
$Mg + H_2O \rightarrow MgO + H_2$ *[2]*
c) (i) Zinc + dilute sulphuric acid → zinc sulphate + hydrogen *[2]*
(ii) $Zn + H_2SO_4 \rightarrow ZnSO_4 + H_2$ *[2]*

Answers to further examination questions (Chapter 15)

21 a) A, good conductivity, low density, no reaction with water (*[2]* for 3 points, *[1]* for 2 points)
b) E, low melting point, low density, high specific heat capacity (*[2]* for 3 points, *[1]* for 2 points)
c) F, low melting point, no reaction with water, good conductivity (*[2]* for 3 points, *[1]* for 2 points)
d) H, very high melting point, low specific heat capacity, poor conductivity (*[2]* for 2 points, *[1]* for 1 point)

22 a) Floats on water/moves/dashes around water surface/melts to form a silvery ball/hissing

noise/steam given off/colour of solution changes from yellow-green to blue (any three for *[3]*)
b) $2Na(s) + 2H_2O(l) \rightarrow 2NaOH(aq) + H_2(g)$ *[2]*
c) Dissolved in water *[1]*
d) Sodium hydroxide solution, *[1]* hydrogen *[1]*
e) Wear safety spectacles, *[1]* work from behind a safety screen *[1]*
f) Storage – below (paraffin) oil *[1]* explanation – the oil prevents the sodium reacting with oxygen or water vapour in the air *[1]*

23 a) An impure substance containing a metal *[1]* and found in rocks or in the Earth. *[1]*
b) Decomposition
c) (i) Filtration/filtering *[1]*
(ii) It fades/becomes paler *[1]*
(iii) $Cu^{2+} + \underline{2e^-} \rightarrow \underline{Cu}$
$\underline{4}OH^- \rightarrow \underline{2}H_2O + O_2 + \underline{4e^-}$ (any four of the underlined points for *[4]*)

24 a) (i) Lead *[1]* – high density, *[1]* excellent resistance to corrosion *[1]*
(ii) To increase the strength *[1]*
(iii) It has a low density *[1]*
b) (i) Copper 85%, tin 15% *[1]*
(ii) A) The atoms of copper can move fairly easily from one position to another when a force is applied. *[1]*
B) The larger tin atoms in the bronze do not allow the atoms to pack so closely together. *[1]*
C) The larger tin atoms break up the regular packing and do not allow the atoms to move so easily from one position to another when a force is applied. *[1]*
c) Galvanising coats the steel with zinc *[1]* so that the steel does not corrode.

Answers to objective questions (Chapter 16)

1	E	**8**	D	**15**	A
2	D	**9**	A	**16**	B
3	B	**10**	D	**17**	A
4	C	**11**	D	**18**	D
5	E	**12**	C	**19**	D
6	B	**13**	B	**20**	C
7	D	**14**	C	**21**	B

Answers to short questions (Chapter 16)

22 Br, *[1]* liquid, *[1]* orange or red/brown *[1]*

23 a) They have similar electron structures, *[1]* with one electron in the outermost shell. *[1]*
b) (i) C *[1]* (ii) N or O or F or Cl *[1]* (iii) O or S *[1]*

24 a) Beryllium or calcium or strontium or barium *[1]*
b) Magnesium + oxygen *[1]* $\rightarrow$ magnesium oxide *[1]*
c) Mg atom has two electrons in outer shell, *[1]* O atom has six electrons in outer shell. *[1]*

25 a) 27, b) 13, c) 10, d) 13, e) positive charge on protons balances negative charge on electrons *[5]*

26 a) Same number of protons/electrons *[1]*
b) Different number of neutrons *[1]*
c) There must be more $^{20}_{10}Ne$ than $^{22}_{10}Ne$, *[1]* because the relative atomic mass is nearer 20 than 22. *[1]*

27 a) D *[1]* b) blue/indigo/violet *[1]* c) copper *[1]*
d) More reactive than Cu *[1]*
e) It reacts *[1]* and forms its ions in solution *[1]*

Answers to further examination questions (Chapter 16)

28 a) (i) 1 *[1]* (ii) rubidium hydroxide *[1]* and hydrogen *[1]* (iii) Rb floats on water/Rb skates here and there on the water surface/Rb gets so hot it starts to burn/bubbles of colourless gas form (any two for *[2]*)
b) (i) Hydrochloric acid *[1]* (ii) magnesium iodide *[1]* and hydrogen *[1]* (iii) Mg sinks to bottom of HI(aq)/Mg eventually disappears/bubbles of colourless gas form (any two for *[2]*)
c) (i) It would turn universal indicator red or orange. *[1]* (ii) The sodium carbonate would remain at the bottom of the mixture/the sodium carbonate eventually disappears/a colourless gas is produced/a white precipitate will form. (any two for *[2]*) (iii) Water or carbon dioxide or calcium selenate *[1]*

29 a) (i) 4 neutrons, 3 protons, 3 electrons *[3]*
(ii) isotopes *[1]* (iii) $^{6}_{3}Li$ *[2]* (iv) Li^+ *[1]*
b) (i) 2,8,7 *[1]*
(ii)

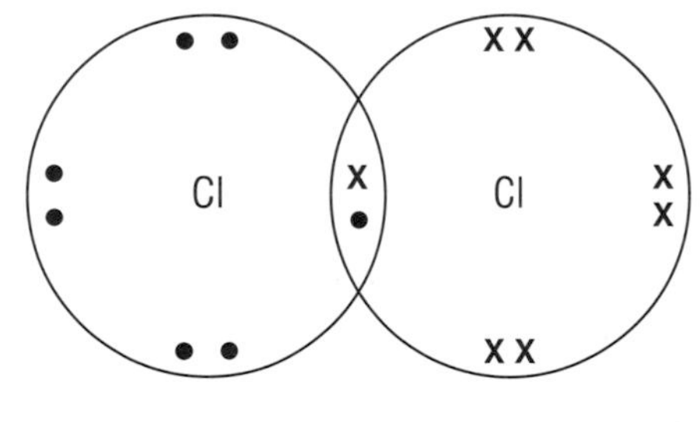

[1]

c)

:Cl• + X → :Cl:ˣ

atom electron ion *[1]*

i.e. Cl + e^- → Cl^-

The atom has been reduced

[1]

30 a) (i) Potassium *[1]* (ii) its outermost electron is further from the nucleus than in Li and Na *[1]* and so is lost most easily. *[1]*
b) Sodium chloride and water *[2]*
c) (i) Neutralisation *[1]* (ii) NaCl(aq) *[1]* + H_2O(l) *[1]*

31 a) (i) 2 *[1]* (ii) 5 *[1]* (iii) 38 *[1]* (iv) 88 *[1]*
b) (i) Strontium, Sr *[2]* (ii) 2^+ *[1]*
(iii) Magnesium or calcium or barium *[1]*
c) Forms positive ions/usually form basic or alkaline oxides/most react with acids to give hydrogen/form ionic compounds/react with non-metals but not with other metals. (any two for *[2]*)

Answers to objective questions (Chapter 17)

1	B	**6**	D	**11**	A
2	D	**7**	A	**12**	C
3	C	**8**	D	**13**	C
4	A	**9**	D	**14**	D
5	B	**10**	B	**15**	B

Answers to short questions (Chapter 17)

16 a) Each Mg atom loses two electrons to form an Mg^{2+} ion. *[1]*
b) Each O atom gains two electrons to form an O^{2-} ion. *[1]*
c) The positive Mg^{2+} ions attract the negative O^{2-} ions. *[1]*
d) (Giant) ionic *[1]*

17 a) (i) Covalent *[1]* (ii) covalent *[1]*

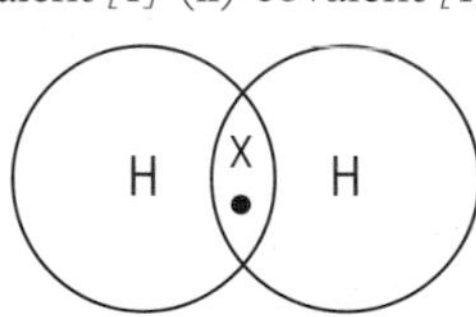

b) shared pair of electrons, *[1]* one electron from each atom *[1]*

18 a) Atoms *[1]*
b) Electrons *[1]*
c) Covalent *[1]*

19 CaO consists of ions/held together by strong ionic bonds/candle wax contains small molecules/held together by weak forces/bonds. *[4]*

20 a) Metals have close packed atoms. *[1]* The outermost electrons *[1]* of the atoms are free to move *[1]* through the structure of positive ions. The negative electrons attract all the positive ions *[1]* and bond the metal atoms strongly. *[1]*
b) The close packing and strong metallic bonds make the iron strong. *[1]* The regular close packing of iron atoms allows the rows/lines/layers of atoms to move over each other (slip), *[1]* when a force is applied. *[1]* After slip has occurred, the close packing is restored *[1]* with strong bonding.

21 a)

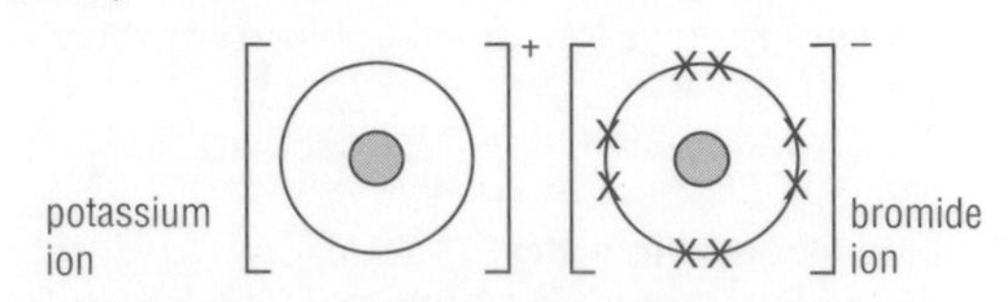

one electron transferred to Br *[1]*
+ and − charges *[1]*
b)

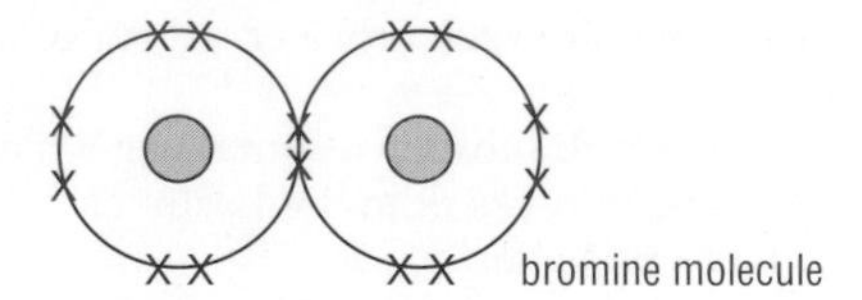

one pair of electrons shared *[1]*
eight electrons around each Br nucleus *[1]*

22 a) 151 + 242 = 393 kJ *[1]*
b) 2 × 202 = 404 kJ *[1]*
c) +393 − 404 = −11 kJ *[1]*
d) The reaction is exothermic or the products are more stable than the reactants. *[1]*

Answers to further examination questions (Chapter 17)

23 a) 2 × 436 + 498 *[1]* = 1370 kJ *[1]*
b) 4 × 464 *[1]* = 1856 kJ *[1]*
c) 486 kJ, *[1]* exothermic *[1]*

24 a) (i) Good ventilation is needed. *[1]*
(ii) The gas is flammable. *[1]*
b) When the temperature rises, the gas particles gain energy and move around faster. *[1]* This means that the particles hit the walls of the cylinder more often *[1]* and when they do so they hit with more force. *[1]* Both these factors cause increased pressure.
c) A shared pair of electrons, *[1]* each atom contributes one electron to the shared pair. *[1]*
d) It has a lower relative molecular mass and is therefore more volatile. *[1]* This means it has a lower boiling point than butane and is less likely to liquefy in cylinders during the winter. *[1]*
e) (i) 2 × 3 × 346 *[1]* + 2 × 10 × 413 *[1]* + 13 × 497 *[1]* = 16 797 kJ
(ii) 8 × 2 × 803 *[1]* + 10 × 2 × 463 *[1]* = 22 108 kJ
(iii) +16 797 − 22 108 = −5311 kJ *[1]*

25 a) (i) Ca has an electronic structure with two outer shell electrons. *[1]* These freely moving electrons allow it to conduct electricity/or these two electrons are lost during reactions to form Ca^{2+} ions. *[1]*
(ii) CaO is basic *[1]* and SiO_2 is acidic. *[1]* They react to form the salt calcium silicate, $CaSiO_3$. *[1]* (any two for *[2]*)

b) (i) A covalent bond *[1]* (ii) an atom *[1]* of oxygen *[1]*
(iii) It has very strong covalent bonds holding all the atoms together. *[1]*
c) (i) It contains two different materials, *[1]* a concrete matrix hardened by gravel. *[1]* The two materials work together to give a superior material. *[1]* (any two for *[2]*)
(ii) By reinforcing it with steel rods *[1]*

26 a) (i) It cannot be broken down into any simpler substance. *[1]*
(ii) It contains two elements chemically combined together. *[1]*
(iii) It contains two different substances *[1]* which are not chemically combined. *[1]*
b) (i) The nucleus *[1]* (ii) NaCl or Na^+Cl^- *[1]*
(iii) One electron from a sodium atom is transferred to a chlorine atom *[1]* to form a positive Na^+ ion and a negative Cl^- ion. *[1]* These ions are held together by the attraction between their opposite charges. *[1]* This is called an ionic bond. *[1]*

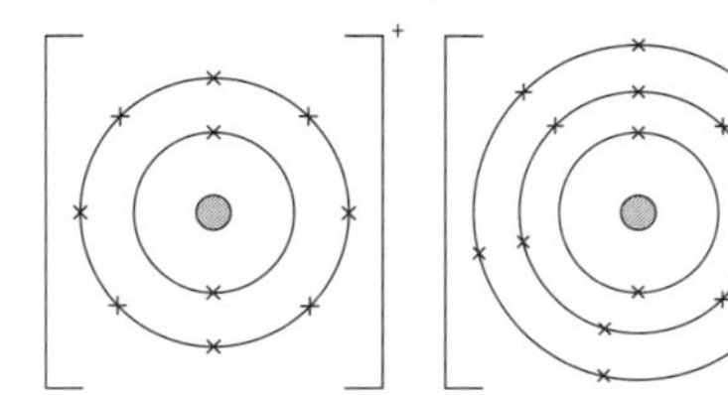

(*[1]* for each correct diagram of the ions, any five for *[5]*)

Answers to objective questions (Chapter 18)

1	D	8	C
2	D	9	A
3	A	10	C
4	B	11	B
5	D	12	A
6	A	13	D
7	B	14	C

Answers to short questions (Chapter 18)

15 a) Hydrogen peroxide → water + oxygen *[1]*
b) The apparatus should be exactly the same for both reactions *[1]*/use the same volume and concentrations of H_2O_2 for both reactions *[1]*/make sure the temperature is the same for both reactions *[1]*/use the same amount (no. of moles of gold and copper) *[1]*/use gold and copper powders of the same fineness *[1]*/mix the powders in exactly the same way with the H_2O_2 solution in both experiments. *[1]* (any three for *[3]*)
c) The glass bottle with rough inner walls will have a larger surface area than that with smooth walls. *[1]* This will cause the decomposition of H_2O_2 to occur faster in the rough-walled bottle. *[1]*

16 a) Axes labelled with units, *[1]* scales on axes, *[1]* points plotted accurately, *[2]* line (curve) of best fit. *[1]*
b) 0.48 ± 0.01 g *[1]*

17 a) It decreases *[1]* as time goes on.
b) It decreases. *[1]*
c) Magnesium is used up *[1]* so its surface area gets less. *[1]*
d) The acid is becoming less concentrated with time and the surface area of the magnesium is getting smaller. This means there are fewer collisions between acid particles and magnesium particles *[1]* per minute *[1]* so the reaction rate decreases.

18 a) Catalysts for biological reactions *[1]*
b) (45 ± 2)°C *[1]*
c) The reaction rate increases as temperature increases at first *[1]* due to increased collisions/more energetic collisions. *[1]* However, temperature changes the enzyme structure *[1]* causing denaturation *[1]* and the enzyme becomes less effective at higher temperatures. *[1]*
d) The enzymes in the biological washing powder act most effectively at about 50°C. *[1]* They will be denatured and much less effective/useless at 100°C. *[1]*

Answers to further examination questions (Chapter 18)

19 a) Carbon dioxide *[1]*
b) (i) Axes labelled with units, *[1]* scales on axes, *[1]* points plotted accurately. *[2]*
(ii) 50 *[1]* (iii) No more bubbles of CO_2 would form. *[1]*
(iv) Some marble was left. *[1]* (v) 40 cm^3 *[1]*
c) (i) More quickly *[1]* (ii) There is a greater surface area of marble to react. *[1]*
(iii) Initial gradient steeper, *[1]* but the curve flattens off earlier to the same final volume (40 cm^3). *[1]*

20 a) Points for reaction A plotted accurately *[1]* with curve of best fit. *[1]* Points for reaction B plotted accurately *[1]* with curve of best fit. *[1]*
b) Curve steeper initially than A, *[1]* flattening off earlier at the same total mass of CO_2 produced (2.2 g). *[1]*
c) Curve steeper initially than B, *[1]* flattening off earlier at the same total mass of CO_2 produced (1.1 g). *[1]*
d) + water + calcium chloride *[1]*
e) (i) 100 *[1]* (ii) 56 *[1]* (iii) 100 g $CaCO_3$ gives 56 g CaO *[1]* so 10 g $CaCO_3$ gives 5.6 g CaO *[1]*
(iv) 100 g $CaCO_3$ gives 44 g CO_2 = 1 mole CO_2 *[1]*

10 g $CaCO_3$ gives 4.4 g CO_2 = $\frac{1}{10}$ mole CO_2 *[1]*
$\frac{1}{10}$ mole CO_2 = $\frac{24}{10}$ = 2.4 dm^3 *[1]*

21 a) (i) Axes labelled with units, *[1]* scales on axes, *[1]* points accurately plotted. *[2]* (ii) Line (curve of best fit) *[1]*
b) (i) and (ii) Iodine molecules must collide with zinc atoms for a reaction to occur. *[1]* If the iodine solution is more concentrated, there are more collisions of iodine molecules with zinc atoms. *[1]* Therefore, the zinc reacts faster *[1]* and the loss of mass of zinc occurs sooner. *[1]*

22 a) (i) X should be on the curve between 0 and 0.5 minutes. *[1]* (ii) Less concentrated solution gives a curve which is less steep initially *[1]* and flattens off below the curve already drawn. *[1]*
b) Heating causes the particles to have more energy and move about faster. *[1]* The particles of H_2O_2 hit the MnO_2 catalyst particles more frequently *[1]* and with more energy. *[1]* This leads to more reactions per minute and therefore an increase in the rate of reaction.
c) (i) Catalytic cracking involves the use of a catalyst *[1]* to cause the breakdown of heavy (large molecular mass) alkanes. *[1]* (ii) Polymerisation involves the addition of smaller units (e.g. ethene) *[1]* to form a polymer (e.g. polythene). i.e. $nC_2H_4 \rightarrow (CH_2)_{2n}$ *[1]*

Answers to objective questions (Chapter 19)

1	A	**6**	C	**11**	D
2	D	**7**	B	**12**	C
3	E	**8**	B	**13**	C
4	B	**9**	E	**14**	B
5	C	**10**	D		

Answers to short questions (Chapter 19)

15 Millions of years ago, deposits of mud, clay and gravel collected on the sea bed. *[1]* These were compacted and cemented by the layers above to form shale and sandstone in stage 1. *[1]* As a result of plate movements in the Earth's crust and convection currents in the mantle *[1]* over thousands/millions of years, the stratified rocks were raised above sea level *[1]* (stage 2). The different rocks have then been weathered by changes in temperature, freezing and thawing, and reaction with water. *[1]* The sandstone and shale have weathered more than the harder more compacted schist which has in turn weathered more than the harder granite. *[1]* (stage 3) (any five for *[5]*)

16 a) Interlocking crystals, *[1]* with different colours, *[1]* no layers, *[1]* no fossils. *[1]* (any two for *[2]*)
b) Basalt will have smaller crystals. *[1]*
c) If the magma cools rapidly, it will crystallise rapidly with small crystals. *[1]* If it cools very slowly, larger crystals will form. *[1]*

17 a) Weathering is the breaking up of rocks by wind, rain, water and temperature changes. *[1]* Erosion is the breaking up of rocks (weathering) and then carrying the weathered material away (transport). *[1]*
b) Temperature changes cause the rock to expand and contract. *[1]* The surface rock expands and contracts more than the rock below and this sets up stresses and strains *[1]* causing the rock to crack.
c) Rainwater reacts with carbon dioxide to form carbonic acid. *[1]*
$H_2O + CO_2 \rightarrow H_2CO_3$. *[1]* Carbonic acid reacts with limestone (calcium carbonate) to form calcium hydrogencarbonate *[1]* which is soluble and so gets washed away. *[1]*
$H_2CO_3 + CaCO_3 \rightarrow Ca(HCO_3)_2$ *[1]* (any four for *[4]*)

Answers to further examination questions (Chapter 19)

18 a) (i) Igneous rocks form when the hot molten magma *[1]* inside the Earth's mantle cools and solidifies. *[1]*
(ii) Metamorphic rocks form when igneous and sedimentary rocks *[1]* are subjected to enormous pressure and/or very high temperatures. *[1]*
b) Igneous rocks can be weathered by temperature changes, freezing and thawing, and sometimes by reaction with water. *[1]* This produces sediments, sand and gravel. *[1]* These deposits may be carried by rivers or ocean currents and deposited elsewhere. *[1]* As the layers of sediment build up over millions of years, the material below is compacted and cemented together as sedimentary rock. *[1]* The whole process will take thousands if not millions of years. *[1]* (any four for *[4]*)
c) As a result of the movement of plates on the Earth's surface, sedimentary rocks are carried deep into the mantle. *[1]* Here the rock is subject to extreme heat forming molten magma. *[1]* At a later stage, the magma may move closer to the Earth's crust or onto the Earth's surface through a volcano. *[1]* The magma therefore cools and crystallises forming igneous rock. *[1]*

19 a) A is sandstone *[1]* – sedimentary *[1]*
B is slate *[1]* – metamorphic *[1]*
C is granite *[1]* – igneous *[1]*
b) A has small grains of sand compacted and cemented together *[1]* from further layers of

sediment built up over thousands/millions of years. *[1]*
B shows sections and layers compacted together. *[1]* This has been caused by subjection of clay and mud to extremes of heat and/or pressure. *[1]*
C has large, interlocking crystals. *[1]* These have formed as magma, in the mantle below the Earth's crust, has cooled very slowly. *[1]*
c) By dating the rocks knowing the half lives of radioactive elements present in them. *[1]*

20 a) 1 is igneous rock, 2 is sediments/erosion deposits, 3 is sedimentary rock, 4 is magma, 5 is metamorphic rock. *[5]*
b) (i) Both may have interlocking crystals. *[1]*
(ii) Igneous – different coloured crystals; sedimentary – similar coloured crystals *[1]*
igneous – no layers; sedimentary – may have layers *[1]*
c) Knowing the half life of a radioactive element in the rock *[1]* and its decay product. *[1]* By finding the relative amounts of these two elements *[1]* it is possible to estimate the age of the rock.

Answers to objective questions (Chapter 20)

1	B	**6**	C	**11**	B
2	D	**7**	A	**12**	A
3	A	**8**	E	**13**	D
4	C	**9**	B	**14**	B
5	E	**10**	A	**15**	C

Answers to short questions (Chapter 20)

16 a) A compound containing only carbon and hydrogen. *[1]*
b) Carbon dioxide *[1]* and water *[1]*
c) (i) Temperature, *[1]* pressure, *[1]* surface area of catalyst. *[1]* (any two for *[2]*)
(ii) The temperature of the engine (cylinders) and the catalyst is low so the benzene does not burn so well. *[1]*

17 a) (i) Cracking *[1]* (ii) polymerisation *[1]*
b) The substance contains one or more carbon–carbon double bonds. *[1]*
c)

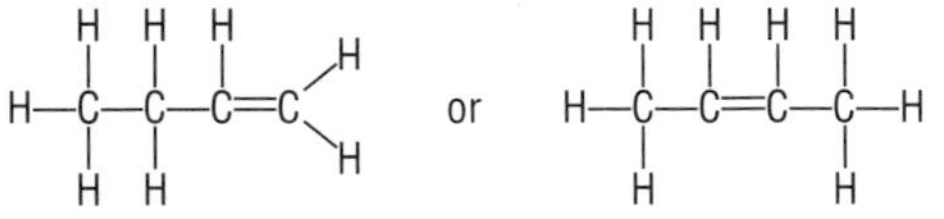

```
    H  H  H                    H  H  H  H
    |  |  |     H              |  |  |  |
 H—C—C—C=C                  H—C—C=C—C—H
    |  |      \              |        |
    H  H       H             H        H
                   or
```

[2]

18 a) Shake the ethene gas *[1]* with bromine water. *[1]*
b) The yellow bromine water goes colourless. *[1]*
c) There would be a greater surface area on which to 'crack' the paraffin vapour. *[1]*
d)

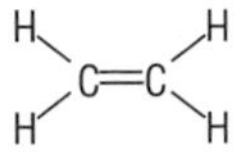

```
H       H
 \     /
  C=C
 /     \
H       H
```

[1]
e) Poly(ethene) *[1]*

19 a) Carbon dioxide *[1]* and water *[1]*
b) $CH_4 + 2O_2 \rightarrow CO_2 + 2H_2O$ correct formulae *[1]*, balanced *[1]*
c) 1 mole methane (CH_4) needs 2 moles oxygen (O_2) *[1]*
$\therefore$ 16 g CH_4 needs 64 g O_2 *[1]* $\therefore$ 10 g CH_4 needs 40 g O_2 *[1]*

20 a) Because they have formed from the remains of dead animals and plants. *[1]*
b) Dead/decaying, *[1]* sea creatures *[1]*
c) Absence of oxygen, *[1]* pressure due to the sea/sediments/additional dead material above, *[1]* high temperatures *[1]* as the dead creatures rotted exothermically

Answers to further examination questions (Chapter 20)

21 As the vapours rise up the column through the holes, *[1]* the temperature falls. *[1]* Different vapours condense at different heights in the tower *[1]* as soon as the temperature falls to just below their boiling point. *[1]* (any three for *[3]*)
b) (i) Splitting/breaking up larger hydrocarbon/alkane molecules into smaller ones *[1]*
(ii) It requires high temperatures to break the strong covalent bond *[1]* between two carbon atoms. *[1]*
(iii) It uses heavier fractions for which there is little direct use. *[1]* It produces additional quantities of petrol. *[1]* It produces important monomers for plastics such as ethene. *[1]* (any two for *[2]*)
c) (i) (Addition) polymerisation *[1]*
(ii) The double bonds in the ethene molecules 'open up' *[1]* and the carbon atoms on separate ethene molecules join together. *[1]*

22 a) Each fraction contains a number of compounds, *[1]* which boil in the temperature range specified. *[1]*
b) Diesel fraction *[1]*
c) Its boiling range is 150°C. That of petrol is only 75°C. *[1]* Its molecules contain fewer C atoms than petrol. *[1]*
d) C_6H_{14} and C_6H_{12} are used to make petrol. *[1]* C_2H_4 is an important monomer for polymers such as poly(ethene). *[1]*

e)

$$\mathrm{H_2C{=}CH_2}$$

[2]

f) (i)

$$\left(\!-\mathrm{CH(H)}-\mathrm{C(H)(CH_3)}-\!\right)_n$$

[1]

(ii) poly(propene) [1]

23 a) It contains a $\mathrm{>C{=}C<}$ bond. [1]

b) $CH_3—CHCl_2$ or $CH_2Cl—CH_2Cl$ [1]

c) (i)

$$\left(\!-\mathrm{C(H)(H)}-\mathrm{C(H)(Cl)}-\!\right)_n$$

[2]

(ii) It does not rust [1] so does not need painting. [1] It is lighter in weight and easier to install. [1] (any one for [1])

d) (i) Combustion of chloroethene will produce chlorine products which are toxic/poisonous [1] (e.g. Cl_2) and acidic [1] (e.g. HCl). These substances will harm or even kill plants and animals. [1]

(ii) PVC is not biodegradeable [1] so it does not harm organisms. [1] However, PVC remains for years and years [1] causing a huge litter problem. [1] (any three for [3])

24 a) Fractional distillation [1]

b) (i) Cracking [1]

(ii)

$$n\,\mathrm{H_2C{=}CH_2} \longrightarrow \left(\!-\mathrm{C(H)(H)}-\mathrm{C(H)(H)}-\!\right)_n$$

[2]

c) (i) $4 \times 415 + 2 \times 495$ [1] $= 2650$ kJ [1]

(ii) $2 \times 740 + 4 \times 465$ [1] $= 3340$ kJ [1]

(iii) Bond breaking in (i) is endothermic + 2650 kJ, bond making in (ii) is exothermic − 3340 kJ [1] ∴ the overall reaction is exothermic. [1]

Answers to objective questions (Chapter 21)

1 A 2 D 3 B 4 E 5 C 6 D 7 E 8 C 9 A 10 C 11 A 12 D 13 A 14 A

Answers to short questions (Chapter 21)

15 a) Soluble/stable/involatile/easily stored/non-toxic/easily transported (any three for [3])

b) Formula mass of NH_4NO_3, 80 g, contains 28 g nitrogen [1]
Formula mass of $(NH_2)_2CO$, 60 g, contains 28 g nitrogen [1]
∴ 1 kg of urea, $(NH_2)_2CO$, will provide more nitrogen. [1]

16 a) $NH_3 + HNO_3 \rightarrow NH_4NO_3$ (two formulae correct [1], all three [2])

b) $NH_4Cl \rightarrow NH_3 + HCl$ (two formulae correct [1], all three [2])

c) $NH_4Cl + NaOH \rightarrow NH_3 + H_2O + NaCl$ (three quarters of formulae correct [1], all five [2])

17 a) $N_2 + 3H_2 \rightleftharpoons 2NH_3$ (three formulae correct [1], correctly balanced [1], equilibrium sign [1])

b) The concentrations of reactants and products do not change. [1]

c) Both the forward and the backward reactions are taking place at the same speed. [1] So, there is no change in the concentration of any substance. [1]

18 a) Essential for synthesis of proteins/chlorophyll [1]

b) Plants are stunted/leaves become yellow (any one for [1])

c) Essential for synthesis of nucleic acids (DNA) [1]

d) Plants grow slowly/small seeds and fruit (any one for [1])

e) May change soil pH/may harm plants and animals in soil/allow elements to accumulate in soil/lead to river pollution (any two for [2])

Answers to further examination questions (Chapter 21)

19 a) Because the reaction involves an equilibrium mixture [1]

b) (i) Special pumps/costly equipment [1] are required to withstand high pressures. [1] Lower pressures avoid these. [1]

(ii) Lower temperatures produce a higher yield, [1] but a slower reaction rate [1] and overall the cost of production (per tonne) will be greater. [1]

c) Some sensible suggestion such as 'NH_3 is more stable at lower temperatures'. [1]

d) Remove the NH_3, [1] by cooling to say

−50°C to liquefy the NH_3, *[1]* and then recycle the N_2 and H_2. *[1]*

20 a) 3 mole $H_2 \rightarrow$ 2 mole NH_3 *[1]*
$\therefore$ 6 g $H_2 \rightarrow$ 34 g NH_3 *[1]*

$\therefore$ 1 g $H_2 \rightarrow \frac{34}{6}$ g NH_3 *[1]*

$\Rightarrow$ 1 tonne $H_2 \rightarrow \frac{34}{6}$ tonne NH_3
= 5.67 tonne NH_3 *[1]*

b) High pressure produces a higher yield of NH_3. *[1]* It also speeds up the reaction *[1]* by increasing the concentration of gases. *[1]* This gives a faster production of NH_3 and a more economical process. *[1]* (any three for *[3]*)

c) (i) So that they can be adsorbed from the soil through the roots of plants.
(ii) The fertilisers are washed out of the soil *[1]* into rivers and lakes where they have detrimental effects on aquatic plants and animals. *[1]*

21 a) (i) One which can go in either direction/be reversed. *[1]*
(ii) N_2, H_2 and NH_3 *[1]*
b) (i) 400 atm *[1]* and 100°C *[1]*
(ii) Lower temperatures give the maximum yield, *[1]* but the reaction rate is slower *[1]* making the manufacture less economic. *[1]* (any two for *[2]*)
c) (i) Atmospheric pressure/high temperature/catalyst (any two for *[2]*)
(ii) $2NO + O_2 \rightarrow 2NO_2$ *[1]*
(iii) $3NO_2 + H_2O$ *[1]* $\rightarrow 2HNO_3 + NO$ *[1]*
d) (i) NH_3 is used to manufacture nitric acid *[1]* so the processes can be continuous *[1]* with no transport costs. *[1]*
(ii) Careful choice of location/reduction in any waste materials/careful disposal of any waste/screening of buildings by trees, etc. (any two for *[2]*)

Answers to objective questions (Chapter 22)

1	C	**6**	A	**11**	D
2	E	**7**	B	**12**	A
3	D	**8**	C	**13**	B
4	A	**9**	C	**14**	C
5	D	**10**	B	**15**	D

Answers to short questions (Chapter 22)

16 a) Speed $= \frac{\text{distance}}{\text{time}}$ *[1]*
$= \frac{200\text{ m}}{19.32\text{ s}} = 10.35$ m/s *[1]*

b) Acceleration $= \frac{\text{increase in speed}}{\text{time taken}}$ *[1]*
$= \frac{11\text{ m/s}}{4\text{ s}} = 3.75$ *[1]* m/s² *[1]*

c) K. E. $= \frac{1}{2}mv^2$ *[1]* $= \frac{1}{2} \times 86 \times 11^2$ *[1]*
= 5203 J *[1]*

17 f = m × a *[1]*, 112 = 7 × a
$\therefore$ acceleration $= \frac{112}{7} = 16$ *[1]* m/s² *[1]*

18 a) Axes labelled with scale *[1]*, points plotted correctly *[1]*, line of best fit *[1]*
b) 8.0 ± 0.5 N *[1]*
c) Extension with 6 N load = 6 cm *[1]*
$\therefore$ length of spring with 6 N load
= 10 + 6 = 16 cm *[1]*
d) 18.0 ± 1.0 N *[1]*

19 a) (i) Thrust and drag *[1]*
(ii) Lift and weight *[1]*
b) (i) Thrust must equal drag *[1]*, lift must equal weight *[1]*
(ii) For the plane to accelerate, thrust > drag, *[1]* for the plane to ascend, lift > weight *[1]*

20 Vehicle travels at 3 m/s for 10 seconds, *[1]* then decelerates *[1]* at $\frac{3}{5}$ m/s² *[1]* between 10 and 15 seconds. Vehicle is stationary from 15 to 30 seconds, *[1]* then accelerates *[1]* at $\frac{9}{10}$ m/s² *[1]* from 30 to 40 seconds and finally travels at 9 m/s from 40 to 60 seconds. *[1]*

21 a) Acceleration $= \frac{\text{increase in speed}}{\text{time taken}}$ *[1]*
$= \frac{3\text{ m/s}}{10\text{ s}} = 0.3$ *[1]* m/s² *[1]*

b) Force = mass × acceleration *[1]*
The force is the same on both cars $\therefore$ as the acceleration of the second car is lower, *[1]* its mass must be greater than that of the first car. *[1]*

22 a) Force = mass × acceleration *[1]*
acceleration $= \frac{\text{increase in speed}}{\text{time}}$ *[1]*
acceleration $= \frac{5\text{ m/s}}{10\text{ s}} = 0.5$ m/s² *[1]*
$\Rightarrow$ Force = 750 × 0.5 = 375 N *[1]*

b) When the driver tries to make the lorry go as fast as possible, the driving force is a maximum. *[1]* This causes the lorry to increase in speed. *[1]* As the lorry increases in speed, friction and drag also increase. *[1]* When friction and drag equal the maximum driving force, the lorry will continue with constant speed. *[1]*

23 a) Work = force × distance *[1]*
90 000 = 6000 × distance
$\therefore$ distance = 15 *[1]* m *[1]*

b) K. E. = $\frac{1}{2}mv^2$ *[1]*
= $\frac{1}{2} \times 800 \times 30^2$ = 360 000 *[1]* J *[1]*
c) Work = force × distance
360 000 = 6000 × distance
∴ distance = 60 m *[1]*
d) The braking distance at a speed of 30 m/s for the car in the question is 60 m. *[1]* In 2 seconds, the car will travel 60 m. *[1]* The advice seems to be sensible. *[1]*

Answers to further examination questions (Chapter 22)

24 a) (i) P. E. = m × g × h *[1]*
= 3100 × 10 × 72 *[1]* = 2 232 000 J *[1]*
(ii) Power = 50 kW = 50 000 J per sec. *[1]*

$$\text{Power} = \frac{\text{work done}}{\text{time taken}} \quad [1]$$

$$\therefore 50\,000 = \frac{2\,232\,000}{\text{time}}$$

$$\therefore \text{time} = \frac{2\,232\,\cancel{000}}{50\,\cancel{000}} = 44.6 \text{ seconds} \quad [1]$$

(iii) The work done by the motor must also overcome friction *[1]* between the carriage wheels and axle and between the wheels and the track. *[1]*
b) (i) P. E. changes to K. E. *[1]*
(ii) Total P. E. at B relative to C = m × g × h
= 3100 × 10 × 65.5 = 2 030 500 J *[1]*
If all this P. E. is converted to K. E. *[1]*,
K. E. = $\frac{1}{2} \times m \times v^2$ = 2 030 500 *[1]*
∴ speed at C, v = 36 m/s *[1]*
(iii) The P. E. at B will be converted to sound energy and heat (friction) as well as K. E. as the carriages move from B to C. *[1]*
∴ they will have no sufficient K. E. at C to carry them to the height of B. *[1]*
Friction must also be overcome as the carriages move from C to D. So, D cannot be as high as B. *[1]* (any two for *[2]*)

25 a) (i) 35 m/s *[1]*
(ii) 300 seconds *[1]*

$$\text{(iii) Deceleration} = \frac{\text{decrease in speed}}{\text{time taken}} \quad [1]$$

$$= \frac{5 \text{ m/s}}{10 \text{ s}} = 0.5 \text{ m/s}^2$$

(iv) Force = mass × acceleration
(deceleration) *[1]*
= 60 × 0.5
= 30 N *[1]*
b) (i) Friction between the tyres and the road had produced heat to warm the tyres. *[1]*
(ii) Heat had warmed up the air in the tyres. Molecules of air in the tyres therefore move faster *[1]* and bombard the walls of the tyre harder and more often. *[1]* This increases the force on the walls of the tyre.

26 a) (i) P. E. *[1]*
(ii) It is converted to K. E. *[1]* and P. E. *[1]* in the metal ball, the plunger and the handle.
b) P. E. = m × g × h *[1]* = 0.1 × 10 × 0.75 *[1]* = 0.75 J *[1]*
c) The energy of the spring is eventually converted to P. E. in the ball, plus P. E. in the plunger and handle *[1]* plus heat *[1]* due to friction between the moving parts. *[1]*

27 a) (i) The horizontal speed is constant. *[1]*
(ii) The vertical speed is increasing. *[1]*

$$\text{b) (i) Speed} = \frac{\text{distance}}{\text{time}} \quad [1]$$

$$= \frac{0.04 \text{ m}}{0.02 \text{ s}} = 0.2 \text{ m/s} \quad [1]$$

(ii) Average speed between P and Q

$$= \frac{\text{distance}}{\text{time}} = \frac{0.2 \text{ m}}{0.2 \text{ s}} = 1 \text{ m/s} \quad [1]$$

(iii) Vertical speed at P = 0 m/s *[1]*
(iv) Vertical speed at Q = 2 m/s *[1]*

28 a) The strength of wood across the grain is roughly as strong as concrete. *[2]* The strength of wood along the grain is roughly as strong as aluminium. *[2]* Wood is very much stronger along the grain than across the grain. *[2]* (any two for *[4]*)
b) There are strong bonds between the atoms in the fibres along the grain. *[1]* There are only weak bonds between the fibres and the material between the fibres across the grain. *[1]*

29 a) The driver's head will be thrown forward on collision. *[1]* The bag must inflate rapidly for protection from serious injury. *[1]*
b) (i) As the gas expands, the pressure falls. *[1]*

(ii) $p \propto \frac{1}{v}$ or p × v = constant *[1]*

(iii) p × v = constant
∴ 6 × 22 = new pressure × 66 *[1]*
∴ new pressure = 2 atmospheres *[1]*
(iv) 4 litres of nitrogen are produced from 44 g
∴ 1 litre of nitrogen is produced from

$$\frac{44}{4} = 11 \text{ g} \quad [1]$$

∴ 22 litres of nitrogen are produced from 11 × 22 = 242 g *[1]*
Twice the amount is normally used to ensure that sufficient pressure is obtained. *[1]*

Answers to objective questions (Chapter 23)

1 C
2 B
3 A
4 C
5 A
6 B
7 C
8 D
9 A
10 A

Answers to short questions (Chapter 23)

11 a) Wind power *[1]* /wave power *[1]* /solar power *[1]* /tidal power *[1]* /hydroelectric power *[1]* /biomass *[1]* (any four for *[4]*)
b) Renewable sources are less reliable depending on winds, waves, sunshine, rainfall, plant growth *[1]* /It is difficult to generate huge amounts of electricity to satisfy large populations from renewable sources. *[1]*/ Significant amounts of electricity require expensive installations initially to use renewable sources – wind farms, tidal barriers, hydroelectric stations and dams. *[1]* (any two for *[2]*)

12 a) (i) 65 (ii) 5 (iii) 30 *[3]*
b) $\text{Efficiency} = \dfrac{\text{useful energy out}}{\text{total energy in}}$ *[1]*
$= \dfrac{1}{30} = 0.03$ *[1]*

13 a) The coldest part inside the refrigerator is the freezer compartment. *[1]* This cools the air near it *[1]* which becomes less dense *[1]* and falls *[1]* to the lower parts of the refrigerator. At the same time, air lower in the refrigerator rises *[1]* towards the freezer compartment. This circulation of air inside the refrigerator (convection currents) keeps the contents cool. (any four for *[4]*)
b) (i) So that convection currents can move up and down the refrigerator unimpeded. *[1]*
(ii) The contents inside the salad box are warmer than those in the refrigerator above. The bottom side of the glass shelf is warmer than its top side. *[1]* Particles (electrons and atoms) on the bottom of the glass shelf collide with particles above them. *[1]* These collisions pass kinetic energy to the atoms above. *[1]* These particles in turn pass their kinetic energy to atoms above them by further collisions. *[1]* So, kinetic energy in the particles (i.e. heat) is conducted through the glass shelf. *[1]* (any four for *[4]*)

14 a) $\text{Efficiency} = \dfrac{\text{useful energy output}}{\text{total energy input}}$ *[1]*
$= \dfrac{10\ 500}{45\ 000} = 0.23$ *[1]*
b) $\text{Efficiency} = \dfrac{490}{700} = 0.7$ *[1]*
c) The electric car *[1]*

Answers to further examination questions (Chapter 23)

15 a) (i) A 9% *[1]* and B 41% *[1]*
(ii) Developing countries cannot afford the technology to develop and use non-renewable sources. *[1]* /Developing countries do not have the technology to discover non-renewable sources. *[1]* /Developing countries have much less dense populations living in rural areas where biomass is readily available. *[1]* /Developing countries cannot afford to buy non-renewable energy sources, etc. *[1]* (any two for *[2]*)
b) (i) A Radiation *[1]* B conduction *[1]*
(ii) Copper painted black *[1]*
(iii) Copper is a good conductor of heat. *[1]* Its black colour will allow efficient absorption of radiation. *[1]*
(iv) Advantage – cheaper/no pollution/saves non-renewable fuel (any one for *[1]*)
Disadvantage – dependent on the Sun shining/temperature rise is limited/water is heated much slower (any one for *[1]*)

16 a) Power of immersion heater
$= 3\text{ kW} = 3000\text{ J/sec}$ *[1]*

$\text{power} = \dfrac{\text{energy}}{\text{time}}\ \therefore \text{energy} = \text{power} \times \text{time}$ *[1]*
$= 3000 \times 1800$
$= 5\ 400\ 000\text{ J}$ *[1]*

b) Heat gained by water
= mass × specific heat capacity × temp. change *[1]*
$= 20 \times 4200 \times 50$ *[1]*
$= 4\ 200\ 000\text{ J}$ *[1]*

c) (i) $\text{Efficiency} = \dfrac{4\ 200\ 000}{5\ 400\ 000} \times 100 = 78\%$ *[1]*
(ii) The energy supplied by the immersion heater must also heat up the copper tank. The energy supplied by the immersion heater must also heat up the heater. (any one for *[1]*)
d) (i) The tank warms the air in contact with it. *[1]* This air expands and becomes less dense. *[1]* The less dense air rises as a convection current *[1]* and cooler air takes its place.
(ii) Paint the copper tank white to reduce radiation *[1]* /Keep the copper tank shiny to reduce radiation *[1]* /Lag the copper tank so that the outside of the lagging is at a lower temperature *[1]* (any one for *[1]*)

17 a) (i) £150 (ii) thick pile carpet/thicker underlay/thicker flooring *[2]*

b) (i) Fibre glass in loft *[1]* because the cost of installing this (£300) is not much more than the cost of the energy lost per year (£250). *[1]* Therefore, the cost of installation will soon be recovered from reduced energy costs per year. *[1]*
(ii) Double glazing. *[1]* The cost of this is very large (£4500) *[1]* and the maximum saved in one year (£100) is only a small proportion of this. *[1]*

18 a) (i) Radiation *[1]* (ii) convection *[1]*
(iii) Air is a very poor conductor. The fleece layer traps pockets of air *[1]* and this prevents convection currents. *[1]*
b) (i) 4200 J raises the temperature of 1 kg of water by 1°C. *[1]*
$\therefore 4200 \times 0.75 \times 90$ *[1]* J raises the temperature of 0.75 kg of water by 90°C.
= 283 500 *[1]* J *[1]*
(ii) Heat will be lost to the air from the flame. *[1]* Heat will be lost to the air from the kettle as it warms up. *[1]* Heat is lost in warming the kettle as well as the water. *[1]* Heat will be lost in warming metal parts of the stove. *[1]* (any three for *[3]*)

Answers to objective questions (Chapter 24)

1 A
2 E
3 B and C
4 C
5 1.5 A
6 1 A
7 C
8 E
9 A and B
10 B
11 D
12 A
13 D
14 C
15 C
16 A

Answers to short questions (Chapter 24)

17 a)

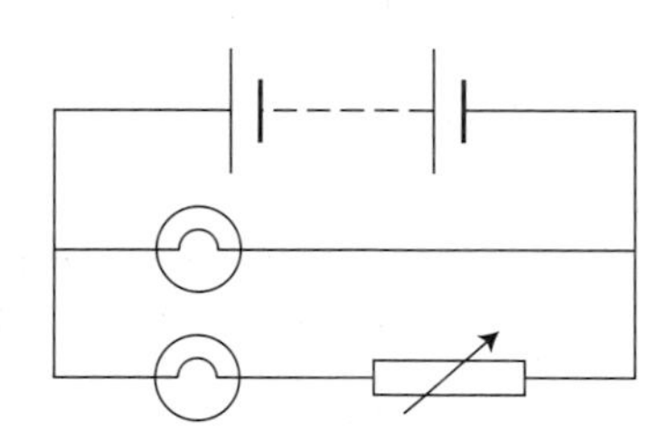

bulbs in parallel, *[1]* variable resistor in series with one of the bulbs *[1]*

b)

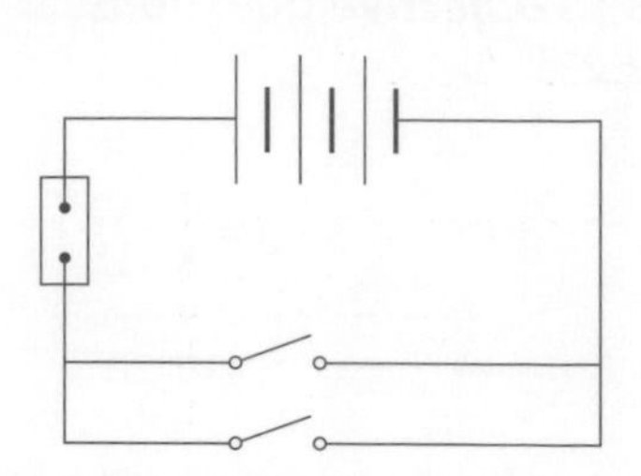

three cells, *[1]* switches in parallel, *[1]* each switch able to complete the circuit. (any two for *[2]*)

18 Chemical energy *[1]* in materials of cell becomes electrical energy *[1]* in the electrons in the electric current which then becomes heat *[1]* and light *[1]* in the bulb.

19 a) (i) $I = \frac{V}{R}$ *[1]* $= \frac{6}{6} = 1$ A *[1]*
(ii) $I = \frac{V}{R} = \frac{6}{2} = 3$ A *[1]*
(iii) 1 A + 3 A = 4 A *[1]*
b) $R = \frac{V}{I}$ *[1]* $= \frac{6}{4} = 1.5\ \Omega$ *[1]*

20 a) Current = 0.40 A *[1]*
b) $R = \frac{V}{I} = \frac{1}{0.40} = 2.5\ \Omega$ *[1]*
c) Current = 0.73 A *[1]*
d) $R = \frac{4}{0.73} = 5.5\ \Omega$ *[1]*

21 As the current increases, the temperature of the light bulb filament increases *[1]* and the atoms in the filament vibrate faster. *[1]* The increased vibration of the atoms hinders/impedes the flow of electrons in the current *[1]* and causes increasing resistance. *[1]*

Answers to further examination questions (Chapter 24)

22 a) B *[1]* b) A *[1]* c) 10 V *[1]*
d) (i) Voltage across A = 10.8 V, *[1]* voltage across B = 5.4 V, *[1]* voltage across C = 3.6 V. *[1]*
(ii) Resistance of A = 27 Ω, *[1]* resistance of B = 13.5 Ω, *[1]* resistance of C = 9 Ω. *[1]*
(iii) Total resistance = 27 + 13.5 + 9 = 49.5 Ω. *[1]*

23 a) Diodes *[1]*
b) (i) L_1 is OFF, L_2 is ON *[1]*
(ii) Diodes will only allow current to pass in one direction *[1]* as

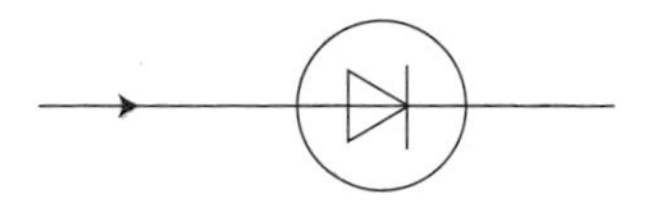

[1]

24 a) (i) Voltmeter across the lamp – i.e.

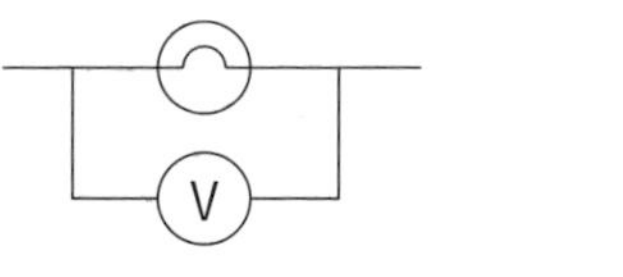

[1]

(ii) Resistance $= \frac{V}{I} \frac{\text{(voltmeter reading)}}{\text{(ammeter reading)}}$ [1]

b) Resistance $= \frac{V}{I}$ [1] $= \frac{6}{2} = 3\ \Omega$ [1]

c) As the p.d. across the lamp increases, the current through it increases and it gets hotter. [1] When the wire gets hotter, the atoms vibrate more and impede the current flow, [1] so resistance is higher.

d) Replace the lamp in the circuit with the LDR. [1] Change the intensity of light on the LDR. [1] (As the light intensity increases, the resistance of the LDR will decrease.)

25 a) The voltage is increased slowly from 0 to +2 volts. The current remains at 0 from 0 to 0.3 volts. [1] From 0.3 to 2.0 volts, the current increases [1] linearly (in a straight line graph) [1] as the voltage increases.
When the current is reversed, the current remains at zero [1] even when the voltage is increased to −2 volts because component X is a diode. [1]

b)

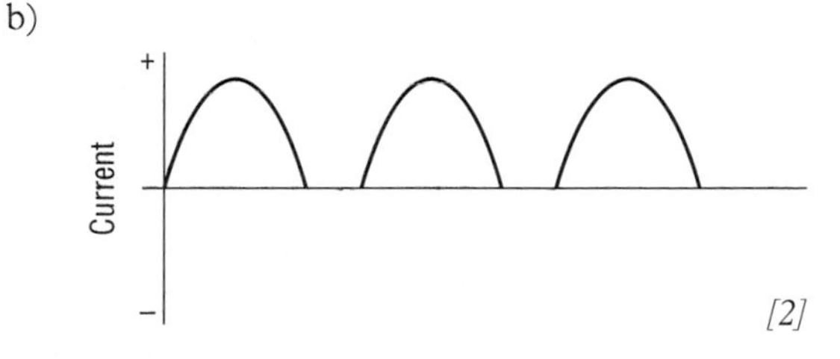

[2]

Answers to objective questions (Chapter 25)

1	B	**6**	D
2	C	**7**	D
3	C	**8**	B
4	D	**9**	A
5	A	**10**	C

Answers to short questions (Chapter 25)

11 a) Positive [1]
b) Positive acetate rod attracts negative charge (electrons) to A [1] ∴ A is negative and B is positive. [1] B remains positive on moving away. [1]

12 a) Prevents electrical fires/protects people using electrical appliances/protects the appliance (any two for [2])
b) Fuse is thin (low melting) wire. [1] If current is too large, wire melts [1] and current ceases. [1]

13 a) 8 units [1]
b) 8 units because charges are not used up (lost, made) [1] and their rate of flow is the same throughout the circuit. [1]
c) If one bulb is removed, the current will be twice as great, [1] i.e. 4 units of charge pass A every second. ∴ 16 units in 4 seconds. [1]

14 a) A is −, B is +, [1] C is + [1]
b) A is −, B is −, [1] C is − [1]
c) Electrons are attracted to bottom of fluff. [1] This makes bottom of fluff negative. [1] So fluff is then attracted to the carpet which is positive. [1]

15 a) The bedside light has both insulated wiring and insulated outer parts. [1]
b) (i) Current × voltage = power [1]
(ii) current × 240 = 80 [1] ∴ current = $\frac{1}{3}$ A [1]
(iii) 3 A fuse [1]

Answers to further examination questions (Chapter 25)

16 a) V × I = P, [1] 240 × I = 25, [1] ∴ I = 0.104 A [1]
b) Energy = V × I × t, [1]
E = 240 × 0.104 × 100 [1] = 2500 J [1]
c) X is the earth wire, protecting the user from an electric shock. [1]
d) The wires in the heating element are thin and carry the whole current. [1] This provides a high resistance to the current and the heating element gets hot. [1] The wires in the cable are thicker and have several strands so the current in each wire is much less and their resistance is much smaller. [1]

17 a) Non-conducting plastic provides insulation for the wires. [1]
b) (i) B [1] (ii) A [1]
c) (i) B [1] (ii) If the current becomes too large, the fuse melts and the current stops. This protects the electric fire/prevents circuit wires becoming overheated/prevents possible harm to someone using the fire. (any two for [2])

18 a) (i) Friction between your hair and the balloon causes the balloon to pick up electrons from your hair. [1] As electrons are negative, the balloon becomes negatively charged. [1]
(ii) Your hair has lost the electrons gained by the balloon ∴ your hair will be positive [1] and the size of its charge will be the same as that on the balloon. [1]
b) As the negative balloon is brought close to

the ceiling, it repels electrons *[1]* from the part of the ceiling nearest to it. This part of the ceiling therefore becomes positive *[1]* and there is an attraction between it and the negative balloon *[1]* sufficient to support the balloon.

19 a) (i) Movements of liquid or gas due to differences in density of the liquid or gas *[1]*
(ii) Part of the liquid/gas is heated and its temperature rises. *[1]* This causes the liquid/gas to expand and become less dense than the surrounding liquid/gas. *[1]* The liquid/gas with a lower density therefore rises, *[1]* whilst cooler, more dense liquid at a lower temperature falls to take its place. *[1]*
b) (i) Power = current × voltage *[1]*
2400 = current × 240 *[1]*
∴ current = 10 *[1]* amps or A *[1]*
(ii) 13 A fuse *[1]*
c) V = R I, *[1]* 240 = R × 10 *[1]* ∴ R = 24 *[1]* ohms or Ω *[1]*
d) Heat supplied = Mass × s.h.c. × temp. rise *[2]*
∴ 588 000 = 2 × 4200 × temp. rise *[1]*
∴ temp. rise = 70°C *[1]*
∴ final temperature of water
= 20 + 70 = 90°C *[1]*

20 a) (i) From the graph, when p.d. (voltage) = 5.0 V, current = 3.8 A *[1]*
Now, V = R I, *[1]* 5 = R × 3.8, *[1]*

$$\therefore R = \frac{5}{3.8} = 1.3 \text{ ohms}$$

(ii) As the p.d. increases, the resistance increases. *[1]*
From the graph, when p.d. = 5.0 V, R = 1.3 ohm

When p.d. = 10 V, $R = \frac{V}{I} = \frac{10}{5} = 2$ ohms *[1]*
b) i) From the graph, when p.d. (voltage) is 10 V, current = 5 A *[1]*
Now power = V × I, *[1]* P = 10 × 5 = 50 *[1]* W *[1]*
(ii) When p.d. (voltage) = 5 V, current = 3.8 A and power = V × I = 5 × 3.8 = 19 W *[1]*
When p.d. (voltage) = 10 V, power = 50 W
The statement is therefore not true. *[1]*

Answers to objective questions (Chapter 26)

1, 2 and **3**

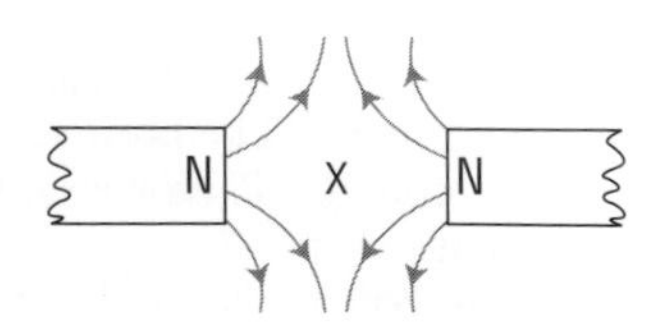

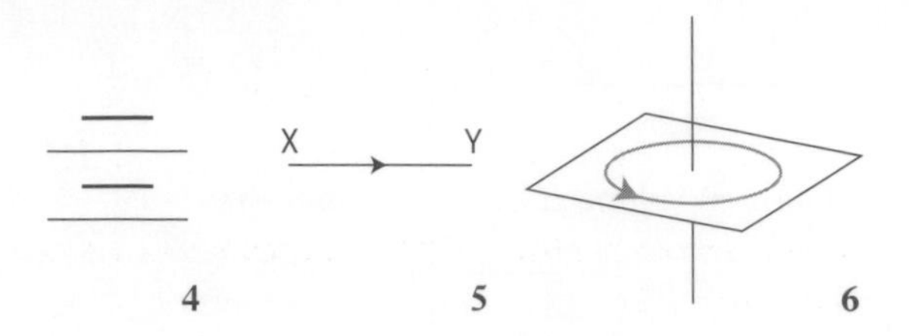

7	D	**11**	B
8	B	**12**	A
9	D	**13**	C
10	B	**14**	B

Answers to short questions (Chapter 26)

15 Current flows in the solenoid (coil around iron core). *[1]* The iron core becomes magnetised *[1]* and attracts the thick iron lever. *[1]* The thick iron lever turns on its pivot *[1]* towards the magnetised iron core and the contact gap is closed. *[1]* A current then flows in the starter motor circuit to start the engine. *[1]* As the wire is thick (low resistance), the current is large. *[1]* (any six for *[6]*)

16 a) An alternating current is one which constantly changes direction. *[1]* A frequency of 50 Hertz means 50 cycles per sec. *[1]*
b) (i) and (ii)

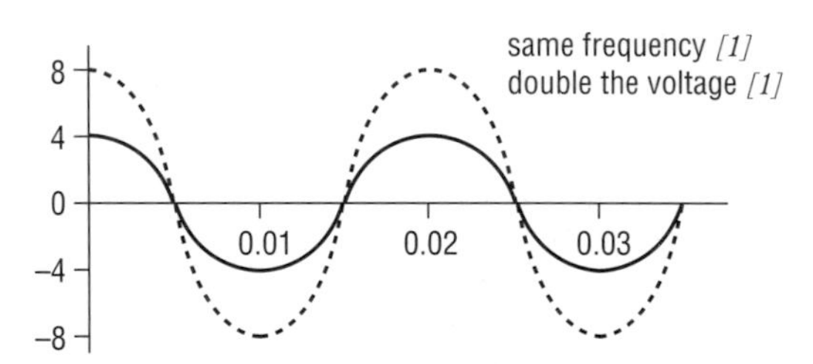

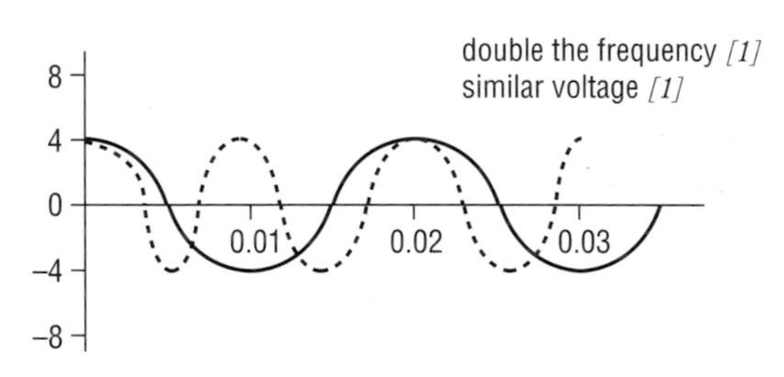

17 a) XY moves along the parallel wires to the right. *[1]*
b) Left-hand motor rule. *[1]* First finger field – downwards, *[1]* second finger – current from X to Y, *[1]* thumb shows movement. *[1]*

18 a) National Grid *[1]*
b) 12 000 V would be very dangerous. *[1]* Household appliances are made to operate at 240 V. *[1]*

c) $\frac{V_p}{V_s} = \frac{n_p}{n_s}$ *[1]*, $\frac{12\,000}{240} = \frac{n_p}{1000}$ *[1]*

$\therefore n_p = 50\,000$ turns *[1]*

Answers to further examination questions (Chapter 26)

19 a) Soft iron *[1]*

b) $\frac{V_s}{V_p} = \frac{n_s}{n_p}$ *[1]*, $\frac{V_s}{240} = \frac{90}{1800}$ *[1]*,

$\therefore V_s = 12$ volts *[1]*

c) The alternating current (voltage) in the primary coil *[1]* causes the magnetic field to continually change (increase, decrease and reverse). *[1]* Therefore, the magnetic field around the secondary coil will change continually. *[1]* This produces a continuous alternating voltage (current) in the secondary coil. *[1]*

d) Step-up transformers allow the transmission of electricity at high voltages and low currents. *[1]* Low currents have a much smaller heating effect in the cables. *[1]* This reduces the energy lost in heating up the power lines.

20 a) One which continually reverses direction. *[1]*

b) (i) As the coil rotates, it cuts through magnetic lines of force to create a voltage. *[1]* When one side of the coil moves up, a voltage is generated in one direction. When that side moves down, a voltage is generated in the other direction. *[1]*

(ii) As the coil rotates, it cuts through the magnetic field (magnetic lines of force) at different rates in different positions. *[1]* When the coil moves through the horizontal, it is cutting through the magnetic field fastest and the voltage is large. *[1]* When the coil moves through the vertical position, it is moving parallel to the field and is not cutting through it so the voltage is zero. *[1]* (any two for *[2]*)

c) Turning the coil faster/having more turns in the coil/using a stronger magnet/using an iron core inside the coil. (any three for *[3]*)

21 a) A current is induced when the magnetic field around the coil is changing. When the magnet is stationary, the magnetic field around the coil does not change, *[1]* so there is no induced current. *[1]*

b) (i) A current surges through the left-hand coil. The sudden increase in current causes a changing magnetic field around both coils. *[1]* The changing magnetic field induces a current in the right-hand coil and a reading on the ammeter. *[1]*

(ii) The magnetic field around both coils stays constant, *[1]* so there is no induced current in the right-hand coil or through the ammeter. *[1]*

(iii) The current in the left-hand coil suddenly falls to zero. This causes a changing magnetic field around both coils. *[1]* A current is induced in the right-hand coil *[1]* opposite in direction to that in part (i). *[1]* (any two for *[2]*)

c) The low voltage a.c. source provides a continually changing alternating current in the left-hand coil. This creates a changing magnetic field around both coils *[1]* which induces a current in the right-hand coil to light the lamp. *[1]* So, electrical energy in the a.c. source becomes light energy in the lamp.

22 a) (i) When the voltage is stepped up, the current is reduced. A smaller current has a much smaller heating effect in the cables. *[1]* Less energy is therefore wasted by heating up the power lines. *[1]*

(ii) High voltages, such as 275 000 V would be far too dangerous for use in our homes. *[1]*

(iii) The most efficient transmission of electricity through the National Grid requires the stepping up and later stepping down of voltages. *[1]* This requires the use of transformers which, in turn, require a.c. rather than d.c. *[1]*

b) (i) The iron core increases the strength of the magnetic field through the coils *[1]* leading to a higher induced voltage in the secondary. *[1]*

(ii) Oil does not conduct electricity, so electrical energy is not lost from bare wire. *[1]* Oil has a large specific heat capacity, so heat lost from the coil causes a smaller increase in temperature. *[1]* (either one)

(iii) $\frac{V_s}{V_p} = \frac{n_s}{n_p}$, $\frac{240}{11\,000} = \frac{n_s}{3000}$ *[1]*,

$n_s = 65$ turns *[1]*

Answers to objective questions (Chapter 27)

1 As his nose is in front *[1]*
2 Laterally inverted, same size, upright, virtual (all four correct *[2]*, three correct *[1]*)
3 Diminished, real, upside-down (any two for *[2]*)
4 D
5 B
6 C
7 C
8 B
9 C
10 B
11 A
12 C

Answers to short questions (Chapter 27)

13 a) Speed = wavelength × frequency *[1]*
b) Speed = 275 × 1089 × 1000 *[1]*
= 2.99×10^8 *[1]* m/s *[1]*
c) 2.99×10^8 m/s *[1]* All e/m waves have the same speed. *[1]*

14 a)

[2]

b)

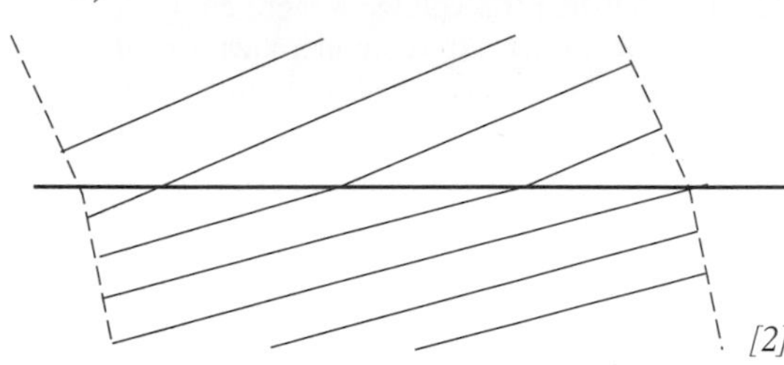

[2]

c) Frequency stays the same, *[1]* wavelength gets less. *[1]*

15 a)

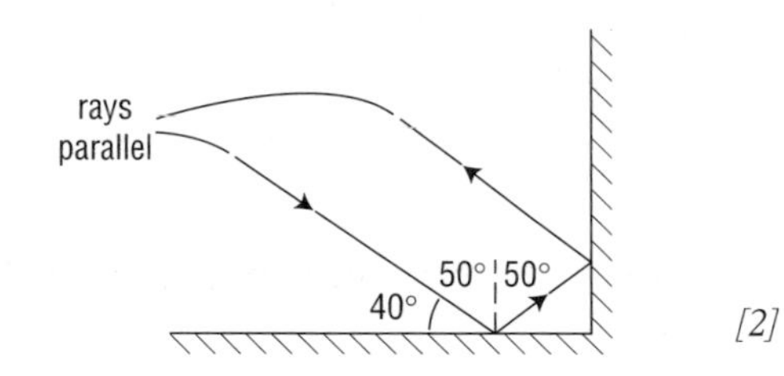

[2]

b)

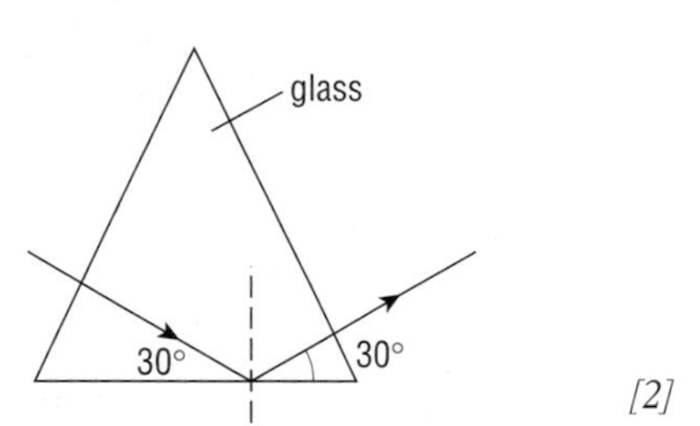

[2]

c)

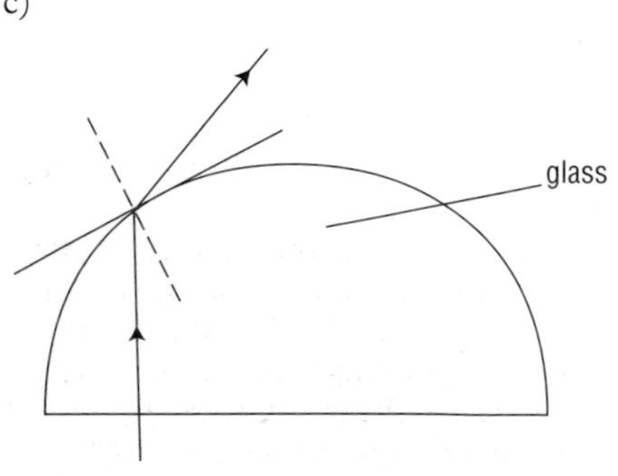

[2]

16 a)

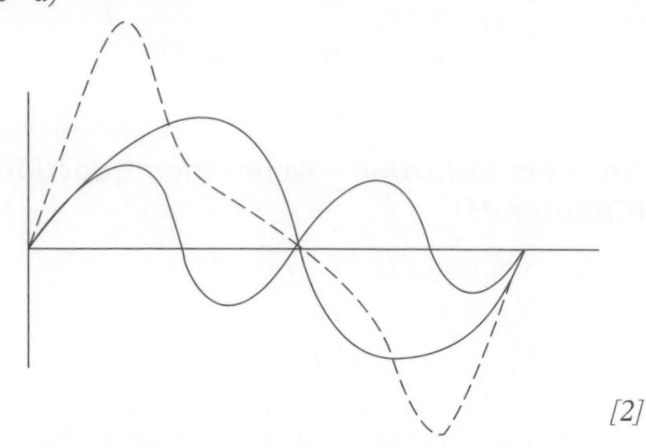

[2]

b) Louder than both X and Y, *[1]* because the amplitude/displacement is greater. *[1]*
c) Same frequency as X, *[1]* half the frequency of Y. *[1]*

Answers to further examination questions (Chapter 27)

17 a) (i) Amplitude *[1]* (ii) Frequency or wavelength *[1]*
b) (i) Distance travelled = 330 × 0.3 *[1]*
= 99 m *[1]* (ii) 49.5 m *[1]*
c) (i) $v = f \times \lambda$ *[1]* (ii) $330 = 33\,000 \times \lambda$ *[1]*,

$\lambda = \frac{1}{100} = 0.01$ *[1]* m *[1]*

18 a) (i) Sound waves with high frequency *[1]* that cannot be heard by human ears. *[1]*
(ii) A Air and bone *[1]* B saline gel and muscle *[1]*
b) Without the saline gel, air would get between the probe and the mother's abdomen. *[1]* This interface would give a major reflection, *[1]* leaving less sound for reflections of the baby within the mother's abdomen. *[1]* When saline gel is used, air is excluded from the space between the probe and the mother's abdomen *[1]* and reflections at the gel/abdomen interface are minimal. *[1]*
(any four for *[4]*)

19 a) Our ears cannot respond to the extremely rapid vibrations. *[1]*
b) Water particles vibrate to and fro *[1]* in the same line/direction as the wave itself. *[1]*
c) Ultrasound does not leave any additional material to pollute the ground like detergent. *[1]*
d) Velocity = wavelength × frequency *[1]*
(speed)
4000 = wavelength × 200 000 *[1]*

$\Rightarrow$ wavelength $= \frac{2}{100} = 0.02$ m *[1]*

20 a) Refraction *[1]*

b)

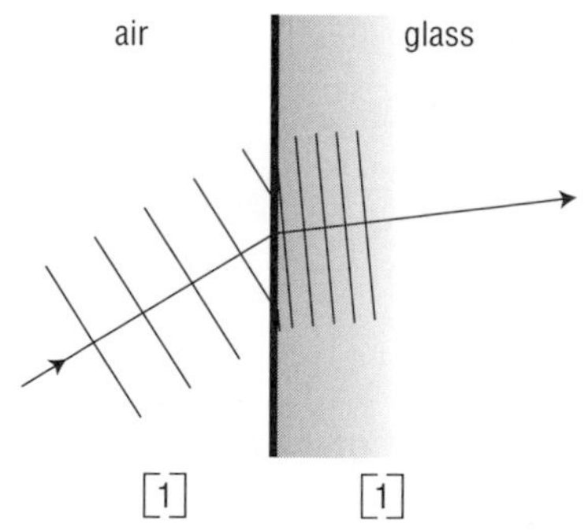

[1] [1]

Light travels slower in glass than in air. *[1]* Therefore, the wavelength (distance between wavefronts) is less in glass. *[1]* This means the wavefronts bend at the interface and the light ray changes direction. *[1]*

c) (i)

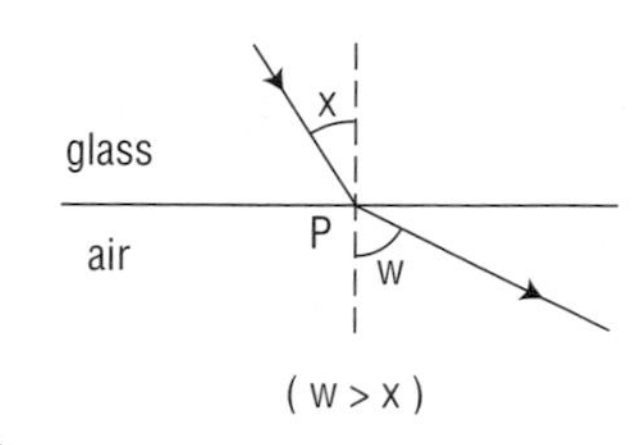

(w > x) *[2]*

(ii)

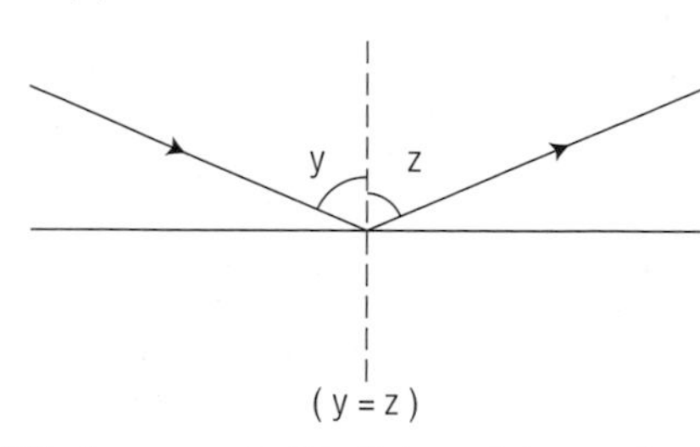

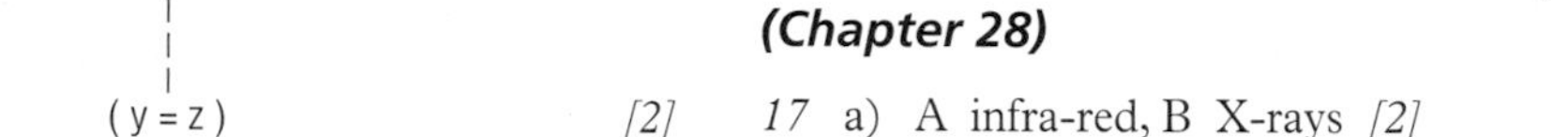

(y = z) *[2]*

(iii) the normal *[1]* (iv) total internal reflection *[1]*

Answers to objective questions (Chapter 28)

1	Infra-red	**7**	C
2	Microwaves	**8**	E
3	Gamma rays	**9**	A
4	Infra-red	**10**	D
5	D	**11**	D
6	A	**12**	B

Answers to short questions (Chapter 28)

13 a) Can be reflected, refracted, diffracted/transverse waves/transfer energy as vibrating electric and magnetic fields/travel through a vacuum/same speed in vacuum (air) (any three for *[3]*)

b) (i) Visible light *[1]* (ii) Infra-red *[1]* (iii) Gamma rays *[1]*

14 a)

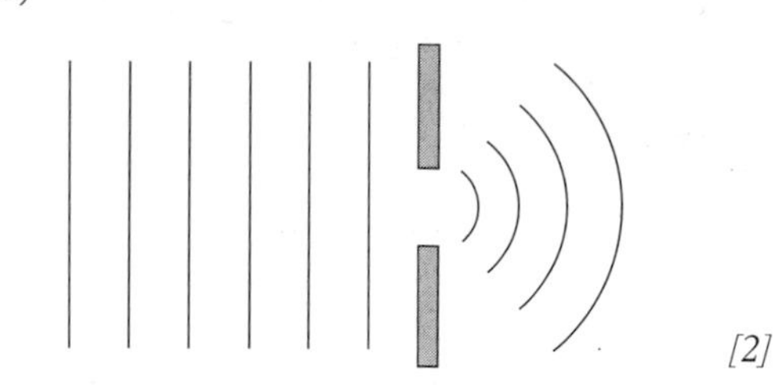

[2]

b) Diffraction *[1]*

c) Because the gap is similar in width to the wavelength *[1]*

d) Stays the same *[1]*

15 a) Speed $= \frac{\text{distance}}{\text{time}}$ *[1]* $= \frac{600}{80}$

$= 7.5$ *[1]* km/s *[1]*

b) 20 seconds *[1]*

c) 250 km *[1]*

16 a) 40 secs *[1]*

b) $\frac{1}{40}$ Hz *[1]*

c) $v = \lambda \times f$ *[1]*, $5 = \lambda \times \frac{1}{40} \therefore \lambda = 200$ km *[1]*

d) (i) Y *[1]* (ii) Z *[1]*

e) Speed = gradient at Y *[1]*

$= \frac{-1}{15}$ m/s *[1]*

Answers to further examination questions (Chapter 28)

17 a) A infra-red, B X-rays *[2]*

b) (i) Speed = wavelength × frequency *[1]*

(ii) $3 \times 10^8 = 10^{-3} \times f$ *[1]*, $\therefore f = 3 \times 10^{11}$ Hz *[1]*

(iii) Microwaves *[1]*

18 a) Radiowaves – 1500 m, *[1]* UV – 3×10^{-8} m, *[1]* visible light – 5×10^{-7} m, *[1]* X-rays – 1×10^{-11} m *[1]*

b) Speed (velocity) = wavelength × frequency *[1]*

$3 \times 10^8 = \frac{3}{100} \times f$

$f = 10^{10}$ *[1]* Hz *[1]*

c) (i) Radiowaves (ii) UV (iii) gamma rays (three for *[3]*)

d) Diffraction is the spreading out of waves, *[1]* when they pass through a gap *[1]* or pass the edge of an obstacle. *[1]* Maximum diffraction (spreading) occurs when the gap is the same size as the wavelength. *[1]* (any three for *[3]*, any suitable example for *[1]*)

19 a) Similarity – both e/m radiation/same speed in vacuum (air) *[1]*
difference – different frequency/different wavelength *[1]*
b) (i) Microwaves are absorbed by water. This heats up the water *[1]* which surrounds the rice. The heat is then transferred by conduction to the rice. *[1]*
(ii) Microwaves will be absorbed by any materials containing water inside a container. This will heat up the water which may boil producing steam. *[1]* Increased pressure inside the container may cause it to explode. *[1]*
c) The infra-red radiation is absorbed by the hot plate. *[1]* Heat is then conducted from the hot plate to the container holding the rice *[1]* and then to the water and the rice. *[1]*

20 a) (i)

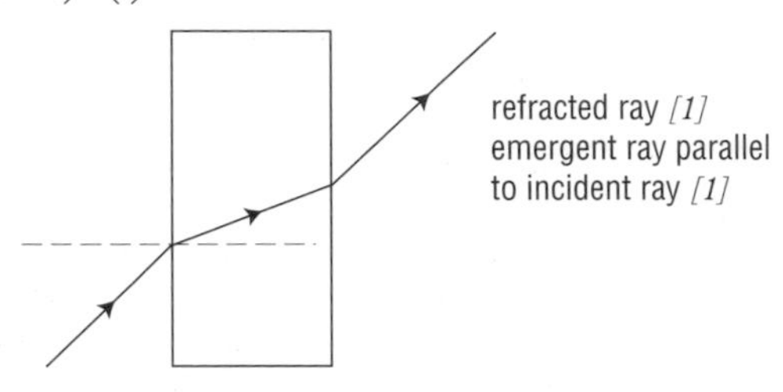

(ii) It is reduced. *[1]*
b) (i) *[2]*

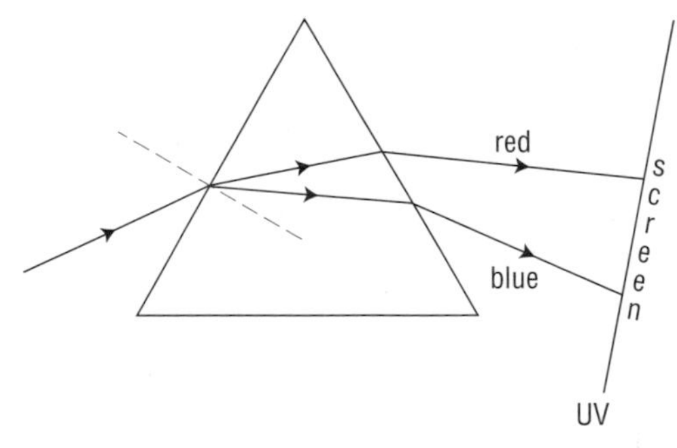

(ii) dispersion *[1]* (iii) see diagram *[1]*
c) (i) To increase the formation of melanin to give the sunbathers a quicker suntan. *[1]*
(ii) UV rays can damage cells and cause skin cancer. *[1]*
d) Infra-red *[1]*
e) (i) X-rays *[1]* (ii) They can destroy and damage cells causing mutations and cancer. *[1]*
f) Can be reflected/refracted/transverse waves/travel through a vacuum/same speed (any one for *[1]*)

Answers to objective questions (Chapter 29)

1	D	**8**	C
2	E	**9**	B
3	B	**10**	B
4	A	**11**	D
5	B	**12**	A
6	A	**13**	A
7	B	**14**	D

Answers to short questions (Chapter 29)

15 a) γ-rays *[1]*
b) It must be able to penetrate through the glass. *[1]*
c) If the bottle is full, the γ-rays give a much weaker signal in the Geiger counter *[1]* than if it is empty because they have to pass through both the glass and the liquid. *[1]*

16 a) Geiger–Müller tube (Geiger counter), *[1]* photographic film *[1]*
b) G–M tube – radiation causes ions to form/current to flow *[1]*
Photographic film – darkens due to radiation *[1]*
c) Harmful effect – nausea, sickness, skin burns, loss of hair, sterility, cancer (any one for *[1]*)
Beneficial use – sterilising medical instruments, tracing leaks, thickness and level gauges, treatment of cancers (any one for *[1]*)

17 a) $^{40}_{20}$Ca – (all three correct *[2]*, two correct *[1]*)
b) 20 g decays to 10 g in 1.3×10^9 years *[1]*
10 g decays to 5 g in another 1.3×10^9 years *[1]*
5 g decays to 2.5 g in another 1.3×10^9 years
∴ age of rock = 3.9×10^9 years *[1]*

18 a) Decays produce heat/radiation *[1]*
b) Cold water piped down to hot rocks, *[1]* later pumped up as hot water *[1]*
c) Advantage – cheap source of energy, *[1]* disadvantage – radioactive materials may get into the water *[1]*

Answers to further examination questions (Chapter 29)

19 a) $^{222}_{86}$Rn *[1]* $^{210}_{83}$Bi *[1]*
b) (i) Count rate is a measure of the number of radioactive decays that have taken place in a given time. *[1]* It is measured using a Geiger–Müller tube. *[1]*
(ii) Points plotted accurately, *[1]* curve of best fit *[1]*
(iii) 22 ± 1 years *[2]*
c) $Pb^{2+}(aq) + CrO_4^{2-}(aq) \rightarrow PbCrO_4(s)$
Ions with charges, *[1]* state symbols *[1]*

d) Age of painting is three half-lives. *[1]* i.e. count rate has fallen from 160 → 80 → 40 → 20 Bq per kg. *[1]* So age of painting is about 66 years – a forgery. *[1]*

20 a) (i) 13n, *[1]* 11p *[1]*
(ii) Inject a solution of sodium-24 chloride into the blood system. Allow time for this to circulate and check the skin graft for radioactivity from time to time. *[1]* If blood is flowing to the new skin, the radioactivity of that area will increase. *[1]*
b) (i) Half-lives indicated on graph, *[1]* 14 or 15 hours *[1]*
(ii) β radiation, *[1]* X = 24, *[1]* Y = 12 *[1]*
(iii) α radiation is least penetrating, *[1]* e.g. only a few cm in air, but it is the most intensely ionising radiation. *[1]* The ions which it produces can attack cells and react with chemicals in cells *[1]* causing nausea and even cancer. *[1]* (any four for *[4]*)

21 a) Proton – relative mass = 1, *[1]* electron – relative charge = −1 *[1]*
b) (i) 19p, 20n, 19e (all three correct *[2]*, two correct *[1]*)
(ii) Potassium-40 has one more neutron. *[1]*
(iii) An atom (isotope) which undergoes radioactive decay. *[1]*
(iv) Calcium/Ca *[1]*
(v) They have the same electron structure. *[1]*
c) (i) ß-radiation *[1]* (ii) sodium/Na *[1]*
(iii) Group VI *[1]*
d) (i) Geiger–Müller tube *[1]* (ii) cancer *[1]*

22 a) (i) Sources of natural radiation *[1]*
(ii) Various rocks/granite/bricks/building material/the Sun (any one for *[1]*)
b) (i) 95p, *[1]* 146n, *[1]* 95e *[1]*
(ii) An atom of the same element *[1]* with a different mass number/number of neutrons *[1]*
c) (i) A helium nucleus/a particle containing two protons and two neutrons *[1]*
(ii) The particle produced will have two fewer protons and two fewer neutrons. *[1]* With two fewer protons it must be a different element. *[1]*
d) (i) γ radiation. *[1]* Penetrating radiation is needed *[1]* to pass through the soil and be detected. *[1]*
(ii) Relatively short half-life, *[1]* so that the radiation disappears fairly quickly. *[1]*

Answers to objective questions (Chapter 30)

1 B
2 D
3 D
4 C
5 A
6 B
7 Nebula
8 Black dwarf
9 Neutron star
10 Red giant

Answers to short questions (Chapter 30)

11 a) An object in a geosynchronous orbit its moving at a speed which makes it appear stationary *[1]* with respect to the Earth's surface. *[1]*
b) It can always transmit messages to or from the places below it on Earth. *[1]*

12 a) As the comet moves from A to B, it slows down *[1]* due to the Sun's gravitational force pulling it back, *[1]* reaching a minimum speed at B. *[1]*
b) From B to C, the comet increases in speed, *[1]* due to the Sun's gravitational force pulling it forward. *[1]*

13 a) Neptune *[1]*
b) Venus *[1]*
c) The smaller planets are denser than the larger planets. *[1]*
d) The planets closest to the Sun (Mercury, Venus, Earth, Mars) are denser than those further from the Sun (Jupiter, Uranus, Neptune). *[1]*

14 a) A huge cloud of dust and gas accumulates in space as a nebula. *[1]* The clouds of dust and gas contract due to gravitational forces. *[1]* As particles are pulled together, they speed up and collide. *[1]* Collisions produce friction and the temperature rises. *[1]* In time, the temperature is hot enough for hydrogen nuclei to fuse. *[1]* Immense heat and light is emitted as a star. *[1]* (any five for *[5]*)
b) star → red giant → white dwarf → black body *[3]*

15 There is no atmosphere on the Moon.
Therefore:
- sun visor to protect from Sun's rays *[1]*
- oxygen supply to maintain life *[1]*
- pressure gauge to maintain pressure on astronaut's body *[1]*
- shiny metal covering on suit to reflect Sun's radiation and reduce absorption. *[1]*

Gravity is only $\frac{1}{6}$th on the Moon relative to the Earth.
Therefore:
- heavy boots to steady movement *[1]*
- rubber pressure suit helps mobility *[1]* (any four for *[4]*)

Answers to further examination questions (Chapter 30)

16 a) Both axes labelled, *[1]* points accurately plotted, *[1]* smooth curve through points. *[1]*
b) The period of orbit increases as the radius of orbit increases *[1]*
c) (i) 24 hours = 24 × 60 × 60 secs = 86 400 s *[1]*
(ii) From the graph, a satellite with a period of

orbit of 24 hours has a radius of orbit = 42 500±1000 km. *[1]* (iii) Satellites such as this are used as communications satellites. *[1]* They move around the Earth at the same rate as it spins so they are always above the same point on the Earth. *[1]*

d) Orbit speed at radius 10 000 km

$$= \frac{\pi \times 20\,000}{10\,000} = 2\,\pi \text{ km/s}$$

Orbit speed at radius 20 000 km

$$= \frac{\pi \times 40\,000}{28\,000} = 1.4\,\pi \text{ km/s} \quad [1]$$

Speed of an orbiting satellite decreases as the radius increases. *[1]*

e) Gravity acts on space craft and pulls it towards the Earth. *[1]* As the spacecraft orbits the Earth, it experiences a centripetal force outwards from the centre of its motion. *[1]* The overall effect of these forces in the direction of the Earth is zero. So, the astronaut appears to be weightless. *[1]*

f) The aeroplane must fly at great speed *[1]* in a path similar to an orbiting satellite *[1]* so that the centripetal force counteracts gravity. *[1]*

17 a) In the Sun the temperature is so incredibly hot *[1]* that hydrogen nuclei fuse to form helium nuclei *[1]* emitting immense amounts of heat and light. *[1]* Make an attempt at a nuclear equation.

e.g. ${}^{1}_{1}H + {}^{1}_{1}H \rightarrow {}^{2}_{1}H + {}^{0}_{1}e$ (positron)
or ${}^{2}_{1}H + {}^{2}_{1}H \rightarrow {}^{4}_{2}He$ *[1]*

b) Hydrogen is used up and fusion processes convert matter into energy. *[1]* The mass of the white hot star gets less. *[1]* This means that the gravitational forces holding the star together are smaller *[1]* so that the star expands *[1]* to form a red giant *[1]* (any four for *[4]*)

c) Nuclei of the heaviest elements in the Sun are also present in the inner planets. *[1]* All the elements identified as being present in the Sun are found on Earth. *[1]*

18 a) Gravitational force *[1]* pulling the probe towards Mercury. *[1]*

b) There is no atmosphere on Mercury to produce drag on the parachutes and slow the probe down. *[1]*

c) The rocket engines could be directed (fired) backwards *[1]* so that the space probe decelerates. *[1]*

d) Mercury's density is significantly greater than the Moon's. *[1]* Mercury has a strong magnetic field. *[1]*

e) Surface features on a planet are largely dictated by atmosphere. There were no winds to cause erosion and movement of material. *[1]* There is no water vapour that might result in rain or snow and surface water in rivers etc. to change the surface features such as craters. *[1]*

f) The parts of Mercury and the Moon in darkness will be well shielded from the Sun. Temperatures in these areas are both roughly the same −180°C and −170°C, respectively. *[1]* Mercury is, however, much closer to the Sun than the Moon. *[1]* So, the maximum temperatures on Mercury at those parts facing the Sun are much higher (430°C) than similar parts of the Moon (120°C). *[1]* (any two for *[2]*)

19 a) (i) Mercury is closer to the Sun than the Earth so its surface temperature is higher. *[1]* (ii) Venus has a higher surface temperature because it is closer to the Sun than the Earth. *[1]* It also has denser gases in its atmosphere than the Earth. The Earth traps radiation *[1]* and raises the temperature even higher than Mercury even though it is further from the Sun than Mercury *[1]* (any two for *[2]*)

b) (i) Colder than −30°C *[1]* (ii) Further than 288×10^6 km *[1]*

c) Fusion *[1]*

d) (i) The mass of the Earth is much greater than that of the Moon so its attraction of any object is greater. *[1]* (ii) The gravitational force from the Earth decreases and that in the opposite direction from the Moon increases. *[1]* The gravitational force increases as it approaches the Moon whilst that from the Earth decreases. *[1]*

e) Red giant *[1]* supernova *[1]*